Shale Gas and Tight Oil 2014

Topical Conference at the 2014 AIChE Spring Meeting and 10th Global Congress on Process Safety

New Orleans, Louisiana, USA
30 March – 3 April 2014

ISBN: 978-1-63439-059-0

Printed from e-media with permission by:

Curran Associates, Inc.
57 Morehouse Lane
Red Hook, NY 12571

Some format issues inherent in the e-media version may also appear in this print version.

Printed by Curran Associates, Inc. (2014)

For permission requests, please contact AIChE
at the address below.

AIChE
3 Park Avenue
New York, NY 10016-5991

Phone: (203) 702-7660
Fax: (203) 775-5177

www.aiche.org

Additional copies of this publication are available from:

Curran Associates, Inc.
57 Morehouse Lane
Red Hook, NY 12571 USA
Phone: 845-758-0400
Fax: 845-758-2634
Email: curran@proceedings.com
Web: www.proceedings.com

TABLE OF CONTENTS

354012 Possible vs. Practical: Engineers Must Lead the Development of Practical Technologies

Abstract:

Chemical Engineering is need today more than ever. There is a strong societal desire for sources of energy and feedstocks that can sustainably support a high standard of living for all. Thermodynamic, material and economic challenges stand in the way. Most of the general public does not have the basic understanding of the energy flows, chemical transformations, and the scale of the systems that support our current way of life. These concepts are fundamental to chemical engineering. Chemical engineers should be on the vanguard of separating hype from truly practical options yet occationally we let society down. This talk will present several examples where the fundamentals of chemical engineering can be used to access the practicality of proposed solutions to energy and water issues, as well as explain why the venture capital model struggles in the materials market.

PROSYS™ Maximize Your Plant.

Operator Errors and What Can be Done to Minimize

Global Congress on Process Safety

March 31, 2014

10TH GLOBAL CONGRESS ON PROCESS SAFETY

PROSYS™ Maximize Your Plant.™

10TH GLOBAL CONGRESS ON PROCESS SAFETY

Presenter

Tom Nolan

- **Graduated from Ohio University with B.S. in Chemical Engineering**
- **24 years experience in chemical process industry in a variety of roles**
- **MAIC and DFSS Six Sigma Black Belt**

PROSYS Maximize Your Plant.™

10th GLOBAL CONGRESS ON PROCESS SAFETY

The Cost of Errors

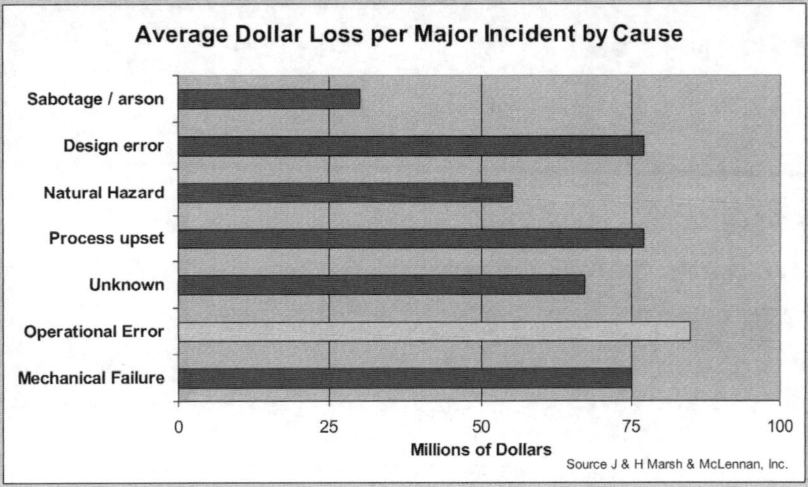

PROSYS Maximize Your Plant.™

10th GLOBAL CONGRESS ON PROCESS SAFETY

The Cost of Operator Errors

- **ASM estimates total loss due to operator error is $8B per year**
- **Chemical Safety Topical Committee - average of one chemical incident per day - cost of over $2 million per incident to <u>comply with requirements</u>**
- **Errors cause 42% of unscheduled shutdowns**
- **70% of process incidents occur during start-up or shutdown**

PROSYS Maximize Your Plant™ 10th GLOBAL CONGRESS ON PROCESS SAFETY

Reasons for Errors

- Lack of Skill
- Lack of Knowledge
- Carelessness
- System Design
  - Operator set up to make errors by inappropriate design – built in errors
  - Errors that are predictable are preventable by better design

PROSYS Maximize Your Plant™ 10th GLOBAL CONGRESS ON PROCESS SAFETY

Focus of this Discussion

- System Design
  - **Alarm Management**
  - **Operator Graphics**
  - Operator set up to make errors by inappropriate design – built in errors
  - Errors that are predictable are preventable by better design

PROSYS™ Maximize Your Plant™

Alarm Floods

In a number of industrial incidents, alarm floods were identified as a significant contributing cause to the incident…

As found by EEMUA in 1999 and CSB

Alarm Flood defined by ISA 18.2 as -
"10 or more annunciated alarms in any 10 minute period per operator"

PROSYS™ Maximize Your Plant™

Why do Alarm Floods Occur?

One reason is not providing dynamic alarm management

- **Alarms need to indicate abnormal situations that require operator action**
- **Processes do not operate in one state**
- **What is normal vs abnormal changes with state**
- **Alarms are typically configured for run therefore many alarms are triggered upon a Change of State: Run to Shutdown**
- **Many of the alarms are not applicable or actionable for the new state impleading the operators ability to act quickly on what is important**

PROSYS™ Maximize Your Plant™

What is Impacted by Alarm Floods?

- **Product quality**
- **Operability or profitability of the process**
- **Loss of equipment**
- **Loss of containment – environmental releases**
- **Injury and loss of life in plant or community**

PROSYS™ Maximize Your Plant™

What Makes Alarm Floods so Dangerous?

Can be a problem for three reasons:

- **A deluge of alarms can cause critical alarms to be missed**
- **Floods can be a significant distraction when dealing with process upsets**
- **Can be an indicator of larger systemic safety issues**

PROSYS Maximize Your Plant.™ 10th GLOBAL CONGRESS ON PROCESS SAFETY

Impacting Alarm Management Design

- Alarm rationalization is not a process to eliminate alarms - its about quality
- Good rationalization will add alarms when appropriate
- Typical (Static) rationalization is only for run mode
- Dynamic rationalization considers all plant modes

PROSYS Maximize Your Plant.™ 10th GLOBAL CONGRESS ON PROCESS SAFETY

Dynamic Alarm Management

- Dynamic rationalization does everything a static rationalization does plus asks "when" for each alarm
- Requires dynamic software to make changes based upon operating mode of the plant
- Eliminates redundant alarms and lowers operator loading during transitions
- Only alarms what is abnormal and actionable for the given state

PROSYS Maximize Your Plant.℠

10th GLOBAL CONGRESS on PROCESS SAFETY

Actual Performance Metrics vs ISA 18.2

Type	Avg Alarm Rate/hr	Low/Hi of Avg Rate/hr	Peak Alarm Rate/hr	Low/Hi of Peak Rate/hr	% Time in Flood	Standing Alarms >24 hrs
Before Rationalization	30	Low=6.2 Hi=61	638	Low=152 Hi=2402	17.2%	9
ISA 18.2 Target Metrics	6	---	<60	---	<1%	<5
After Dynamic Alarm Management	2	Low=0.09 Hi=2.5	25	Low=13 Hi=42	0.25%	4

PROSYS Maximize Your Plant.℠

10th GLOBAL CONGRESS on PROCESS SAFETY

Results After Dynamic Alarm Management

Annunciated Alarms Per Hour (Monthly) **Area: G1-G5-G6** **June 2013**

Average Counts | Peak Counts | Total Counts

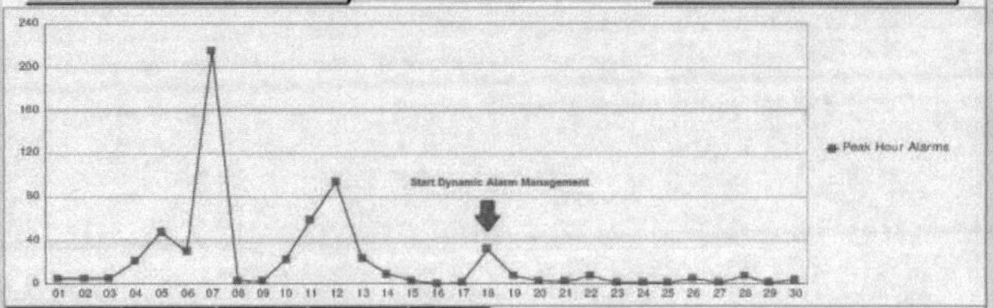

Day	Min Hour Alarms	Avg Hour Alarms	Peak Hour Alarms	Total Alarms
01-Jun-2013	0	0.3	5	6
02-Jun-2013	0	0.4	5	9
03-Jun-2013	0	0.5	5	13
04-Jun-2013	0	1.1	21	27
05-Jun-2013	0	13.3	48	319
06-Jun-2013	0	3.6	30	86
07-Jun-2013	0	11.9	215	285
08-Jun-2013	0	0.4	3	10
09-Jun-2013	0	0.3	2	7
10-Jun-2013	0	1.5	23	35
11-Jun-2013	0	4.7	59	112
12-Jun-2013	0	11.9	94	285
13-Jun-2013	0	3.8	24	90
14-Jun-2013	0	1.3	9	32
15-Jun-2013	0	0.3	3	8

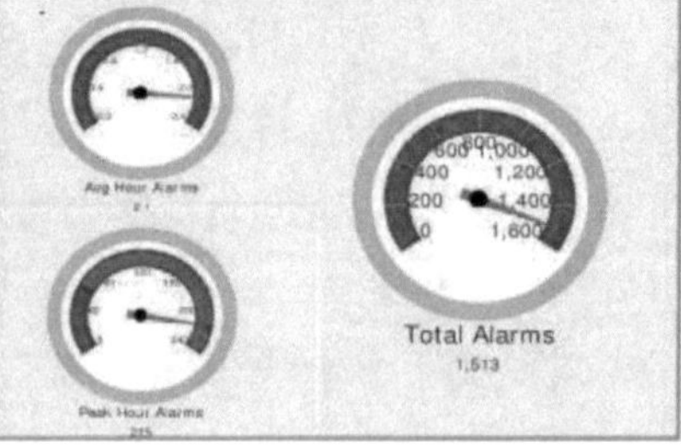

PROSYS™ Maximize Your Plant.™

10TH GLOBAL CONGRESS ON PROCESS SAFETY

Operator Graphics – The Risk of Changing Focus

- **Distraction of changing focus from process graphic to faceplate window**
- **Added workload from managing multiple open windows**
- **Increased probability of errors when changing values for un-intended tag with multiple faceplates open**
- **All of above become multiplied and more complex when process is transitioning from one state to another**

PROSYS™ Maximize Your Plant.™

10TH GLOBAL CONGRESS ON PROCESS SAFETY

Faceplate Operation

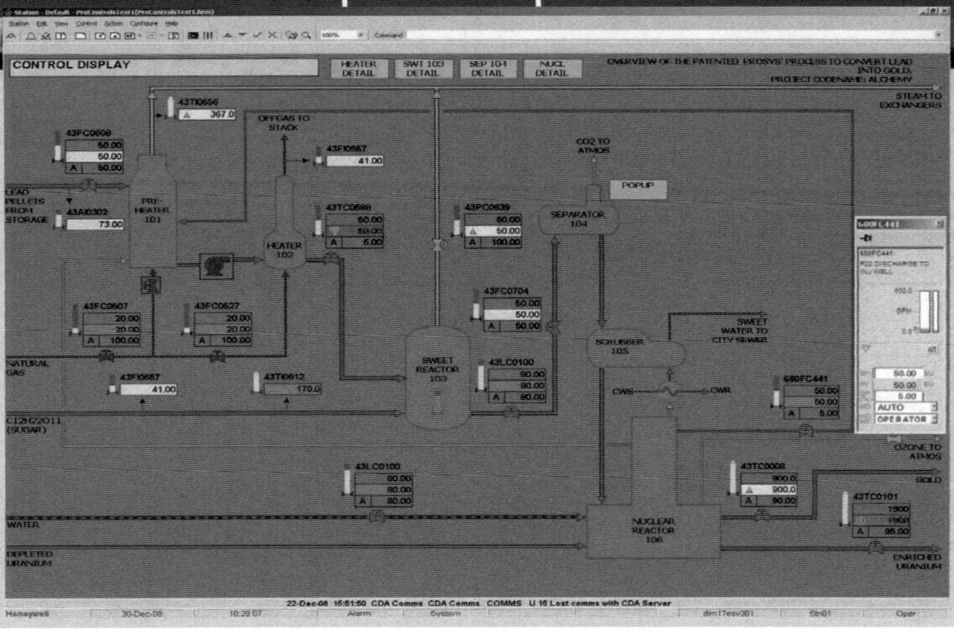

PROSYS Maximize Your Plant.™ 10th GLOBAL CONGRESS ON PROCESS SAFETY

Potential for Errors

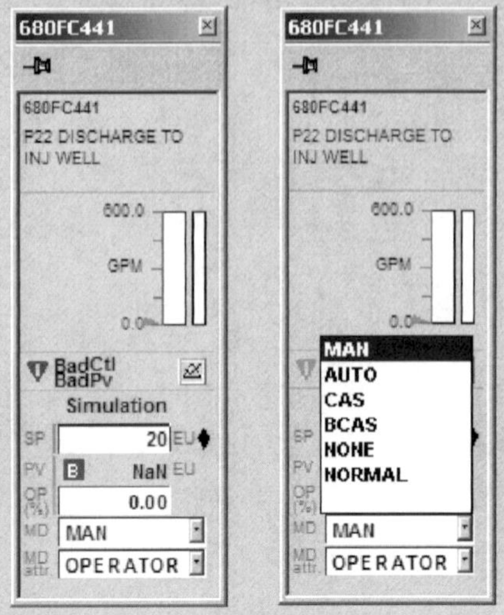

Faceplate Design -can introduce additional errors

- Mode drop down list covers SP, PV and OP values
- Operator may select wrong Mode if SP, PV and OP values are hidden from view
- Mode list offers more options than necessary

PROSYS Maximize Your Plant.™ 10th GLOBAL CONGRESS ON PROCESS SAFETY

Direct Entry Operation

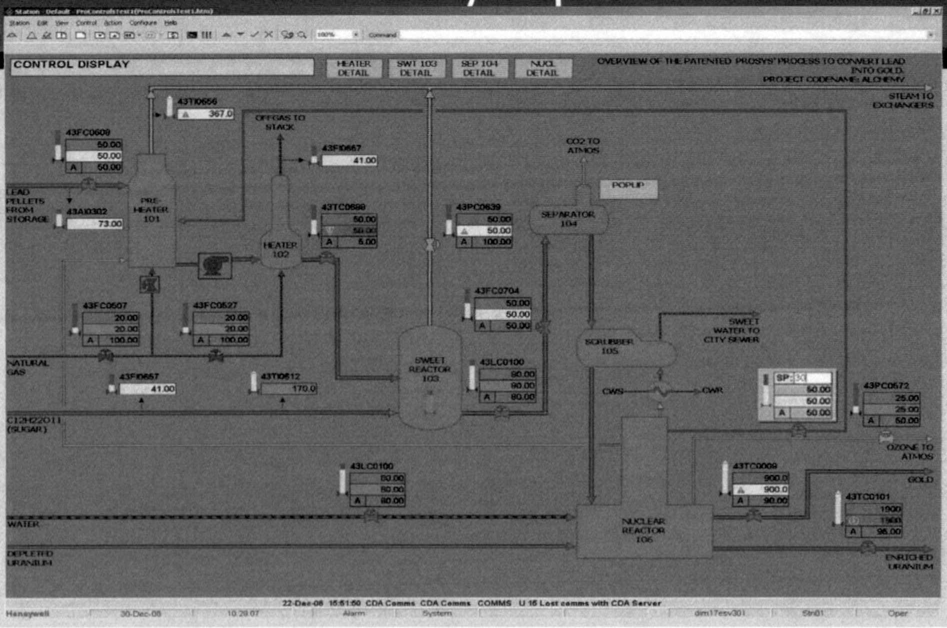

PROSYS Maximize Your Plant.™

Direct Entry Fields

Values can be entered by:

- Select point
- Typing value via keypad
- Pressing enter

Key Factors:

- By maintaining focus of points after value change, it improves safety for input errors, multiple clicking functions and is more time efficient
- Easily recognizable operator enterable fields

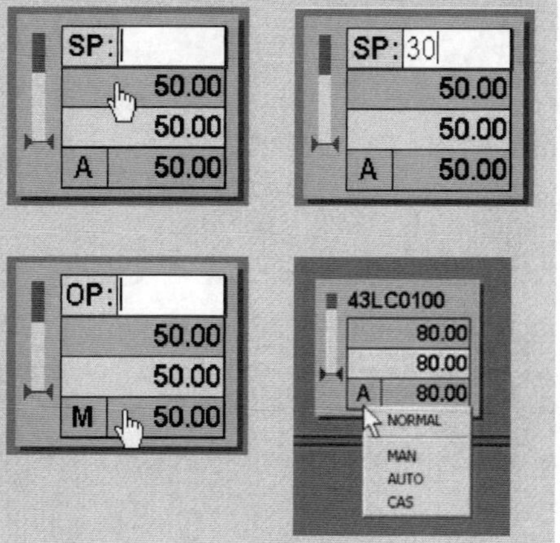

PROSYS Maximize Your Plant.™

Evaluation of Methods

Comparison of Faceplate to Direct Entry

- **Keystroke Level Analysis**
- **Operator Loading Analysis**
- **Risk Analysis**

PROSYS Maximize Your Plant™

10th GLOBAL CONGRESS on PROCESS SAFETY

Keystroke Level Model

Symbol	Time (s)	Description
K	0.28	Keystroke
P	1.1	Point to object
BB	0.2	Click on object
H	0.4	Home hands on keyboard or mouse
M	1.2+	Mental act or routine thinking

PROSYS Maximize Your Plant™

10th GLOBAL CONGRESS on PROCESS SAFETY

Estimate Execution Time Analysis

- **Evaluates execution time by an operator comparing faceplates to direct entry for SP/OP and Mode Changes**

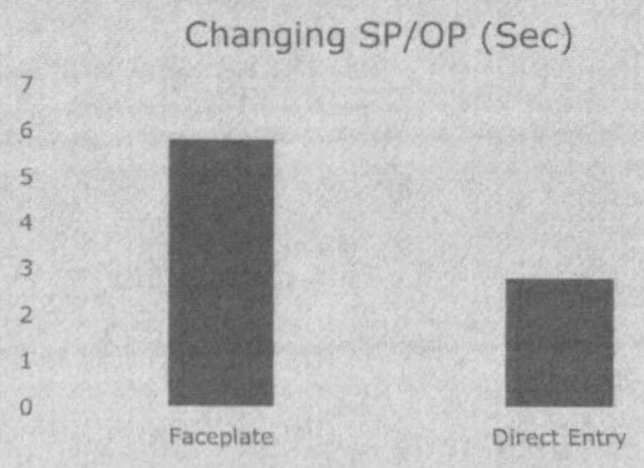

- **50 % Reduction**

PROSYS Maximize Your Plant.™

Estimated Operator Loading Analysis

- **Evaluates Mental and Physical Load on Operator**
  - Tasks that require very little thought are removed – such as (BB), (H)

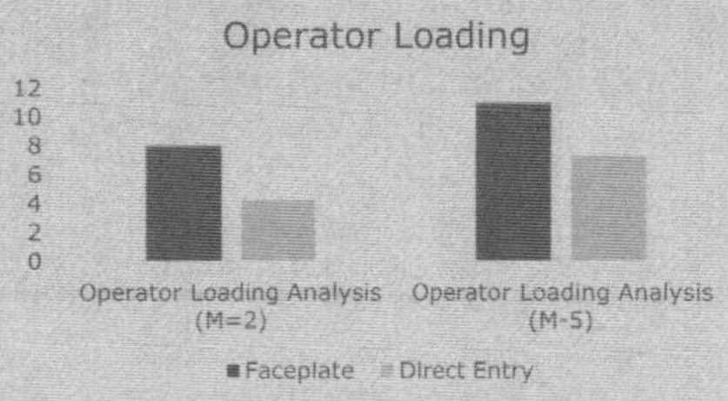

  - Mental and Physical tasks are weighted based on the amount of fatigue they cause – assigned a weight of M=2.0 to 5.0

PROSYS Maximize Your Plant.™

Risk Analysis

- **Identifies actions where operator entry errors can occur**

PROSYS Maximize Your Plant.™

10th Global Congress on Process Safety

Risk Analysis

Faceplate operation KLM for changing SP/OP	Potential Error
Move to shape (P)	Select wrong point
Click on shape (BB)	
Move to faceplate (P)	Point in faceplate is previous point
Click on SP/OP field (BB)	Change wrong parameter
Move hands to keyboard (H)	
Type in value and press enter (4K)	Mistype value and press enter
Move hands to mouse (H)	
Move to close faceplate (P)	
Click to close faceplate (BB)	
Potential error count	**4**

PROSYS Maximize Your Plant.™

10th Global Congress on Process Safety

Risk Analysis

Direct Entry operation KLM for changing SP/OP	Potential Error
Move to SP/OP (P)	Select wrong parameter or point
Click on SP/OP field (BB)	
Move hands to keyboard (H)	
Type in value and press enter (4K)	Mistype value and press enter
Potential error count	**2**

PROSYS Maximize Your Plant.™

10th GLOBAL CONGRESS ON PROCESS SAFETY

Direct Entry Vs Faceplate

Summary of Benefits

- **Execution Time SP/OP Changes - 51.5% less**
- **Execution Time for Mode Changes – 50% less**
- **Operator Loading – 33 % to 47% less**
- **Risk of Errors – 50% less**

PROSYS Maximize Your Plant.™

10th GLOBAL CONGRESS ON PROCESS SAFETY

Conclusion

- **Many industrial errors are a result of operators using control systems with flawed designs**
  - Poor design and performance by alarm management distracts operators and/or occludes critical alarms
  - Operator graphics with multiple open faceplates can cause a change intended for one controller to be entered into the faceplate of another
- **The two mechanisms listed above are predictable and therefore preventable through better design available today**

PROSYS™ Maximize Your Plant.
Questions?
Comments?
10TH
GLOBAL
CONGRESS
ON PROCESS
SAFETY

Problem Statement

visium KMS
Process Safety Management Solution

- Any organisation can improve by leveraging knowledge and lessons learned.
- Knowledge can be externally or hidden locally, in a global enterprise, where lessons learned are difficult to share beyond their initial impact.
- There is so much data, information and knowledge (opportunities) coming at your company that can make you stronger or make you look like your not paying attention.
- Not paying attention or managing the available knowledge, can sink your ship… nobody wants that!

Expert Insight Delivered 2

Osys
Part of the Rolls-Royce Group

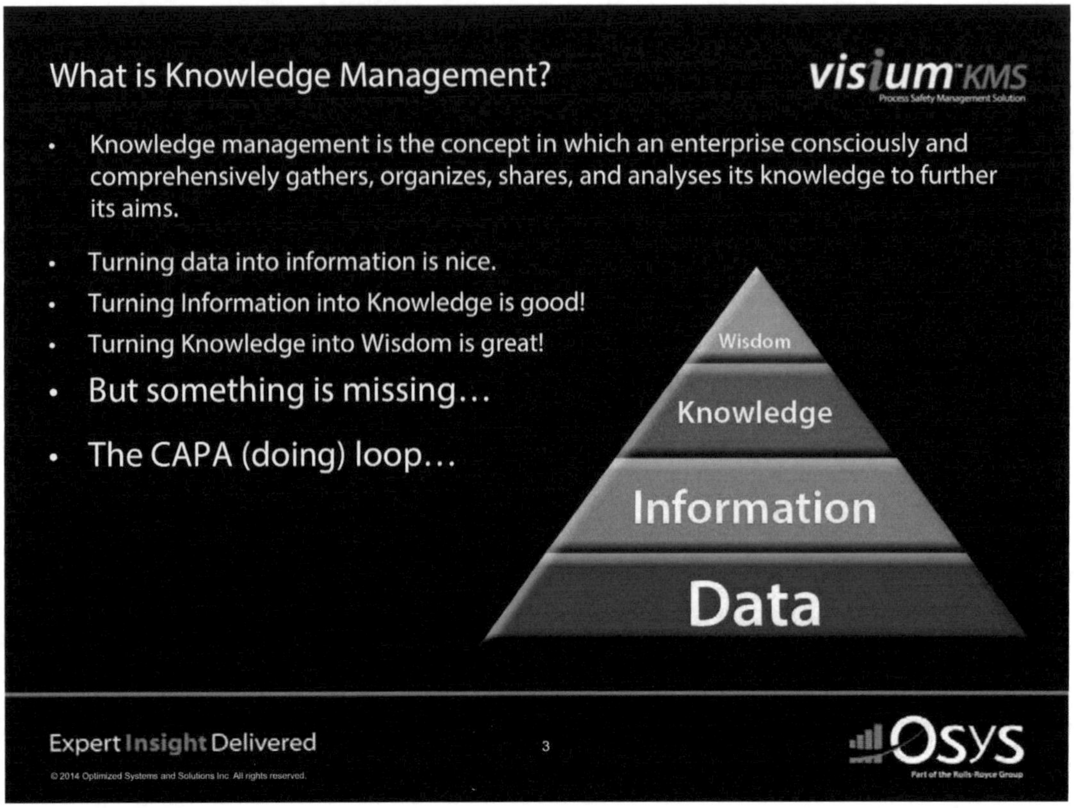
What is Knowledge Management?
visium KMS
Process Safety Management Solution
• Knowledge management is the concept in which an enterprise consciously and comprehensively gathers, organizes, shares, and analyses its knowledge to further its aims.
• Turning data into information is nice.
• Turning Information into Knowledge is good!
• Turning Knowledge into Wisdom is great!
• But something is missing…
• The CAPA (doing) loop…
Wisdom
Knowledge
Information
Data
Expert Insight Delivered
3
© 2014 Optimized Systems and Solutions Inc. All rights reserved.
Osys
Part of the Rolls-Royce Group

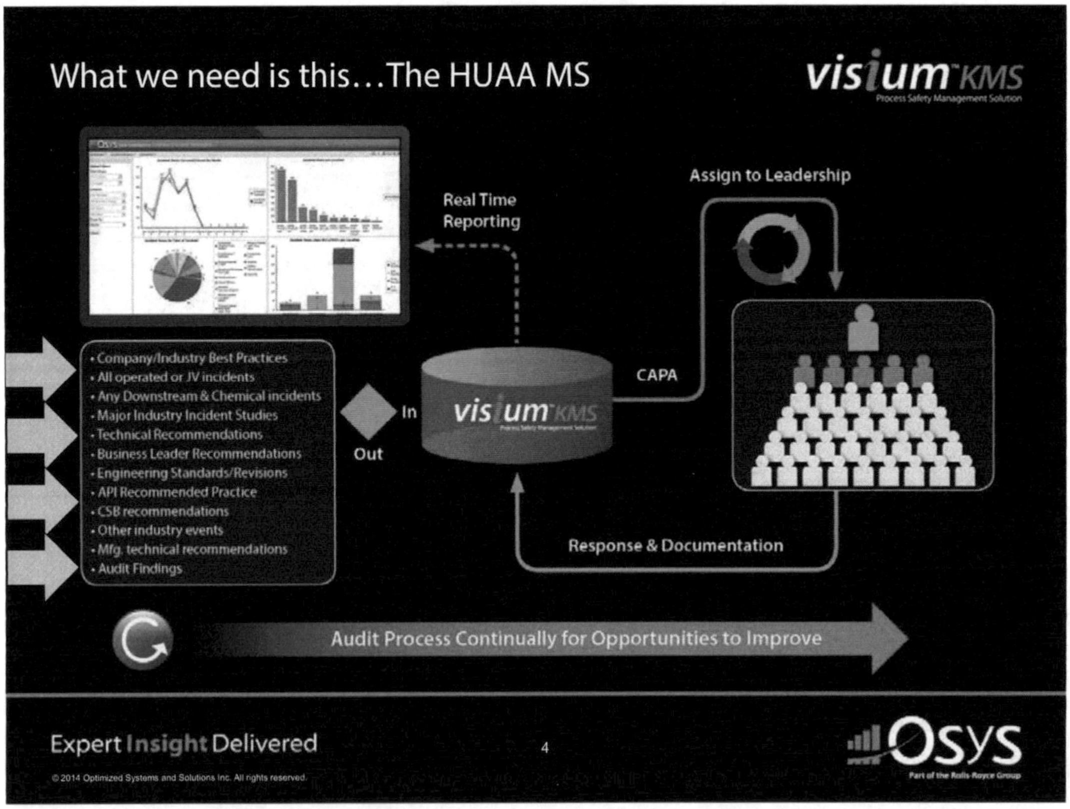
What we need is this…The HUAA MS
visium KMS
Process Safety Management Solution
Real Time Reporting
Assign to Leadership
• Company/Industry Best Practices
• All operated or JV incidents
• Any Downstream & Chemical incidents
• Major Industry Incident Studies
• Technical Recommendations
• Business Leader Recommendations
• Engineering Standards/Revisions
• API Recommended Practice
• CSB recommendations
• Other industry events
• Mfg. technical recommendations
• Audit Findings
In
Out
visium KMS
CAPA
Response & Documentation
Audit Process Continually for Opportunities to Improve
Expert Insight Delivered
4
© 2014 Optimized Systems and Solutions Inc. All rights reserved.
Osys
Part of the Rolls-Royce Group

What is the HUAA Management System…

visium KMS
Process Safety Management Solution

HUAA – Heard, Understood, Acknowledged, Actioned.

1. Opportunities identified (heard) by your experts,
2. Captured by an electronic system,
3. Expertly Reviewed (understood),
4. Accepted or rejected (acknowledged) and
5. Assigned (actioned) to Sr. Managers for resolution and tracked to closure.

Expert Insight Delivered 5

OSYS Part of the Rolls-Royce Group

HUAA – Management Systems - Detail

visium KMS
Process Safety Management Solution

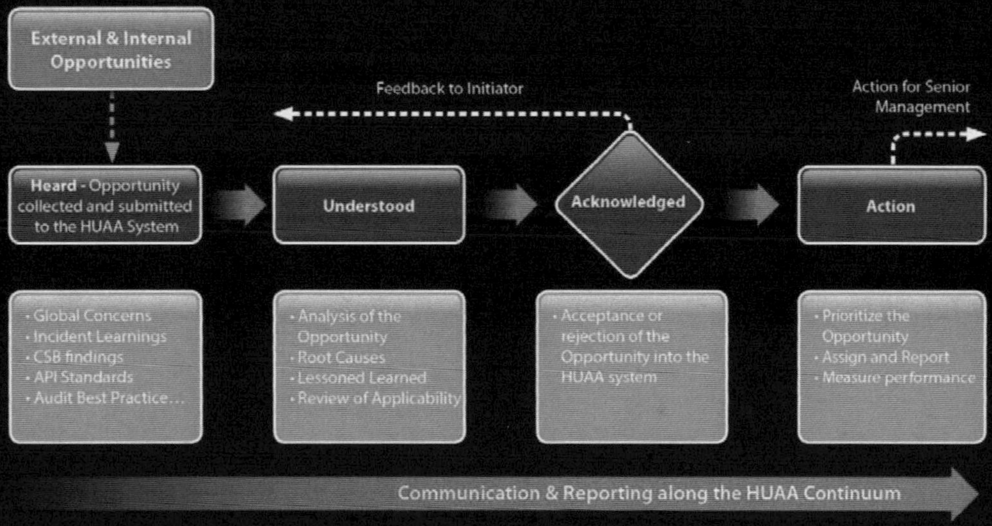

Expert Insight Delivered 6

OSYS Part of the Rolls-Royce Group

HUAA Results

visium KMS
Process Safety Management Solution

- Efficient Collection - Important Opportunities are identified and collected from many sources (internal and external), and entered into the HUAA management system.
- Expert Vetting - These opportunities are vetted, parsed and resolved for inclusion or exclusion; documenting & communicating both.
- Accountable Assignment - Included opportunities are systematically assigned to Sr. Leaders and managed across the enterprise at SVP and C levels.
- Action - Improvement will be assured through the accountable actions of these leaders.
- Results - The organization will gain competitive advantage by active listening, learning from their experience, as well as the experience of others, and by taking ACTION!

Expert Insight Delivered 7

OSYS
Part of the Rolls-Royce Group

Sharing is not enough to keep your ship afloat!

visium KMS
Process Safety Management Solution

Expert Insight Delivered 8

OSYS
Part of the Rolls-Royce Group

352631 Autocad - Smart Grid – Agriculture - Architecture and the United States Government

Abstract:

The computational methods, process control, process design, process safety, public policy of the smart chip for agriculture industry, and architecture industry in the United States utilizing the computational methods of AutoCAD in the structural development stages of the smart grid for device/component/equipment and geographical development and public policy for grant funding.

352747 Production, Characterization and Catalytic Studies of Biobased Carbon Materials

Abstract:

Carbon-based nano materials were produced from wood chips, lignin and bio-char. The influence of reaction conditions on the production of nano-carbons was investigated. Carbon-based nanomaterials were characterized using scanning electron microscopy (SEM), transmission electron microscopy (TEM), high resolution transmission electron microscopy (HRTEM), and X-ray diffraction (XRD). The catalytic performance of these nanomaterials as syngas cleaning adsorbents and chemical conversion catalysts for bio-fuel processes has been evaluated.

347835 Uncertainty in Sour Gas Viscosities Estimation, What Is the Effect on Your Reservoir Inflow and Tubing Performance

Abstract:

In light of the need to find an applicable, consistent and promising correlation to accurately calculating sour gases viscosity under several reservoir conditions, this study aims to compare the existing correlations and other industry standard methods, check their accuracy and range of applicability for sour/acidic gases. Moreover, investigating the impact of uncertainties in estimating sour gas viscosities on reservoir flow performance relationship (IPR), reserves and resources estimations, tubing performance, pipeline and facilities design, the design of compressors and more importantly the need to find such correlation to be used in further advancements of reservoir simulation.

354052 Chromonic Nanocarriers for Chemotherapeutics: Size Distribution and Control Release Studies

Abstract:

Conventional chemotherapy fails to release the drug in the controlled manner; rather they release the drug instantly due to which re-administration is required as therapeutic index falls. Moreover, they target both the cancer cells and non-cancer cells which make the treatment of the cancer cells highly ineffective due to excessive toxicities. Recently, the nanoparticle mediated drug delivery for chemotherapeutics has gained much attention. The present study reports the use of a novel class of materials – *Chromonics* in cancer drug delivery. Folic acid as a chromonic material has been explored to develop nanocarrier for sustained release of chemotherapeutic drugs *in-vitro*. With the help of a biocompatible polymer as a tool, folate nano-domains were formed from self-assembled liquid crystalline folate solutions. These nano-domains were crosslinked to form a stable nanoparticle system. The effect of different factors like crosslinking agent, relative polymer and folate concentration, on the size distribution was investigated in detail. Based on the observations, we have reported that not only the synthesis but also the size of the developed nanoparticles can be controlled by varying the folate: HPMC weight ratios. The present study then explored the release of drugs from these nanocarriers. We have investigated the release of folic acid as a model drug from these nanoparticles as well as release of methotrexate and cytarabine as an anticancer drug for the period of 60 days. It was observed that release of drug was nanoparticle's size-dependent, release medium dependent, pH of the release medium dependent, crosslinking agent and its concentration dependent. The significant part of the current study is the synthesis of chromonic nanoparticles, how the size can be controlled, how controlled release can be obtained as well as how the release profiles can be modified depending on the dosage forms. Therefore, we can say that folate nanoparticles can be a potent nanocarrier for the sustained release of cancer drugs.

IPEX

Committed to Excellence

Double Containment Piping Systems: A Fail-Safe, Leak-Free Solution

Darin Johnson, P.Eng.

Applications Engineer

IPEX USA LLC

AIChE – 2014 Spring Meeting

New Orleans, LA

Monday, March 31, 2014

IPEX

Committed to Excellence

Why Double Containment Piping Systems?

- December 1988 – law change within the ***Resource, Conversation, and Recovery Act*** in the United States regarding underground piping.
- U.S. Environmental Protection Agency (EPA) requirement: Code of Federal Regulations (CFR), Title 40, Part 280.
  - Petroleum Products
  - Hazardous Substances (over 3200, CFR Title 40 Part 302.4)
  - Hazardous Wastes

Why Double Containment Piping Systems?

- Buried tank - all buried piping connected to it must be double contained.
- Above ground tank – associated buried piping consists of 10% or more of the tank volume, it also must be double contained.

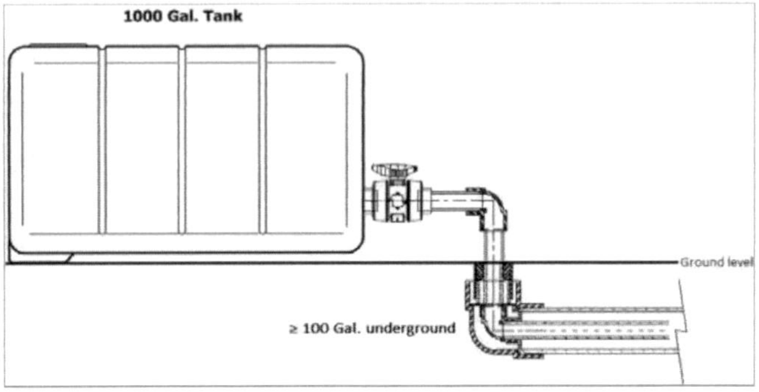

Why Double Containment Piping Systems?

- If the piping is less than 10% of the volume of the aboveground tank, and you choose not to double contain the buried piping, ***responsibility for any clean-up or remediation resulting from a release will still fall on the Client.***

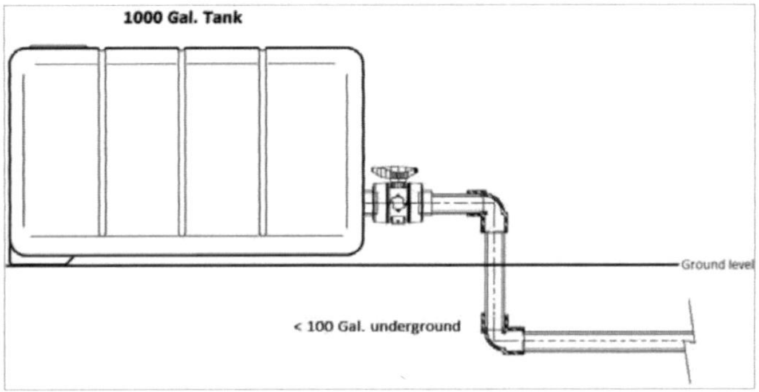

Why Double Containment Piping Systems?

- **Above Ground Piping Systems**
  - Different jurisdictions have to ability to require chemicals be double contained.

Nomenclature of Double Containment Piping

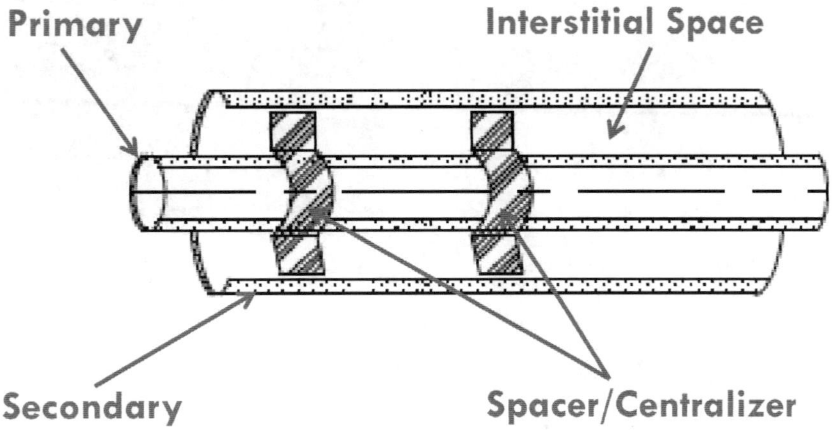

IPEX
Committed to Excellence

Double Containment Piping Systems: Material Combinations

SYSTEM Carrier	Containment	Carrier SIZE inches
PVC 80	PVC 80	1/2 - 12
PVC 80	PVC 40	1/2 - 12
PVC 40	PVC 40	1/2 - 12
PVC DWV	PVC DWV	1-1/2 - 12
PVC 80	CS SCH 10	1/2 - 12
CPVC 80	CPVC 80	1/2 - 8
CPVC 80	PVC 80	1/2 - 12
CPVC 80	PVC 40	1/2 - 12
CPVC 80	CS SCH 10	1/2 - 12
NAT PP 80	PVC 80	1/2 - 2
NAT PP 80	PVC 40	1/2 - 2
RED PVDF 80	CPVC 80	1/2 - 2
RED PVDF 80	PVC 80	1/2 - 2
NAT PVDF 80	PVC 80	1/2 - 2
ENFUSION PP	PVC DWV	1/2 - 6
S40 A53B CS BW	PVC 80	1/2 - 8
S40 A106B CS BW	PVC 80	1/2 - 8

SYSTEM Carrier	Containment	Carrier SIZE inches
S40 A53 BLK SW	FRP	1/2 - 2
S40 A53 BLK SW	PVC 80	1/2 - 2
S80 A53B CS SW	PVC 80	1/2 - 2
S10 T304L SS BW	PVC 80	1/2 - 8
S10 T304L SS BW	PVC 40	1/2 - 8
S10 T0304L SS BW	CPVC 80	1/2 - 8
S10 T304L SS BW	FRP	1/2 - 12
S10 T304L SS BW	S10 T304L	1/2 - 12
S40 T316L SS BW	PVC 80	1/2 - 8
S40 T316L SS BW	CPVC 80	1/2 - 8
S40 T316L SS BW	FRP	1/2 - 12
S40 T304L SS SW	PVC 80	1/2 - 2
S40 T304L SS SW	PVC 40	1/2 - 2
S40 T304L SS SW	CPVC 80	1/2 - 2
S40 T304L SS SW	FRP	1/2 - 2
S10 T316L SS SW	PVC 80	1/2 - 2
S10 T316L SS SW	CPVC 80	1/2 - 2

IPEX
Committed to Excellence

Double Containment Piping Systems: Design Considerations

- **Material Selection**
  - Primary Pipe
    - Chemical Compatibility
    - Pressure Capabilities
    - Thermal Properties
  - Secondary Pipe
    - Chemical Compatibility
    - Loading (Static & Live) for buried applications
    - Hanging requirements for above ground
    - Leak detection requirements
    - UV requirements

Double Containment Piping Systems: Design Considerations

- **Thermal Expansion & Contraction**
  - Two basic thermal relationships may be present:
    - When only one of the two systems experiences thermal expansion/contraction.
    - When both systems experience different magnitudes of thermal expansion/contraction.
  - The Floating System
  - The Restrained System

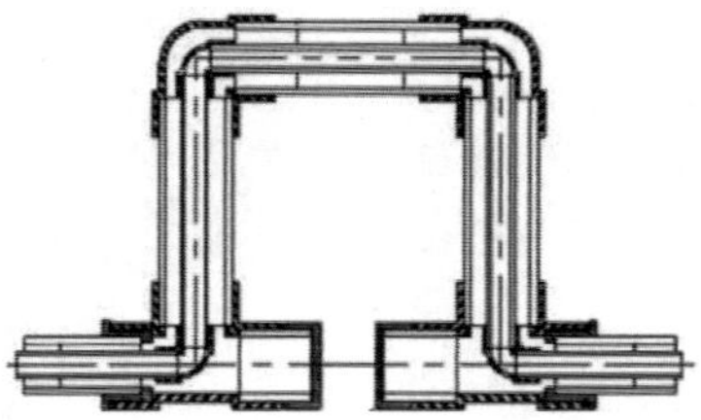

Double Containment Piping Systems: Leak Detection Requirements

- EPA requirement: (CFR), Title 40, Part 280
  - Pressurized systems require electronic leak detection that must be checked once per year.
  - Drainage systems (gravity flow) require visual leak detection with monthly manual inspection for a media leak.

Double Containment Piping Systems: Leak Detection Requirements

- The following forms of leak detection are acceptable as per regulations
  - Visual (drainage applications only)
  - Electronic low point probes
  - Continuous cable

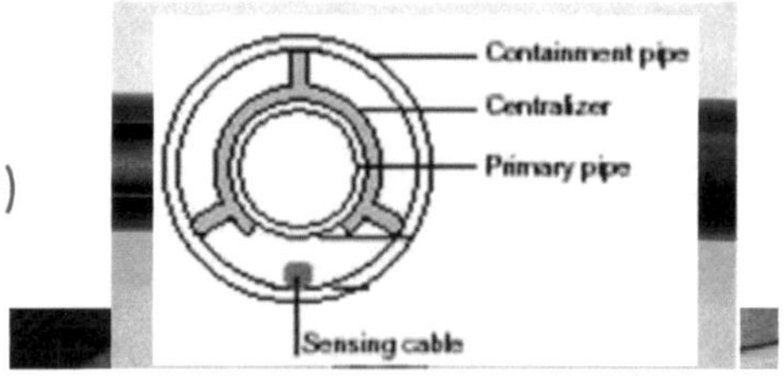

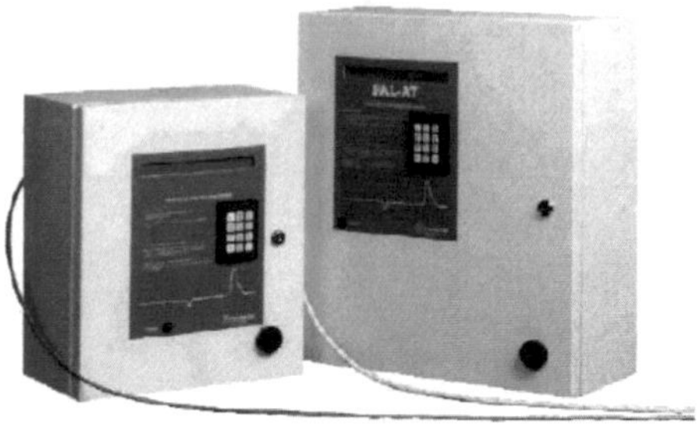

Double Containment Piping Systems: Applications

- Underground Applications
- WTP & WWTP
- Oil & Gas
- Hospitals
- Museums
- Food and Beverage
- Pharmaceuticals
- Laboratories
- Data Centers

IPEX
Committed to Excellence

Double Containment Piping Systems: A Fail-Safe, Leak-Free Solution

Questions?

Thank you!

DTP Process

"Now Ready to License"

JGC Corporation
Mitsuo Morita

2014

JGC CORPORATION

MITSUBISHI CHEMICAL

What is DTP Process ?

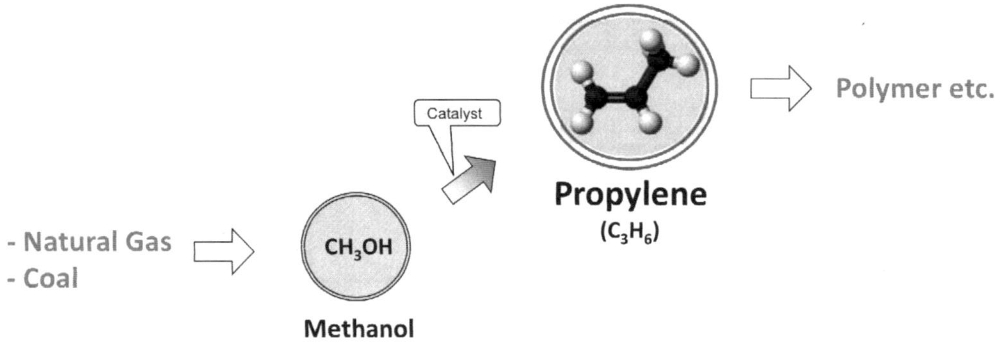

-On-Purpose Propylene Production from Methanol

JGC CORPORATION

- 2 -

MITSUBISHI CHEMICAL

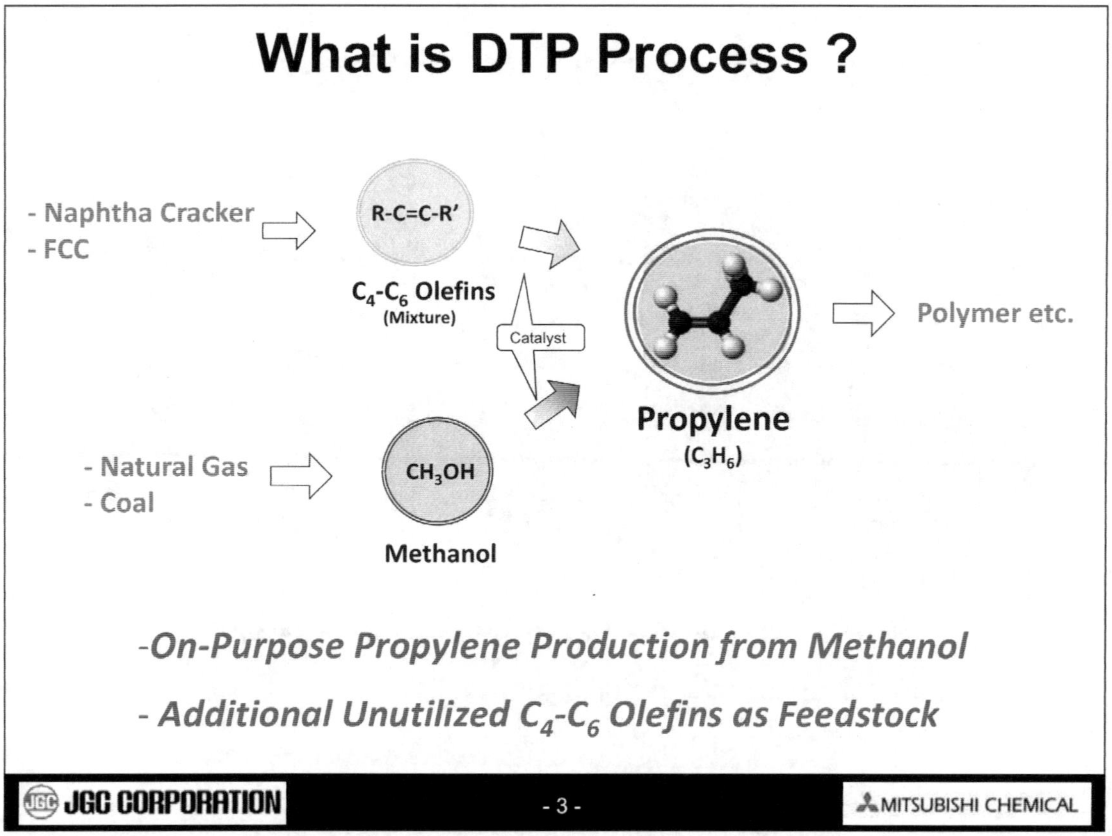
What is DTP Process ?
- Naphtha Cracker
- FCC
R-C=C-R'
C_4-C_6 Olefins
(Mixture)
Catalyst
Propylene
(C_3H_6)
Polymer etc.
- Natural Gas
- Coal
CH_3OH
Methanol
-On-Purpose Propylene Production from Methanol
- Additional Unutilized C_4-C_6 Olefins as Feedstock
JGC CORPORATION
- 3 -
MITSUBISHI CHEMICAL

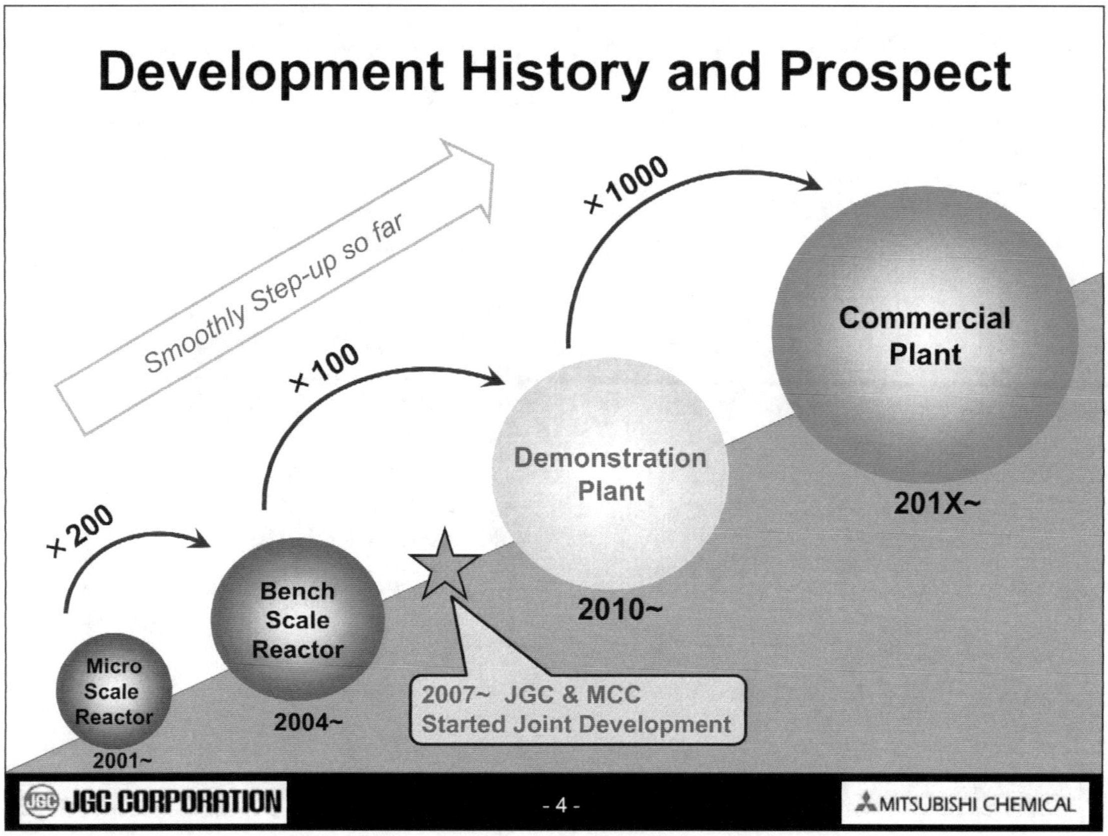
Development History and Prospect
Smoothly Step-up so far
×1000
×100
×200
Commercial Plant
201X~
Demonstration Plant
2010~
Bench Scale Reactor
2004~
Micro Scale Reactor
2001~
2007~ JGC & MCC
Started Joint Development
JGC CORPORATION
- 4 -
MITSUBISHI CHEMICAL

Demonstration Plant Overview

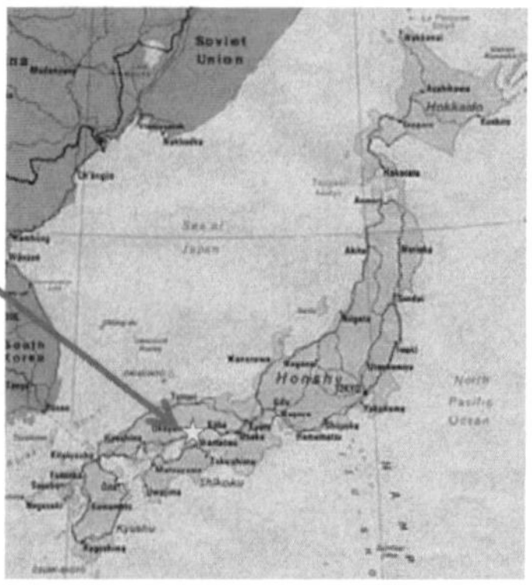

- **Located in Mitsubishi Chemical Mizushima Plant**
- **Successfully Start-up in August 2010**
- **Continuous Operation for Two Years**

Process Flow (Standalone Case)

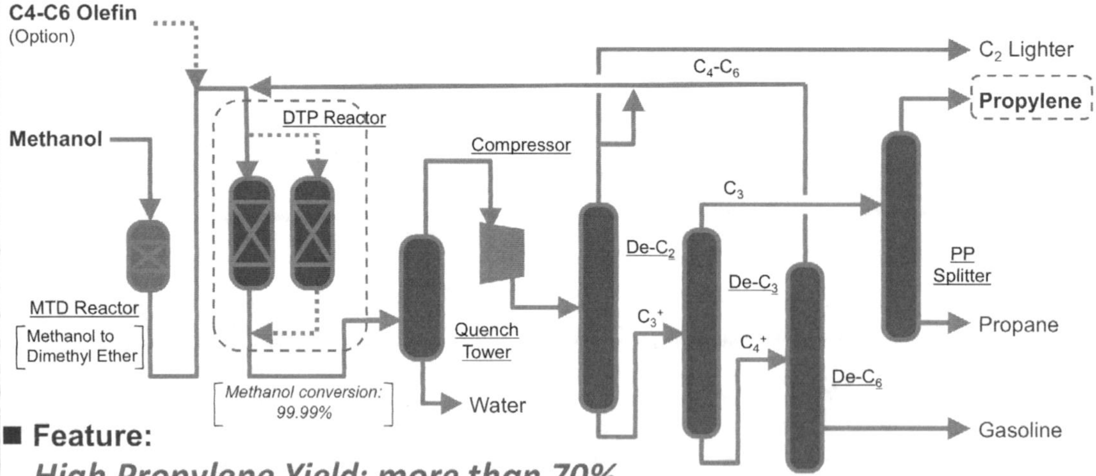

■ **Feature:**

- ***High Propylene Yield: more than 70%***
- ***Applied Existing Proven Technologies***

Process Flow (Integration Case)

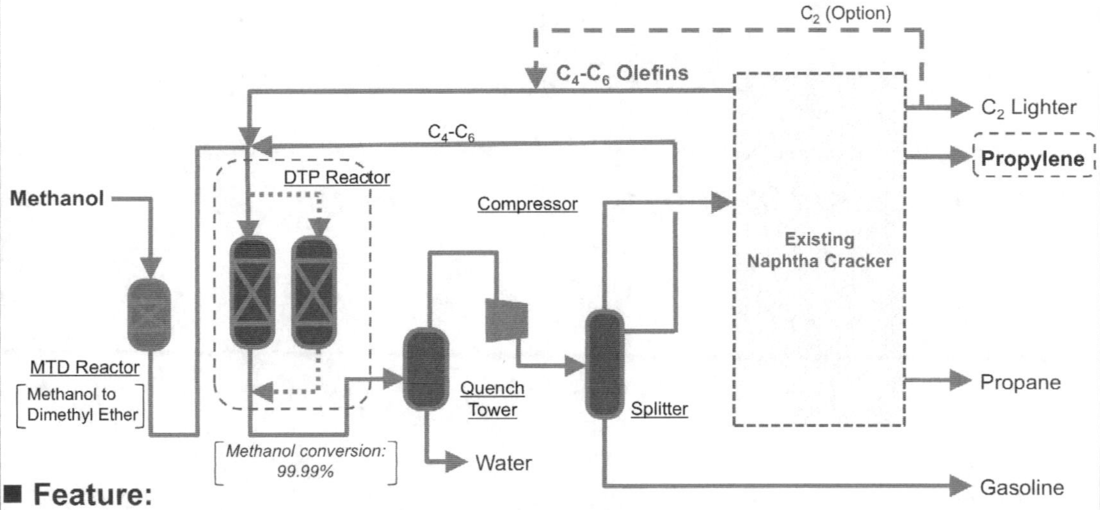

■ **Feature:**

- ***High Propylene Yield: more than 70%***
- ***Applied Existing Proven Technologies***
- ***Low Investment Cost in case of Integration with Naphtha Cracker***

JGC CORPORATION - 7 - MITSUBISHI CHEMICAL

Application

(1) Naphtha Cracker Integration Case (Propylene Capacity: 20KTA – 200KTA)

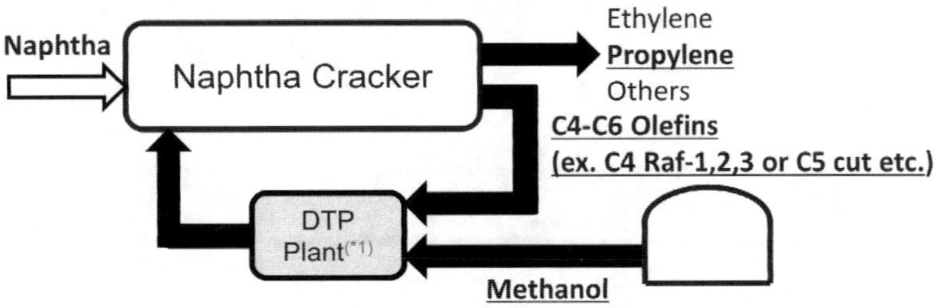

(*1) Utilizing facilities in Naphtha Cracker.

(2) Standalone Case (Propylene Capacity: 250KTA – 500KTA)

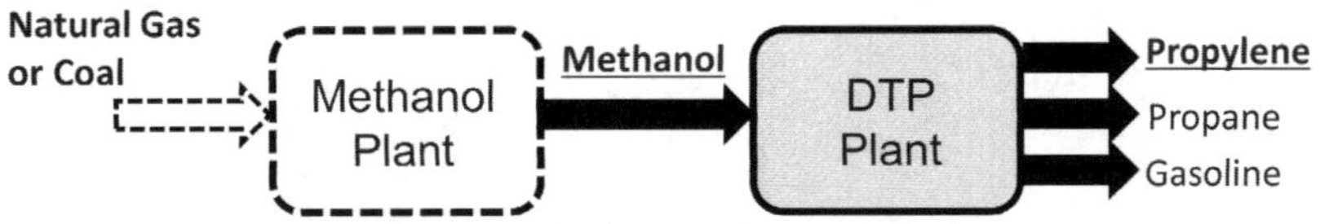

- ***Two Cases available depending on Client requirements !!***

JGC CORPORATION - 8 - MITSUBISHI CHEMICAL

DTP is...

- *High Yield Propylene Production Process*
- *Flexible Feed Methanol to Olefin Process*
- *Process that increase P/E Ratio for Cracker*

Thank you very much !!

Contact:
Mitsuo Morita
Market Development Dept.
Global Marketing Division
2-3-1, Minato Mirai, Nishi-ku,
Yokohama 220-6001, Japan

Phone: 81-45-682-8371
FAX: 81-45-682-8183
E-Mail: morita.mitsuo@jgc.co.jp

352533 Self-Cleaning "Bernoulli" Type Filters Used in Onshore and Offshore Applications

Abstract:

Georg Schuenemann GmbH (SAB-BREMEN) has developed a new Filter Type F-480, dubbed the "Compact Giant" for applications in onshore water intake and cooling water systems, sea water desalination plants, and in offshore projects, such as FPSO's, Drillships, and Production Platforms.

This paper will describe the essential design features of the filters, based on the "Bernoulli principle" which result in a fully automated self-cleaning filter system, and choice of materials in a sea water service with high corrosion rates.

A case study will be presented in which stringent project requirements such as substantial flow rates, adaptibility to space limitations and pipe connections had to be observed. Reliability of the system, extended run times without the need for frequent replacement of filter elements, as well as ease of monitoring and operator involvement have been achieved.

352653 Effects of Support on Sulfur Tolerance and Regeneration of Pt Catalysts Measured By Ethylene Hydrogenation and EXAFS

Abstract:

We investigated the effect of support on sulfur tolerance and regenerability under reducing environments by measuring the rate of ethylene hydrogenation, combined with hydrogen chemisorption, and extended x-ray absorption fine structure (EXAFS). Alumina and titania (P25) supports were used to prepare Pt catalysts at 1 wt.%. The catalysts were tested after treatment in H_2S/H_2 at 250 °C followed by regeneration treatments in H_2 at 250 °C, 350 °C and 450 °C. The sulfided catalysts exhibited a 20-27 times decrease in the rate of ethylene hydrogenation and a 4-6 times drop in the Pt surface area. Regenerations up to 450 °C were unable to remove all the sulfur, as shown by the presence of Pt-S bonds by EXAFS. However, a partial recovery of the hydrogenation rate per mole of Pt was observed on sulfided Pt/Al_2O_3 after reduction at 450 °C, while the induction of strong metal support interactions (SMSI) at reduction temperature above 350 °C was observed on Pt/P25, regardless of the presence of sulfur. For Pt/P25, the reversal of the SMSI state together with sulfur removal by mild oxidation suggests that sequential reduction/oxidation treatments may be more effective in restoring the S-free state of TiO_2-supported catalysts.

352799 Improved Operational Efficiency and Reliability through Insulation Materials Selection

Abstract:

This paper will explore the implications of insulation materials selection on the energy efficiency and reliability for Ethylene production units. Considerations will include insulation properties, structural support, installation methods, installation and maintenance costs.

BOILER MACT COMPLIANCE: MERCURY IN FUEL GAS

PATRICK LAINE, WES YONGE, & VIC VICKERY
PEI, HOUSTON, TX 77043

Patrick Laine, Ph.D.
Chief Scientist
Houston, TX
850-866-6686
plaine@pei-tx.com

Wes Yonge, PE
Chief Engineer
Houston, TX
713-806-8396
wyonge@pei-tx.com

James 'Vic' Vickery
Chief Technology Officer
Houston, TX
281-536-0899
vvickery@pei-tx.com

1 Introduction & Objectives

For most ethylene producers the primary objective of this sampling and analysis effort will be to verify and "establish a mercury fuel specification through which gas-fired boilers that use a fuel other than natural gas or refinery gas may be considered Gas 1 unit" as identified in 40 CFR Part 63.7521.(f).

Measurement and Monitoring Requirements: Decision Rules

- ❑If Hg concentration ≤ 50% of Gas 1 specification* then no further testing required
- ❑If Hg concentration > 50% and ≤ 75% of the specification* then semi-annual sampling required
- ❑If Hg concentration > 75% of the specification* then monthly sampling required

*Specification is **40 µg/Sm³**

The goal will be to execute a sampling and analysis campaign that will produce compliance quality data. The key to producing and validating compliance quality data is a well-designed and executed QA/QC protocol. At a minimum, a performance based QA/QC protocol is recommended that includes the following components:

- Sorbent Trap Breakthrough Bed:
  - ❑ Sorbent traps consisting of a minimum of two sorbent beds in series so that the trailing bed can be analyzed to verify that no breakthrough occurred from the leading bed.
- Duplicate Samples:
  - ❑ The sampling system should be configured with two trains to accommodate the collection of simultaneous duplicate sorbent trap samples, in parallel, providing a means of verifying precision in the sample collection process.
- NIST (National Institute of Standards and Technology) Traceable Spike:
  - ❑ Sorbent traps with a third bed (downstream) that is spiked with a NIST traceable spiked mass of mercury serves as a matrix spike and provides a means of verifying that components in the matrix (the fuel gas) are not affecting the trapping efficiency of the sorbent trap.

2 Methods

Sorbent Trap Mercury Sampling Methods Comparison

Method	Description	Method Type: S = Sampling A = Analysis	Evaluation	Sample Collection Phase QA/QC
ISO 6978	Hg in Natural Gas by Thermal Desorption (CVAFS)	S & A	Method calls for low sample volumes can lead to non representative results.	No
EPA Method 30B	Determination of Gaseous Hg Emissions from Coal Fired Combustion Sources using Carbon Sorbent Traps	S	Requires modification (Modified M30B)	Yes
ASTM D6350	Hg S&A in Natural Gas (AFS)	S & A	Method calls for short term samples (i.e., low sample volume) which leads to poor accuracy.	No
ASTM D5954	Hg S&A in Natural Gas (AAS)	S & A	AAS is less sensitive than AFS.	No
EPA Method 1631	Measurement of Hg in Water	A	No	Yes

3 EPA Method 30B QA/QC Performance Criteria

QA/QC Test or Specification	Acceptance Criteria	Frequency
Analyzer calibration	Correlation Coefficient: 0.999	Daily
Transfer efficiency of analyzer	%Transfer Efficiency: 90%–110%	Once per sorbent trap
Initial calibration blank	< 0.5 ng	Once before each calibration
Sorbent Trap Section B breakthrough	≤ 10% of Section A Hg mass, for Hg concentrations > 1µg/dscm ≤ 20% of Section A Hg mass, for Hg concentrations ≤ 1µg/dscm	At each sample point
Spike Recovery Test	Average recovery between 75% and 125 % for Hg_0	Once per sample point
Paired Sorbent Traps Agreement	≤ 10% RD for Hg concentration > 1µg/dscm ≤ 20% RD or 0.2 µg/dscm absolute difference for Hg concentrations ≤ 1µg/dscm	At each sample point

4 Equipment

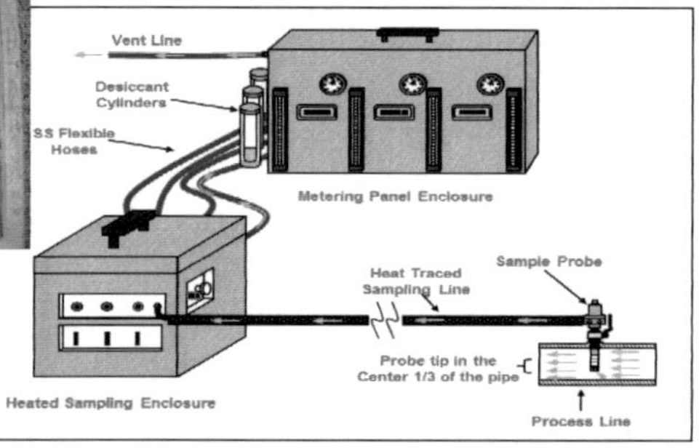

Flow

QA/QC Acceptance Criteria:	99% of Total Collected Hg	≤ 10% of Primary Mass	Recovery between 75% and 125%

Gold Trap:

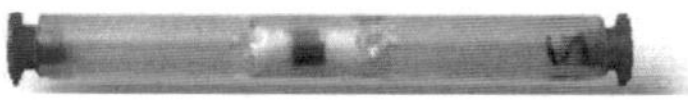

5 Conclusions/PEI Recomendations

- ❑ There are five methods accepted by the Boiler MACT rule. PEI has extensive experience and expertise employing **all five** of these methods.
- ❑ Operators can achieve compliance without submitting a Fuel Analysis Plan by using ASTM D6350, however, we have noticed variable results with this method.
- ❑ The **PEI recommendation** for obtaining the most accurate compliance quality data is to use Modified EPA Method 30B in conjunction with EPA 1631 where traps should be analyzed in a certified ISO 17025 laboratory. The recommended pathway includes a rigorous, performance based QA/QC protocol that validates accuracy and precision of all sample collection procedures.

351376 Young Professional Simulation Tutorial

Abstract:

18C00 Young Professional Simulation Tutorial

Authors: Naomi Hua and Mike Donahue, Invensys

Process simulation is a complex skill to master. Even with years of experience, a typical user is constantly facing new problems that need to be addressed. There are many issues that can lead to unexpected and unnecessary errors with the use of process simulation tools. Regardless of what simulation program you are using, these issues are common from one program to another. In this presentation, targeted for the young professional, we will discuss the "Most Common Pitfalls in Process Simulation". For each topic, we will discuss various techniques and methods these troublesome problems and prevent their occurrence.

Topics to be discussed include:

1. Erroneous data entry
2. Simulation defaults
3. Bad plant data
4. Thermodynamic method selection

5. Simulation tolerances and convergence

6. Recycle identification and handling

7. Model complexity

8. Difficult or impossible specifications

9. Assay pseudo-component development

10. Assay property blending

1. Erroneous data entry is the most common mistake made when building a process simulation model. This presentation will cover methods that can help ensure proper data entry in your process simulation model, such as checking the data you entered, particularly the units of measure, using "alternate forms" of data input such as copy/paste and data import, and reviewing all warning and error messages.

2. Defaults are methods, values, and conditions that are automatically selected for you by the program when you begin a new simulation model. These defaults will affect the results calculated by your model and every simulation program has them. Most defaults within simulation programs fall under categories such as convergence criteria, thermodynamic methods, tolerance and units of measure. The problem with defaults is unintended results from not fully understanding the effects a particular default has on the model. Take the time to understand the defaults of the program and how these defaults will affect the results. If you do not agree with the defaults selected by the program, most simulation programs will allow the user to change the default.

3. Plant data changes over time. The feed information and process conditions of the unit operations may be different. Not only does plant data change over time, but measurement information is often very inaccurate. This could be due to faulty or poorly calibrated measurement equipment, leaks in lines, sensor failures, and other issues. Because of these changes and inaccuracies over time, you always have to validate the information that is entered in your model to ensure that you are obtaining the proper results. To do so, you can utilize programs that can help with data validation and reconciliation like SimSci's ROMeo online optimization program. With it, you can connect your model to your data historian to constantly import current data measurements into your model and then reconcile these measurements to determine when problems arise within your plant and where those problems are occurring. You will also need to use your engineering judgment to determine if the data is valid. More details will be provided during the presentation.

4. The proper selection of thermodynamic methods for your simulation is the most important decision that you make when modeling a process. The method you select could have a drastic effect upon your results. Often, any number of thermodynamic methods is suitable for a given application. It is up to the user to determine the best method to use based upon knowledge, personal experience, and best judgment. Some important factors that need to be considered are the components being used and the operating conditions.

Oftentimes, the selected thermodynamic method still needs to have additional inputs such as binary interaction parameters to simulate the process accurately.

5. Many of the calculations in process simulation are iterative. When these calculations iterate, the program compares the results calculated from one iteration to the results calculated in the next iteration in order to determine

whether it is converged to a solution or needs to continue calculating. This method of comparing results is where tolerances come into play and simulation programs use tolerances to determine when a solution is reached. One of the major mistakes when modeling a process is not understanding or using the correct tolerance values. A looser tolerance may make the file easier and quicker to converge, but it also reduces the accuracy of the model. A tighter tolerance may be more accurate, but might make it impossible for the model to solve or increase the convergence time exponentially. A common practice is that when you have reached a converged solution with your model, tighten the tolerances within the model and solve it again. If you notice that your answers have changed significantly, you need to continue tightening the tolerance values until your results change little from one iteration to another. You will then know that you have reached an accurate solution for your model.

6. Recycles in process simulation occur when you return to a previously calculated unit operation to recalculate the unit due to a change made elsewhere in the model. They are very common and can often be difficult to converge. Some suggestions on troubleshooting recycle convergence include initially solving your model without the recycles to see how it behaves, providing initial estimates for recycle streams, using tighter tolerances on unit operations within recycle loops than the total recycle tolerances themselves, and ensuring that all components have a way in and out of a recycle by performing mass balances and using purge or makeup streams where necessary.

7. The general rule of thumb when modeling is to start simple and add complexity as you go. Each time you add a few streams or unit operations converge and save the model before moving on. Make sure you justify any additions that you make to the model. You can most likely model your system without including every unit from your PFD in the simulation. When modeling, think about the goal of your simulation and not about literally translating every real-world piece of equipment into the simulation model. Make the judgment call on what is most important and what the purpose is of every unit you add to your simulation. For example, if your real-life process shows a heat exchanger, a pump and a splitter, but all that is necessary is that the temperature, pressure, and separation are achieved properly, consider using a flash drum to accomplish all three.

8. Specifications within process simulation are everywhere and every stream and unit operation you place in your model contains specifications. In its simplest form, a specification is a constraint that you place on the model that the model needs to meet in order to successfully converge. For streams, these are usually temperatures, pressures, flowrates and compositions. For unit operations, they range from process conditions, like temperatures and pressures, to product conditions, like an overhead composition in a column. There are some pitfalls, however, that you can run into with specifications such as azeotropes in columns, high purity specifications, and the possibility of multiple solutions.

9. Petroleum assays are sets of distillation data that are used to characterize all of the components that reside in a given crude sample. When you enter distillation data into a process simulator, the program converts it into components that it can use when modeling your process. The components that are generated are called pseudo-components, and each pseudo-component represents all of the unknown components that would be within a given boiling point range in the crude. Some of the things you will want to ensure when properly developing your assay components are did you use the correct distillation curve basis, did you enter the data correctly, did you use enough cuts and are you using the correct characterization methods .

10. When you have more than 1 assay in a simulation model, you need to be concerned with assay blending and how it will affect the results of your file. Assay blending is the mixing of 2 or more groups of components (i.e. ¨C sets of cuts/pseudo-components) in order to calculate averaged component properties for cuts that are in the same boiling point range as one another. Once these new averaged component properties are determined, the program will use these averages from this point on in the simulation model any time it needs to access component

properties for a given pseudo-component. All component properties are blended, including normal boiling points, specific gravity, molecular weight and refinery inspection properties, and usually the blended component properties are calculated using a weighted average

These topics are based on experience and feedback from end-users to technical support representatives for process simulators. They often represent the most commonly asked questions or areas that most users find to be problematic. While the specific examples highlighted in each topic come from SimSci software, the intent of this tutorial presentation is to be broad-based and is particularly applicable to young professionals seeking to develop their experience with simulation.

Polymeric Sulfonic Acid Composites for Fuel Cell Applications: Effects of Casting Solvent and Sonication Time on Membrane Performance

*Jimoh K. Adewole[1], Abdullah S. Sultan[1, *], Amir Al-Ahmed[2], and S. M. Javid Zaidi[3]*

[1]*Center for Petroleum & Minerals, King Fahd University of Petroleum & Minerals, Dhahran 31261, Saudi Arabia*

[2]*Center of Research Excellence in Renewable Energy, King Fahd University of Petroleum & Minerals, Dhahran 31261, Saudi Arabia*

[3]*School of Chemical Engineering, The University of Queensland, Australia*

Abstract

Polymer blends comprising sulfonated poly (ether ether ketone) (SPEEK) and 3wt% (2-acrylamido-2-methyl-1-propanesulfonic acid) (PAMPS) were prepared by solution casting method. Two solvents (N,N-Dimethylacetamide (DMAc) and N-Methyl-2-pyrrolidone (NMP)) were uesd separately. Residual solvents in the membrane samples after drying were quantified using thermo gravimetric analysis. Effects of periods of sonication (0 to 90 minutes) and types of solvent on the membrane performance were evaluated interms of morphology and water uptake. An attempt was made to correlate membrane performance with the time of sonication in order to select the most suitable solvent and membrane casting conditions for subsequent membrane preparations. For both solvents, the water uptake was observed to first decrease at 10 minutes of sonication and then increased as the sonication time was increased. The minimum water uptake recorded for DMAc samples was 19.23% at 20 minutes sonication time while that of NMP sample was 24.24% at 10 minutes. However membrane morphology was found to deteriorate with increase in sonication time. No correlation was obtained between the membrane properties and the sonication time. In general, membrane performance was better for membranes samples using DMAc than NMP solvent. Therefore, DMAc was selected as the casting solvent for membrane preparation on which further studies will be carried out.

Keywords: morphology, water uptake, residual solvent, thermogravimetric analysis

Introduction

Water soluble polymers have received academic and industrial interests due to their potential applications in fuel cells and other electrochemical devices (Ye et al. 2012). One of the water soluble polymers that have attracted publication interest is poly (2-acrylamido-2-methyl-1-propanesulfonic acid) (PAMPS) (Florjańczyk et al. 2003). Generally, some of the major performance criteria for fuel cell materials include proton conductivity, water and methanol uptake, good mechanical properties, chemical stability, thermal stability and methanol permeability (Gil et al. 2004; Swier et al. 2005; Ye et al. 2012). PAMPS was reported to posses higher proton conductivity than partially hydrated Nafion. In addition, it has been found to be more tolerant to water fluctuation in high temperature operations. Some of the major drawbacks in the use of PAMPS are its poor mechanical properties, thermal stability, water and methanol uptake and solubility in many solvents (Florjańczyk et al. 2003). One of the approaches that have been proposed to solve these problems is development of polymer composites.

This research paper investigates the synergistic effect of combining water soluble PAMPS with sulfonated poly (ether ether ketone) (SPEEK) which is non - water soluble polymer to prepare composite membranes for application in fuel cells. Specifically, preliminary results on the performnance of the composite were presented. Performance were evaluated based on the composite morphology and water uptake. Effects of sonication time and types of casting solvent on membrane properties were also evaluated.

Material and Method

Material

Sulfonated poly (ether ether ketone) (SPEEK) purcahsed from FuMA-Tech, Germany, and poly (2-acrylamido-2-methyl-1-propanesulfonic acid) (PAMPS) purchased from Aldrich Chemical Co., are the two major polymers used in this research. The SPEEK has a DS of 0.533 and solubility of 99% in DMAc. SPEEK is derived by sulfonation of Poly (ether ether ketone) (PEEK). PEEK is a commercially successful polymer due to its excellent thermal stability and solvent resistance (Mohr et al. 1991). For sometimes, further application of PEEK have been stalled due this solubility issues. By sulfonation, PEEK can be made soluble in most commercial solvents. Importantly, the solubility of SPEEK also depends on its degree of sulfonation (DS). Since, this work involves the use of two polymers, a solvent which will disolve both the PAMPS and SPEEK must be selected. Literature review showed that PAMPS dissolves in N,N-Dimethylformamide (DMF), N,N-Dimethylacetamide (DMAc), propylene carbonate (PC), and water (Florjańczyk et al. 2003). Initial solubility studies that was carried out in our laboratory was done for N,N-Dimethylacetamide (DMAc) and N-Methyl-2-pyrrolidone (NMP). The choice of NMP was due its unique solvent properties and its structtural distiction from DMF and DMAc (Figure 1). Moreover, no literature was found where NMP was used as casting solvent for PAMPS. Thus, DMAc and NMP were used as solvents in this work. All materials were used as received from the manufacturers.

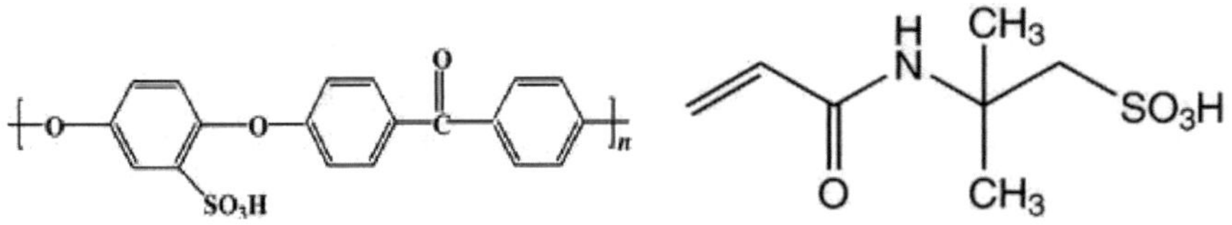

Figure 1 Structure of SPEEK and PAMPS

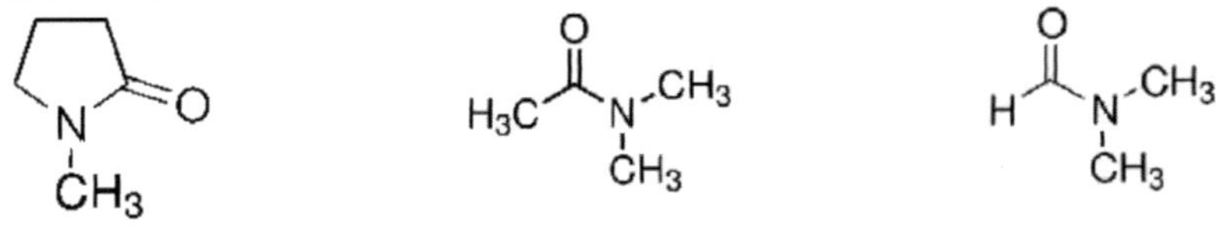

N-Methyl-2-pyrrolidone   N,N-Dimethylacetamide   N,N-Dimethylformamide

Figure 2 Structure of Solvents

Experimental Procedure

Membrane Preparation

SPEEK was first dissolved in solvents (DMAc and NMP) and stirred for 2hours at 70°C. PAMPS was then added and the blend was stirred for 24hours. The solution was sonicated for various periods of time ranging from 10 to 90minutes using an Ultrasonic bath. In untralsonic bath, the final temperature achieved for a given time of sonication depends on the tempearutre of the laboratory (summer or winter). It is therefoe genrally advisable that maximum temperature of the bath is determined and maintains (as the so - called equilibrium tempearture), when operating continuously under ambient conditions (Santos et al. 2009). During the sonication, equilibrium temperature was maitained by simply filling the bath with water that was preheated to sample temperature.

Membrane Characterization

Water uptake and morphological tests were performed. Water uptake was carried out by first vacuum drying all membrane samples at 100°C until constant weights were obtained. Dried samples were then immersed in deionized water for 24h at room temperature (25°C). Samples were taken out of the water after 24h, wipped to remove surface water and then weighed immediately. Water uptake was calculated using Eq. 1.

$$water - uptake = \frac{W_{wet} - W_{dry}}{W_{dry}} \quad \text{Eq. 1}$$

Surface morphology of the membrane was done using scanning electron microscopy (SEM). Thermogravimetric analysis (TGA) was done from room temperature to 350°C using TA Q600 analyzer at heating rate of 20°C/min under nitrogen atmosphere with flow rate of 20mL/min using ~ 8 -10mg membranes samples.

Results and Discussion

Membrane Morphology by Scanning Electron Microspcopy (SEM)

Results of surface morphological analysis performed on membrane were shown in Figure 3 and 4. The overall pictures of this figure depicts a dense membrane with some surface defects observed for most of the sonicated samples. For example, Figure 3A showed some kind of layered arrangement with no visible pores at the surface. Also, the PAMPS seems to be well distributed throughout the SPEEK. The micrograph in Figure 3B can be observed to be rough and possesses some uneven ramdom distribution of PAMPS agglomerates indicating that the distribution of PAMPS have been disturbed by the 10minutes sonication. An increase in sonication time to 20 minutes was observed to improve the distribution of PAMPS. This was however accompanied by some cracks believed to be due to the slight increase in the viscousity of the blend solution. The vicousity increase was brought about by the excape of some fraction of the solvent during sonication. As the sonication time was increased, PAMPS was observed to become more evenly distributed and the cracks become more visible and larger in size (Figure 3D to F). SEM images of samples at sonication time of 10 and 30 minutes were reproduced in higher magnification of 10μm (Figure 4) in order to further revealed the effect of sonication on the morphology of the somposites. Similar morphological behaviour has been reported in literature, though for inorganic/organic clusters of nanoclay/epoxy composites membranes (Lam et al. 2005).

Water Uptake

Water uptake have been reported to be affected by the degree of sulfonation, membrane pretreatment, hydration, ambient relative humidity and water temperature (Xing et al. 2004; Wu et al. 2006). Figure 5 showed the effect of sonication time and solvent type on water uptake that was measured at room temperature (25°C). For both solvents, water uptake is same at zero sonication time. There was however a further decrease in the water uptake for NMP at 20 minutes sonication time before it beging to increase. Thus, the best sonication time for both solvent lied between 10 and 20 minutes with respect to water uptake. Interestingly, this result correspond to that reported by Lam (2005) on the effect of ultrasonication on microhardness of composites.

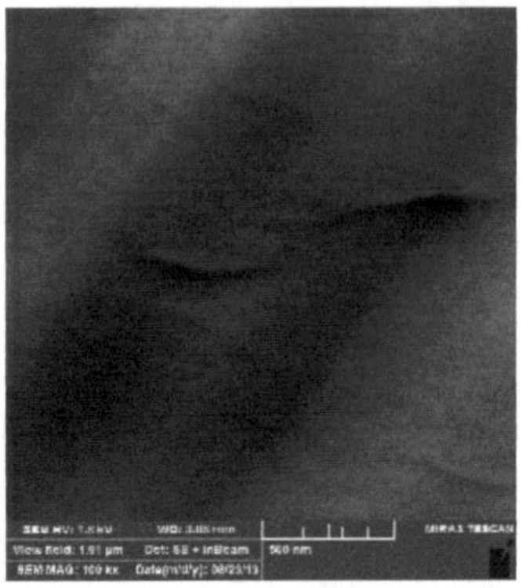

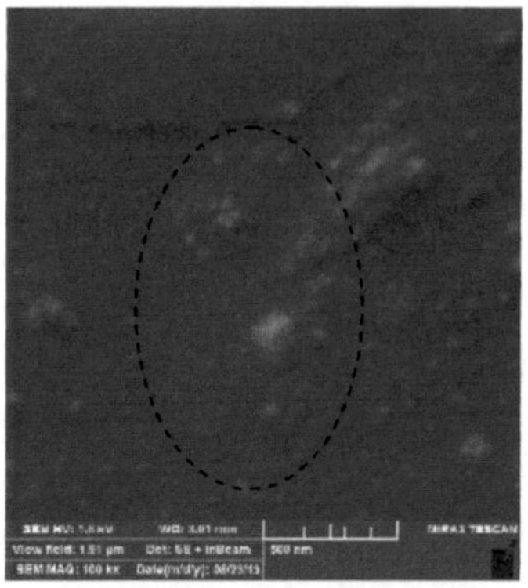

A) No sonication

B) 10 minutes sonication

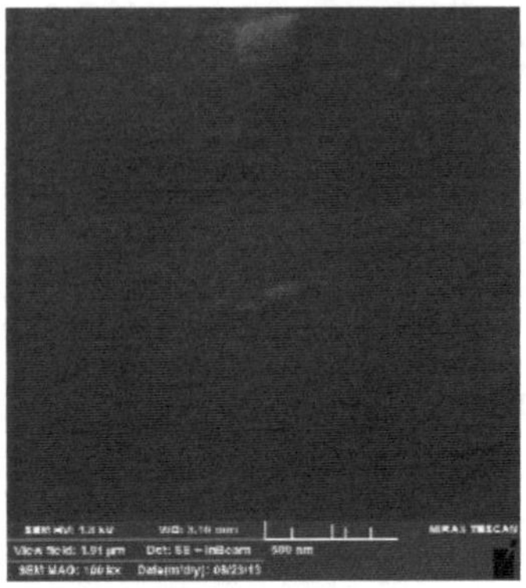

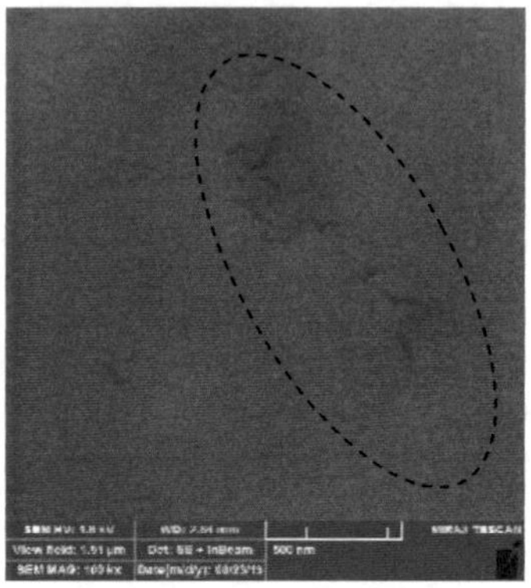

C) 20 minutes sonication

D) 30 minutes sonication

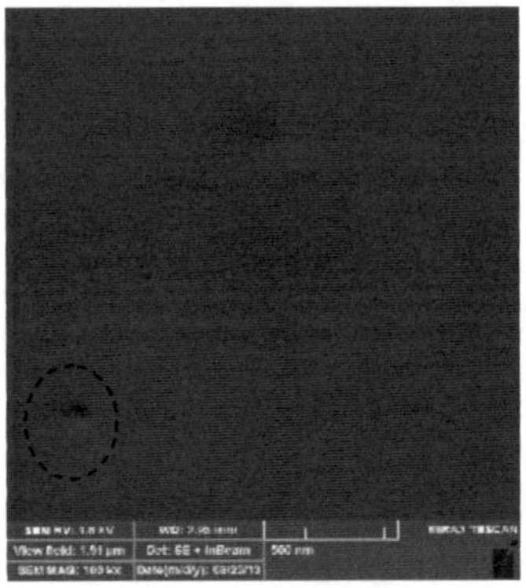

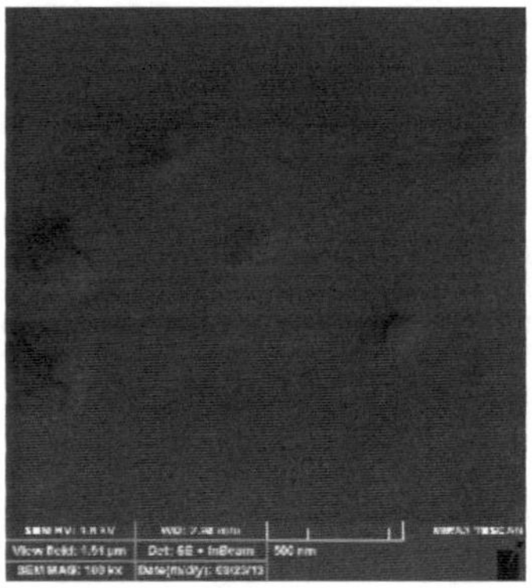

E) 60 minutes sonication

F) 90 minutes sonication

Figure 3 SEM micrograph of SPEEK-PAMS composite with various times of sonication with magnification of 500nm

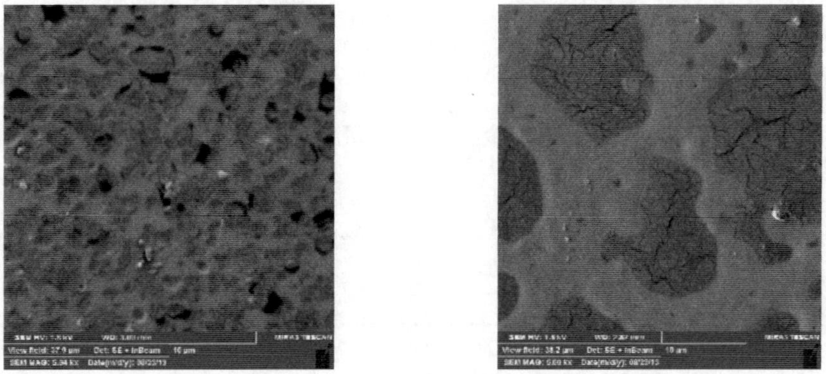

A) 10 minutes sonication B) 30 minutes sonication

Figure 4 SEM micrograph of SPEEK-PAMS composite with various times of sonication with magnification of 10μm

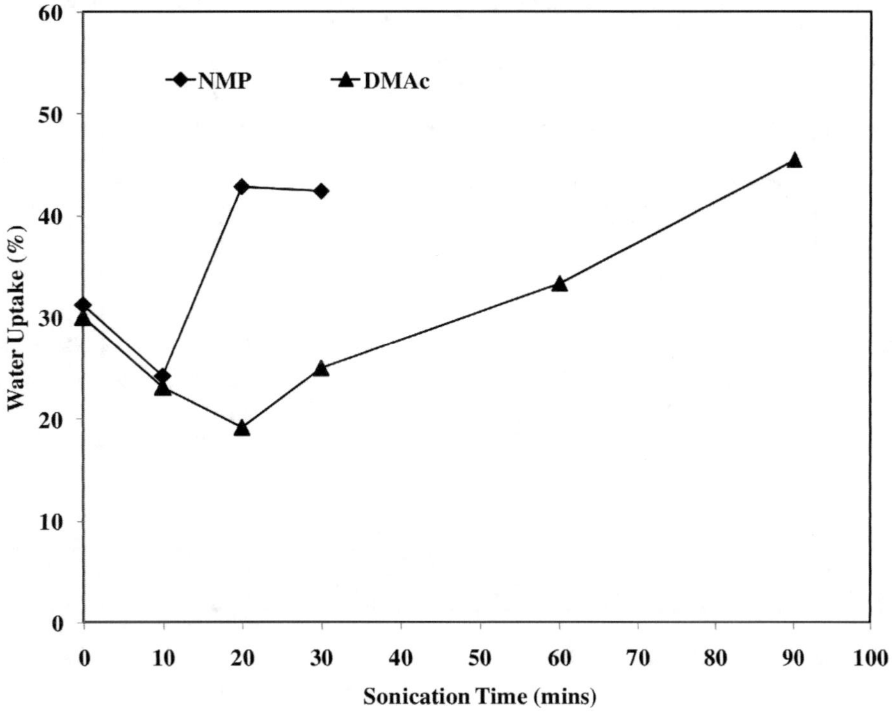

Figure 5 Water uptake of SPEEK-PAMS (3wt%) composite with variuos sonication times

Thermogravimetric analysis (TGA)

The difference in membrane performance resulting from the use different casting solvents have been attributed to the presence of residual solvents in the membrane samples after evaporation (Joly et al.

1999). TGA was thus used to quantify the residual solvent and evaluate its effect on the membrane performance as has been previously demonstrated in lieterature (Recio et al. 2007; Iqbal et al. 2008). Figure 6 shows the TGA curve for samples prepared from NMP and DMAc with different post vacuum heat treatment conditions. From this figure, weight loss can be attributed to evaporation of variety of substances, including solvents, from the samples (Rich et al. 1997). Table 2 contains the solvent and its boiling points, and weigth loss at the boiling point range that were extracted from TGA results. For sample heat treated for 24hrs, the overall percentage weight loss up to 204°C was about 8wt% of membrane sample for NMP while it is about 6.3wt% for DMAc. These will definitely include compnenents such as water, solvents and other unidentified substances. To quantify residual solvents, weight loss was estimated between 192 -214°C for SPEEK-PAMPS-NMP and between 154 and 176 for SPEEK-PAMPS-DMAc. The results obtained showed that the amount of the residual solvent in NMP is larger than that of DMAc by 1.25wt% (Table 1). To further investigate if this difference is responsible for higher value of water uptake for the NMP sample, membrane samples made from both solvents were further heat treated under vacuum for 72hrs. Residual solvent after this treatment was slightly higher in DMAc than NMP. Despite this, the value of water uptake of the NMP membrane is still higher than that of DMAc by approximately the same value as before the heat treatment (1.71wt%).

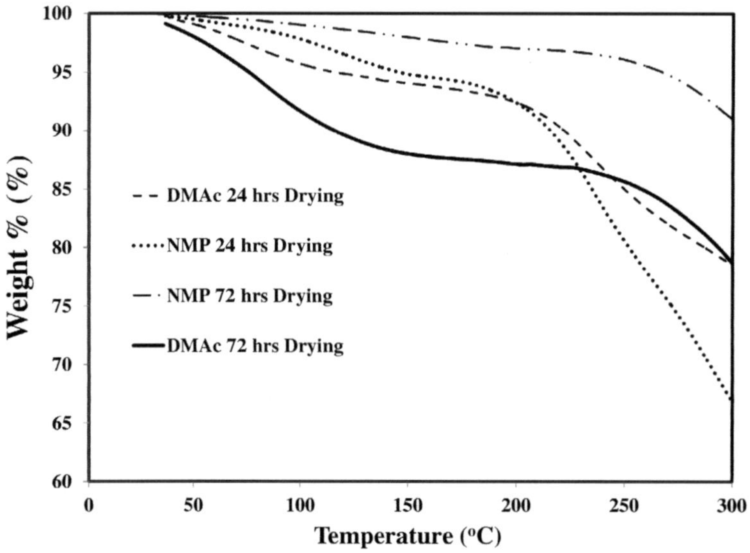

Figure 6 TGA of SPEEK-PAMPS-NMP and SPEEK-PAMPS-DMAc

Table 1 Residual solvent in SPEEK-PAMPS composites membrane samples

Membrane Sample	Solvent	B. Pt. (°C)*	Residual Solvent (wt%)	Water Uptake
DMAc 24 hrs Drying	DMAc	164.5 – 166	0.57	30.00
NMP 24 hrs Drying	NMP	202 – 204	2.85	31.25
DMAc 72 hrs Drying	DMAc	164.5 – 166	0.367	21.21
NMP 72 hrs Drying	NMP	202 – 204	0.217	22.92

*Boiling Points of Solvents

On a general note, results from Figire 1 – 5 showed that the best perfomance can be obtained for variety of sonication time depending the property of preference. Also, performance of DMAc solvent was observed to be better than that of NMP on the basis of water uptake. For this reason, DMAc was selected as the casting solvent for membrane preparation on which further studies will be carried out. Water uptake of DMAc membrane, prepared without sonication, was comapred with those reported in previuos literature. Table 2 illustrates the comparison between SPEEK-PAMPS and other relevant membranes previuosly reported in literature. In addition, this table serve to provide insight into the possible potentials of the composites. From this table, it was observed that the water uptake for SPEEK-PAMPS is within the range of values of other membranes on the table. Its water uptake is less than that of Nafion 112 which is a commercial membrane. The differences observed between the water uptake values of SPEEK – PAMPS and other membranes from the table can be explained based on three important factors - SO_3H group, the degree of sulfonation and experimental temperature (Gil et al. 2004; Shan et al. 2006; Wu et al. 2006). Water uptake of SPEEK – PAMPS is higher due to the presence of SO_3H group in the PAMPS molecules (Figure 2). SO_3H group have been reported to enhance membrane acidity and hydrophilic properties (Shan et al. 2006). Infact, the higher the SO_3H group the larger the water uptake. The degree of sulfonation has been reported to affect the water uptake of membranes. The higher the degree of sulfonation, the larger the value of water uptake (Gil et al. 2004). Thus, water uptake of SPEEK – PAMPS is lower than SPEEK -3 due their different degree of sulfonation, 53.3 and 72% respectively and the experimental temperature 24 and 60°C, respectively. The experimental temperature of SPEEK – 2 is close to that of the composite. However, the results cannot be compared in this case since the DS of SPEEK – 2 was not known. Above explanation is also true for the values of water uptake for SPEEK – 1.

Table 2 Comparison of water uptake values obtained with those from Literature (Xing et al. 2005)

Membrane	Water Uptake (wt%)	Experiment Temperature (°C)	**D S (%)	Reference
Nafion 112	22.85	N/A	-	(Xing et al. 2005)
SPEEK – 1*	16.00	30	45	(Ren et al. 2005)
SPEEK – 2*	44.90	25	N/A	(Zhong et al. 2008)
SPEEK – 3*	85.00	60	72	(Wu et al. 2006)
SPEEK– 4*	15.00	N/A	N/A	(Handayani and Dewi 2011)
SPEEK/PAMPS (3wt%)	21.21			This work

*The numbering have been included for explanation purposes.
**DS means Degree of Sulfonation

Conclusion and Recommendations

Blend of SPEEK and PAMPS was prepared by solvent casting method and its performance was evaluated with respect to its morphology and water uptake. Effect of sonication time (10 – 90minutes) and solvent types (DMAc and NMP) were also investigated. Performance of membrane prepared using DMAc was observed to be better than that of NMP. There was no correlation between the sonication time and membrane properties, but it was observed that the sonication deteriorate the membrane morphology. Membrane prepared without sonication shows better surface morphology, whereas the best water uptake was observed between 10 and 20 minutes. Thus, other properties need to be investigated in order decide the best sonication time. Therefore, further investigation such as thermal stability, methanol permeability, ion exchange capacity and gas (O_2 and H_2) permeability of the membrane are recommended for a complete evaluation of the membrane performance for direct methanol fuel cell (DMFC) application. All these investigations are in progress in our laboratory.

Acknowledgement

Authors would like to thank the National Science, Technology and Innovation Program (NSTIP) of KACST for funding this research under project number 10-ENE1374-04. Also, authors extend their appreciation to Center of Petroleum & Minearls, Center of Excellence in Renewable Energy, Mr. Dan Jerry D.Cortes of the Nanotechnology Laboratory, Chemical Engineering Department, King Fahd University of Petroeleum & Minerals, Saudi Arabis for their supports.

REFERENCE

Florjańczyk, Z., et al. (2003). "Proton conducting electrolytes based on poly(2-acrylamido-2-methyl-1-propanesulfonic acid)." Electrochimica Acta **48**(14–16): 2201-2206.

Gil, M., et al. (2004). "Direct synthesis of sulfonated aromatic poly(ether ether ketone) proton exchange membranes for fuel cell applications." Journal of Membrane Science **234**(1–2): 75-81.

Handayani, S. and E. L. Dewi (2011). "Influence of Silica/Sulfonated Polyether-Ether Ketone as Polymer Electrolyte Membrane for Hydrogen Fueled Proton Exchange Membrane Fuel Cells." Internat. J. of Sci. and Eng. **2**(2): 26-30.

Iqbal, M., et al. (2008). "Solvent effect on morphology and CO2/CH4 separation performance of asymmetric polycarbonate membranes." J. Membr. Sci. **318**(1–2): 167-175.

Joly, C., et al. (1999). "Residual solvent effect on the permeation properties of fluorinated polyimide films." Separation and Purification Technology **16**(1): 47-54.

Lam, C.-k., et al. (2005). "Effect of ultrasound sonication in nanoclay clusters of nanoclay/epoxy composites." Materials Letters **59**(11): 1369-1372.

Mohr, J. M., et al. (1991). "Gas transport properties of a series of poly(ether ketone) polymers." Polymer **32, 13**: 2387-2394

Recio, R., et al. (2007). "Gas separation of 6FDA–6FpDA membranes: Effect of the solvent on polymer surfaces and permselectivity." J. Membr. Sci. **293**(1–2): 22-28.

Ren, S., et al. (2005). "Surface modification of sulfonated poly(ether ether ketone) membranes using Nafion solution for direct methanol fuel cells." Journal of Membrane Science **247**(1–2): 59-63.

Rich, D. C., et al. (1997). "Effect of cure conditions on Probimide 32 polyamide-imide." Journal of Applied Polymer Science **63**(9): 1113-1126.

Santos, H. M., et al. (2009). The Power of Ultrasound. Ultrasound in Chemistry: Analytical Applications. J.-L. Capelo-Martinez. Weinheim, Wiley-VCH.

Shan, J., et al. (2006). "Sulfonated polyether ether ketone (PEEK-WC)/phosphotungstic acid composite: Preparation and characterization of the fuel cell membranes." Pure Appl. Chem **78**(9): 1781 - 1791.

Swier, S., et al. (2005). "Sulfonated poly(ether ketone ketone) ionomers as proton exchange membranes." Polymer Engineering & Science **45**(8): 1081-1091.

Wu, H.-L., et al. (2006). "Preparation and characterization of poly(ether sulfone)/sulfonated poly(ether ether ketone) blend membranes." European Polymer Journal **42**(7): 1688-1695.

Xing, D. M., et al. (2005). "Characterization of Sulfonated Poly(Ether Ether Ketone)/Polytetrafluoroethylene Composite Membranes for Fuel Cell Applications." Fuel Cells **5**(3): 406-411.

Xing, P., et al. (2004). "Synthesis and characterization of sulfonated poly(ether ether ketone) for proton exchange membranes." Journal of Membrane Science **229**(1–2): 95-106.

Ye, Y.-S., et al. (2012). "Water Soluble Polymers as Proton Exchange Membranes for Fuel Cells." Polymers **4**(2): 913-963.

Zhong, S., et al. (2008). "Modification of sulfonated poly(ether ether ketone) proton exchange membrane for reducing methanol crossover." Journal of Power Sources **180**(1): 23-28.

10TH GLOBAL CONGRESS ON PROCESS SAFETY

Fixed Plant Facility Siting – A Mega Project Lifecycle Framework

M.F. Haider
mech_haider@yahoo.com
mhaider@suncor.com

Prepared for Presentation at
American Institute of Chemical Engineers
2014 Spring Meeting
10th Global Congress on Process Safety
New Orleans, LA
March 30 – April 2, 2014

Fixed Plant Facility Siting – A Mega Project Lifecycle Framework

M.F. Haider
mhaider@suncor.com
mech_haider@yahoo.com

Keywords: Facility Siting, Project Management, Mega Project, Consequence Analysis (CA), Quantitative Risk Assessment (QRA), Adaptive, Change driven, Agile methodology

Abstract

Facility Siting based on tools such as Consequence Analysis (CA) and Quantitative Risk Assessment (QRA) is often times successfully applied for Brownfield operating unit projects, projects with relatively smaller footprints and for projects lasting relatively short time-periods. Challenge remains to apply and realize the full potential and value of Facility Siting for Mega-Projects. This is mostly due to project complexity and length – this paper will introduce a Management Framework to successfully apply CA and QRA as part of Facility Siting for Mega-Projects. The paper will also outline key strategies to ensure successful (i.e. value adding) implementation of facility siting, suggest a list of activities, associated sequence and pitfalls to avoid. Project Management team (e.g. Project Sponsor, Project Engineers, Managers) of both owner/operator and contractor/vendor may find this paper more useful as it provides a broad overview (i.e. when to initiate, how to plan, execute and close such assignment to maximize value) including real life examples of facility siting process over the project lifecycle.

1. Introduction

The driving force for process safety has been primarily based on catastrophic events [1]. With an increasing number of incidents, the process industry and governments started to take measures such as Facility Siting initiatives with hope to minimize loss of life and property. In the 1970s and 1980s, some of the world's most tragic industrial accidents took place. The Flixborough explosion in 1974 was reinforced by that of the Seveso tragedy in 1976 [2]. The Macondo blowout in the Gulf of Mexico on April 20, 2010, killed 11 employees and led to an uncontrolled oil spill lasting 87 days. This blowout was the most significant offshore incident in the US to-date. Finally, the Fukushima Daiichi nuclear plant incident in Japan in March 2011 drew the attention of the global process and power industries and encouraged them to incorporate natural disaster risks in a hazard analysis study [3]. As part of inherently safer design, many companies now-a-days apply CA and QRA as part of facility siting for their new projects. Due to the scarcity of information at the beginning, the studies in many cases start later in the project lifecycle for them to be effective and value adding. This paper will focus more on how to manage such initiatives for new Mega-Projects over the technical intricacies of the CA/QRA processes. The intention of the paper is to enhance the Body of Knowledge publicly available for management team in as much non technical languaga as possible to make their greenfiled Mega-Project initatives a success i.e. safer for people, environment and property.

So, what is different for industrial Mega-Projects?

- Industrial Megaprojects, as the name implies, are very large. To provide a simple and simply applied definition, we are defining a megaproject as any project with a total capital cost of more than $1 billion (U.S. dollars) as measured on January 1, 2003 [4].
- Most often engineering and construction activities run in parallel and not in sequence
- Multiple sub-projects under the 'Mega-Project umbrella' run in parallel
- Complex – design is almost 'never' complete in early stages – goes through many iterations
- Progressive elaboration is evident
- Scarcity of critical workforce with mega-project experience
- Scarcity of information – requires project team to make assumptions for CA/QRA
- Timeline: 2-5 years

2. Key Strategies

2.1 Facility Siting Charter

It is not only important or good to do but critical that the stakeholders need to identify, understand and agree on the purpose and value to conduct facility siting for their projects. The Author's personal experience over the years has been in order to maximize value of Facility Siting the project team has to start 'early' to develop project Charter and start with the end in mind. As can be seen from the following generalized Figure 1 that as the company starts to appraise opportunities, it is

Figure 1: Stages and Milestones to apply CA/QRA in a Mega-Project

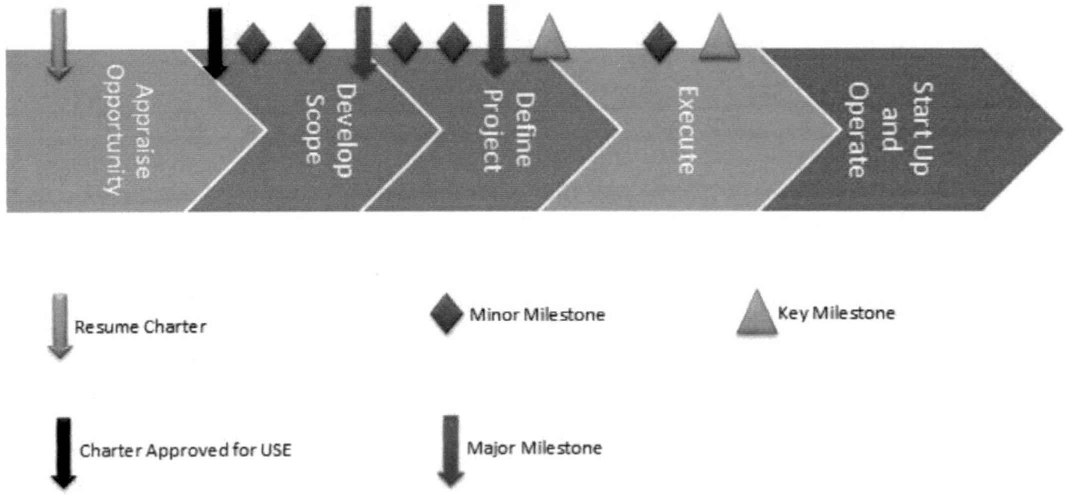

prudent to resume developing the Facility Siting Charter by bringing in the stakeholders together. Figure 1 shows a generalized project lifecycle for a typical Petro-Chemical Mega-Project. The

Figure also shows that the Key Deliverable is developed through a successive series of repeated cycles over the project life-cycle. The Milestones will be defined later in the paper. Part of starting with the end in mind is to clearly outlining ultimate goal/s, e.g. vision is to design a safer facility and utilize Mitigated QRA Report to formally document and communicate that facility is within the Risk Threshold or justification for accepting risk profile as is by Senior Management. The Vision in the Charter also provides opportunity to strike a balance between business goals and climbing the Risk Mountain – at each stage of increasing safety (decreasing risk). As part of striking a balance, the Project Engineer has to Identify options up-front if and/or when it becomes apparent that aggregate Residual Risk will not be within the Risk Threshold or acceptable so that team can put more effort in developing justification for either accepting or rejecting the risk profile by Management

Figure 2: Strike a balance between Business goal and Climbing the *expensive* Risk Mountain

2.2 Management Strategy

In order to successfully manage Facility Siting, CA and QRA for a Mega-Project, the stakeholders have to align on a management strategy. There are two primary management approaches for the purpose of managing such work:

1. Predictive or Plan driven
2. Adaptive or Change driven

The Author's past experience has been if the organization undertook very 'similar' projects in the near past (<5 years), critical experienced workforce is retained, Lessons Learned is readily available, and most importantly, the past project is still considered to be a success, it is advisable to use Plan driven approach. On the other hand, if the organization is introducing new technology, has limited past experience with similar project, involves relatively high number of stakeholders, capital cost > $1B, project life spans >2 years, it is recommended to utilize Change driven approach to maximize value. It is to be noted that it is not essential to meet all the suggested criteria mentioned above to pick one management strategy over the other.

2.2.1 Benefits of Adaptive or Change Driven Approach

In Adaptive/Change Driven Planning, project team intentionally repeats one or more project activities as the project team's understanding of the product increases ……….Also known as …agile methods …..intended to respond to high levels of change and ongoing stakeholder involvement [5]. The heart of Facility Siting involves making timely but informed decision. Some of the characteristics of this framework:

1. To add small incremental improvements that will deliver value to stakeholders
2. Milestones develop the product (e.g. Mitigated QRA Report) through a series of repeated cycles, while increments successively add to the functionality of the product
3. For each Milestone, a deliverable (read 'model') or set of deliverables will be completed. Future iterations may enhance those deliverables or create new ones. Each iteration incrementally builds the deliverables until the exit criteria for the phase are met, allowing the project team to incorporate feedback.

The biggest challenge with Mega-Project is scarcity of key information needed as inputs to develop scenarios and models. Some of the critical benefits, a project may gain by applying Change Driven approach:

1. The approach acts as Leading Indicator on Process Safety for Plant Layout - allows for multiple opportunities to make critical decisions on plant layout over the project life-cycle capitalizing on what is known at a given time. This lets the project team to use "what they know", to ensure "what they don't know" falls into the right place
2. Opportunity to make decisions over the project lifecycle not only works to save life but also helps with striking a balance between business needs and climbing the risk mountain
3. Aides design-development with progressive elaboration while producing final CA/QRA deliverable for the project

2.3 Execution Strategy

The main idea of Change Driven process starts with decomposing the known preliminary layout and process information into a set of requirements, work to be performed and their priority sequence, sometimes referred to as a *Product Backlog.*

It is advisable that the owner/operator forms a multi-use agreement (contract) with the chosen vendor. This way the plant areas under the Mega-Project umbrella can get their specific work done while the CA/QRA Project Team provides overall oversight and guidance.

At the beginning of an iteration, the team will work to determine how many of the highest priority items on the backlog list can be delivered within the next iteration. The time-length of each iteration should be less than 4 weeks. As shown in Figure 1, during "Develop Scope" and "Define Project", the length of each iteration could be even less than 2 weeks. This iterations (otherwise known as "Minor Milestone" as shown on Figure 1) will develop scenarios, assumptions and preliminary CA/QRA models. Assumptions are required due to lack of available quality information in the early stages of a project.

At the end of each iteration, the Minor Milestone/deliverable should be ready for review by the customers/stakeholders. This does not mean that the customer is required to accept delivery, just that the product should not include unfinished and/or incomplete features.

The sponsor and customer representatives should be continuously engaged with the other project stakeholders to provide feedback to Contractor on deliverables as they are created and to ensure that the *Product Backlog* reflects Sponsor/stakeholders' current needs. Below, you will find two examples using two different management strategies. For confidentiality purpose, the specific name/location is kept generic and few items were modified to suit.

2.3.1 Project X Facility Siting Using Change Driven Approach

A CA/QRA team was given the assignment to support design development during "Develop Scope" phase (Figure 1). The Sponsors already selected the site for the project. But the fixed plants were yet to be located on the site. The Vision for the initiative was for the QRA result to come under the organization Risk Threshold. Multiple EPC's were given to engineer different plant areas. The CA/QRA team selected a vendor to support them with defining the *Product Backlog* based on the preliminary Process Design Basis, intended product the facility is expected to sell, the feedstock necessary etc. The CA/QRA team had prioritized the Product Backlog based on hazard level of feedstock, preliminary PHA schedule and project schedule. This also involved obtaining feedback from different Process Engineering and Operations teams to collect necessary inputs to put together the assumptions and scenarios to be modeled within the 1st iteration.

Out of 25+ scenarios, only 7 perceived to be the most hazardous were modeled for CA that gave ample input (e.g. explosion, toxic and fire contours) for the CA/QRA team to work with the stakeholders to recommend location for specific process units on the given site. Note that, CA/QRA team integrated the potential consequence to understand local incident and global consequence events. At this time QRA was not performed. This was used as input for minor modification of preliminary plant layout. It is to be noted that the site plan was still preliminary after the first iteration but assisted different EPC accountable for their specific area to recognize where to locate the specific plants i.e. finalized key plans for the overall site plan.

With further design-development, more refined information became available from each EPC. This new information led to 130+ scenarios and was used for further CA/QRA iterations that created either new model and/or embellished the past ones. Each EPC was accountable for facility siting within their boundary while the CA/QRA team provided technical oversight and integration support for the overall project. The Sponsor/stakeholders were engaged on a regular basis to ensure that the Product Backlog reflects future modeling requirements to properly site facilities e.g. Fire Water Pump house, Tank Farms, site Medical Centre etc. Early engagement of CA/QRA team and higher management willingness to incorporate their feedback with the progressive elaboration of the project led to QRA result meeting the organization risk threshold and ultimately safer facility. Changes suggested based on CA/QRA work in earlier stage (i.e. Develop Scope, Define Project) were easier to incorporate into the design.

2.3.2 Project Y Facility Siting Using Plan Driven Approach

The sponsors for project Y selected the site for the project and decided to hold on to not perform CA/QRA work with the progressive elaboration of the project activities. This was mostly decided based on the fact that not all information was available to properly conduct the 'studies'. Fundamentally, the sponsors looked CA/QRA as deliverables requiring a more 'complete' plan. Once the Develop Scope (as referred in Figure 1) is almost complete, the sponsors decided to engage CA/QRA team and a vendor. By this time, the project had significantly high number of stakeholders. Many of the stakeholders were not very familiar with CA/QRA and the value it could bring to them. There was not much time to schedule and train the stakeholders as well to bring them up to speed. As a result, the stakeholders were unable to align on a Vision. The Vendor was asked to provide explosion, toxic and fire contour maps for specific scenarios and locations that were almost 'fixed' on the site plan at that point in time. The resultant contour maps showed significant consequence on nearby facility with over 700+ people. It was also noted that the plant area consuming the most hazardous substance would impact the nearby facilities significantly from consequence perspective. It should be noted that the result was only based on a selected few scenarios. By the time the project reached the "Execution" phase, more scenarios were added as part of the scope. But it was already too late to incorporate most of the recommended changes resulting from CA/QRA work. The project went through several area by area risk reduction workshops only to accept (by the company executives per organization risk matrix) the aggregate residual risk level which was also found to be not within the organization risk threshold.

2.3.3 Importance of Starting Early

Figure 3: Impact of variable based on project time [5]

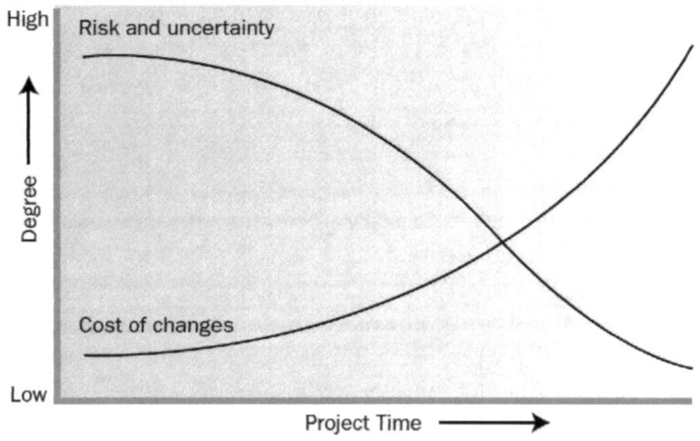

The above two examples highlight the importance of starting CA/QRA work early enough to realize and maximize the value CA/QRA brings to a Mega-Project. The high-level view provides common frame of reference for comparing projects for CA/QRA initiative, even if the projects were dissimilar. Change driven methodology allows for applying the known information at early stage in a way that supports laying out plant facilities to reduce risk and thus maximize potential value. While these characteristics remain present to some extent in almost all project life cycles, they are not always present to the same degree [5]. Change Driven framework, in particular, keeps stakeholder influences higher and the costs of changes lower especially in the early stages of the life cycle than in predictive/plan life cycles. The Figure 3 highlights a well-known fact of project management that even though risk and uncertainty remains high in the early stages of a project, cost of incorporating changes remains much lower. This translates into a key opportunity for project stakeholders to align on laying out the key areas on a site plan intelligently to reduce risk and maximize value by applying tools such as CA and QRA.

2.4 Key Decision Points

Facility Siting, from a management point of view, is all about bringing the stakeholders closer and find the 'right' spot for the facilities. Finding the 'right' spots requires making timely decisions. The challenge has always been to make timely decisions as the stakeholders more often find themselves forced to make those decisions based on information that are frequently of poor quality and at times uncertain, especially in the early stages of a project. At times, due to the scarcity of information, stakeholders rely on their past experience to make those decisions on plant layout. As a result, depending on the individual experience, two different teams for a project area may come up with two very different plant layout. CA/QRA work, if started early enough, will also depend on input that most often requires embellishment through assumptions to create needed scenarios and models. Even with this limitation, CA/QRA provides a key 'input' in allocating real estate for key facilities on a plant site at early stage of a project lifecycle. With each iteration and deliverable 'Change driven' methodology allows for multiple opportunities to make critical decisions on plant layout over the project life-cycle. It allows the stakeholders to decide when to 'sharpen the pencil' (e.g. PHAST to Computational Fluid Dynamics model), relocate facility or accept the consequence and/or risk, 'pull the plug' on further modeling/refinement and start developing justification, if required, to recommend management to accept/reject the aggregate risk profile.

2.5 Alignment with Process Hazard Analysis (PHA) Planning

CA/QRA results are a useful input to PHA (e.g. HAZOP, LOPA) activities. This is particularly true as projects move forward after establishing the plant layout on a high level and starts to conduct "Define Project" (Figure 1). PHA activities may also result in recommendations (i.e. new scenarios) that goes into the *Product Backlog* and requires project team to develop new models or refine old ones. Mega-projects have multiple sub-projects running in parallel most of the time, having an Integrated Schedule for all the reviews is a good strategy so that the right stakeholders are booked for the sessions. The integrated PHA schedule also supports prioritization of *Product Backlog* for CA/QRA activities over the lifecycle of the project.

2.6 Integration Strategy

It is critical for a Mega-Project involving multiple plant areas/silos to understand the BIG picture so that site risk can be dealt with on an aggregate basis. There is strong possibility of one area to impact other plants in the vicinity. It is recommended to review integration aspect and aggregated results on a regular basis. The following Figure 4 shows that the origin of an incident may reside in one particular plant area, while it may impact other areas in the vicinity.

Figure 4: Consequence of an incident in one area impacting multiple areas

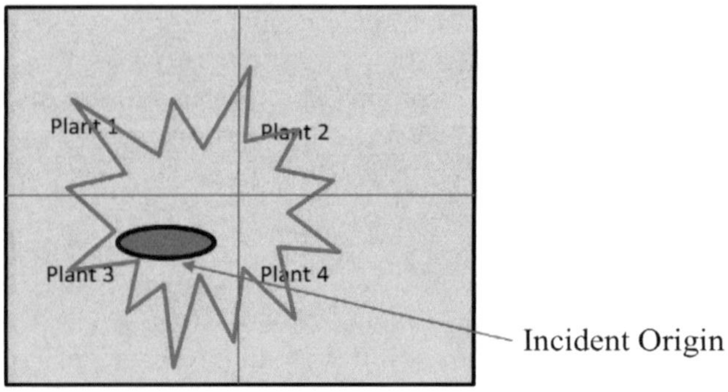

Recommended Strategy:

- Setup Strong Center - Task Force Single Point of Contact (SPOC)
- Collaborate – SPOC will promote Collaboration among Plants/Areas

3. List of Activities (suggested)

3.1 Overall Sequence

The CA/QRA work typically should start with the approved Charter as shown on Figure 1. A project Task Force i.e. SPOC is formed comprising of Project Engineer, Operations rep, Process rep and CA/QRA Technical Expert. The SPOC is ultimately accountable to produce the CA/QRA deliverable studies but also acts to support the design development and site plan finalization efforts in the early stages of the project.

The team collects 'input' such as Preliminary site Layout and design information, works with area process and operations team to identify scenarios, develop assumptions for known-unknowns and unknown-unknowns, develop modeling 'Product backlog', prioritize backlog based on PHA schedule and produce modeling output (i.e. deliverable) for stakeholder use. The stakeholder usually has the option of accepting or rejecting consequence. If the result of an iteration is acceptable within area causing the incident – communicate to outside stakeholders to ensure there is no cross boundary impact. If the result of the iteration is NOT acceptable to outside of the area – use the result as input to PHA for the areas to mitigate. Collaboration recommended between origin area and impacted area/s to identify appropriate mitigations. There is also a possibility that the consequence resulting from the iteration is not acceptable to the area causing the incident, in that case the area will initiate a PHA to mitigate the consequence. As part of developing mitigations, consider the following:

- Develop QRA if appropriate
- Perform Building Assessments for Critical Buildings
- Apply Cost-Benefit analysis to optimize risks/opportunities

- Determine Safe Haven locations and their characteristics

If Mitigation is not possible from Business perspective – perform As Low As Reasonably Practicable (ALARP) justification for Senior Management to formally accept the posed risk

- To apply mitigations on a site wide basis, review Milestone deliverables by taking on a BIG picture perspective
- Future iterations may enhance those deliverables or create new ones as new information is available. Each iteration incrementally builds the deliverables to produce the Mitigated CA/QRA report to support Facility Siting for the stage

The overall effort is expected to produce a maximum of two single reports for the overall site:

1. Mitigated version of QRA Report
2. and Consolidated CA Report

It is recommended to develop as-built, pre-start up CA/QRA reports prior to start-up. Table 1 and 2 outlines typical activities performed by SPOC and plant areas respectively once key process areas are established/finalized on Site Plan. Please note that the activities are project dependent and may warrant modifications to suit specific situation.

Table 1. SPOC Sequence of Activities to conduct Facility Siting

Initiative	Key Steps
Setup Facility Siting Team	Sponsor to support to setup a team (i.e. Project Engineer, Technical Expert, Operation, Process representative) to act as SPOC
Integration	Provide functional (technical and management) guidance to Areas
	Assess cross boundary impact/opportunity
	Communicate incremental outputs after each Area PHA/Risk Reduction
IFA versions (prior to "Define Project" stage)	Consolidated CA Report
	Mitigated and Integrated QRA Report
IFD CA/QRA (prior to "Execution" stage)	Re-validate IFA CA/QRA post PHA and vendor PHA
As-built versions	Post-oil version

3.2 Typical Area Activities

Table 2. Area (Typical) Sequence of Activities to conduct Facility Siting

Initiative	Key Steps
Plant 1 (Typical)	Perform Modeling and Calculations
	Conduct Risk Reduction Workshop (if appropriate) – Apply mitigations
	Incremental output for use based on reduced risk level
	Assess Buildings/facilities
	Identify Safe Haven/shelter Locations (if required)
	Estimate Area Project Scope impact
	Perform area scope updates/changes e.g. passive fire protection requirement
	Carry out QRA (if appropriate)
	Update Product Backlog
	Develop/update contract to support area specific activities

4. Pitfalls to Avoid

The following observations were noted by the Author during his time to manage Facility Siting scope:

1. Facility Siting does not replace good and common sense engineering practice e.g. locating a plant that uses high volume of hazardous material will significantly increase risks to other plants/areas (e.g. camp) within the vicinity
2. Evaluate effectiveness of 'Cost Reduction' philosophies as opposed to added risk e.g. different modularization techniques may decrease the plant footprint by reducing pipe lengths but makes the plant more compact and likely to increase severity of explosion consequence
3. Avoid developing models with hope to realize risk reduction
4. Sharpening the pencil for models most likely may not reduce risk
5. Avoid jumping into CFD or other similar methods without evaluating cost-benefit for undertaking such effort
6. Utilization of QRA results is much more controversial than the methodology. Watson [6] has suggested that QRA should be considered as an argument, rather than a declaration of truth. In this view – focus on accuracy over precision
7. CA/QRA Task Force should be part of Management of Change (MOC) process
8. Evaluate carefully before locating facilities e.g. tank farms close to main entrance/egress as part of Facility Siting

9. Keep the focus on VISION for the overall effort to avoid scope creep
10. Note the difference between Facility Siting as a Deliverable and tool for design-development activities

5. Conclusion

This paper presented Adaptive/Change driven framework to manage initiatives like facility siting/CA/QRA as part of Mega-Projects. Change driven framework does not mean "lack of planning or no planning". As demonstrated in the paper, this methodology starts with a vision and then develops the ultimate product both iteratively and incrementally when it is possible to define small incremental improvements that will deliver value to stakeholders. This allows making key decisions on laying out the facilities in early stages of a project rather than waiting for design to develop to a significant extent. Often times, project team wait too long for design to mature enough before applying CA/QRA to validate consequence impact and/or risk level. Most often, especially for Mega Projects, this leads to delays in executing the CA/QRA work properly at the right time to influence site layout. This delay in execution leads to scenarios when scope changes cannot be incorporated anymore. This translates into value erosion as project was unable to reap the full benefit of Facility Siting. The management framework recommended in this paper i.e. Change driven/adaptive framework allows for the stakeholders to start early in the project lifecycle to apply CA/QRA in laying out the plant facilities. The key is to start EARLY but START with the END in mind. Alignment among stakeholder on a VISION up-front is also important as this is expected to save life while striking a balance between business needs and climbing the Risk Mountain. Vision also allows avoiding scope creep while applying Adaptive/Change driven approach. It is highly recommended that stakeholders focus more on accuracy over precision. In the early stages of project lifecycle, quality information is scarce. Change driven/Adaptive framework gives you as the project stakeholder the opportunity to use "what you know" intelligently to ensure "what you don't know" falls into the right place.

6. References

[1] M. Mannan, A. Chowdhury and O. Reyes-Valdes, "A portrait of process safety: From its start to present day," College Station, 2012.

[2] M. Mannan, Lees' Loss Prevention in the Process Industries, Volumes 1–3 (3rd Edition), , 2005, Elsevier, 2005.

[3] K. McAndrews, "Consequences of Macondo: A Summary of Recently Proposed and Enacted Changes to US Offshore Drilling Safety and Environmental Regulation," in *Society of Petroleum Engineers, Americas E&P Health, Safety, Security and Environmental Conference*, 2012.

[4] E. W. Merrow, Industrial Megaprojects Concepts, Strategies and Practices for Success, New Jersey.: John Wiley & Sons, Inc., 2011.

[5] A Guide to the Project Management Body of Knowledge (PMBOK® Guide) – Fifth Edition, Newtown Square: Project Management Institute, Inc., 2013.

[6] S. Watson, "The Meaning of Probability in Probabilistic Risk Assessment," *Reliability Engineering and System Safety,* vol. 45, pp. 261-269, 1994.

348480 Global Energy and Transportation and Mobil Oil

Abstract:

Biofuels, Computational Methods, Engineering Economics, Polymers, Thermodynamics. Polymers-plastic, Thermodynamics- oil, Engineering Economics – the money it takes to finish the method of Global Engineering Research, Computational Methods – numerical input of analytical data, Biofuels – the transport fluids that utilized in more than one method from industrial usage to everyday living which enters the atmosphere. Exxon has completed a $600 million dollar deal for production of algae biofuel, amongst Mobil Oil has performed major research Global Energy and the effects of oil. Reading a most recent article on Mobil website.

348748 Characterization of Iron Phthalocyanine As the Active Material for Lithium Batteries

Abstract:

Results on the determination of the lithium-ion diffusion coefficient in the iron phthalocyanine (FePc) active material as a function of intercalation level, using a modified electrochemical impedance spectroscopy technique, will be presented. Also, developed thermodynamic functions such as the Gibbs free energy, entropy, and enthalpy changes for the lithium insertion reaction with FePc, derived from cell open-circuit voltage measurements as a function of intercalation level, are presented. Also presented are estimates of the maximum amount of lithium that can be intercalated into the FePc based on constant volume and temperature conditions.

349086 Design of a Free-Fall Reactor for Fast Pyrolysis of Waste Plastics

Abstract:

Thermal pyrolysis has earned significant reputation in the recovery of valuable chemicals, especially petrochemical based fuels from various materials, including biomass and plastics. Of many available technologies, *fast* pyrolysis was preferred in his study owing to its minimal heat and mass transfer limitations and better control on process conditions. The focus of this study is towards efficient conversion of waste plastics into gaseous products consisting primarily of alkanes, alkenes and aromatics. Lab scale experiments were performed using a micro reactor to determine favorable conditions, including reaction temperature and particle size that would result in maximum yield of ethylene. Using these results, a simple industrial size radiatively heated free-fall reactor was designed for the purpose of pyrolyzing waste plastics. The goal is to design a *fast* pyrolysis reactor that improves upon inefficiencies in traditional reactors while resulting in high yield of ethylene.

349732 Biosorptive Dehydration of Ethanol/Water Azeotropes Using Compound Starch-Based Adsorbent

Abstract:

Abstract: Recently, the interest in the production of bioethanol has greatly increased as a result of the increasing fossil fuel prices and associated environmental issues. Bioethanol is made from either grains or biomass and the broth contains 6 to 12 percent ethanol. In addition, water removal represents a serious problem as ethanol form azeotrope with water at atmospheric pressure. Distillation to water-free ethanol consumes 50 to 80 percent of energy used in the fermentation ethanol manufacturing process. Various techniques have been developed to break the azeotropic point, such as salting-out method, azeotropic distillation, extractive distillation, reactive distillation, adsorptive distillation and pervaporation. Among the methods, the adsorptive option is the most attractive, especially the biosorption, which has distinct advantages over conventional methods as it is environmentally friendly, highly selective, easily available, easy to operate, cost effective and reusable in repeated cycles in the treatment of aqueous organics.

In this work, a specially formulated compound starch-based adsorbent (CSA), which consists of corn, sweet potato and foaming agent, was developed. As an exploratory study, this research was focused on the investigation of CSA for biosorptive dehydration of aqueous ethanol and optimization of conditions for its maximum adsorption under various conditions of vapor feed flow rate, bed temperature and bed height through an orthogonal design method. Furthermore, to investigate the adsorption mechanism, dynamic saturated adsorbance of CSA was studied and characterization of the adsorbent before and after adsorption, as well as after regeneration, was performed.

In the first step, the net retention time and separation factor of ethanol and water were obtained using inverse gas chromatography (IGC), the results indicated that the net retention time of water on CSA is significantly longer than that of ethanol which attested the feasibility of this adsorptive separation purpose and low temperature could be propitious to the process which was consistent with the traditional theory of adsorption.

Then, through a batch adsorption experiment, the optimum adsorption condition was determined under vapor feed flow rate, bed temperature and bed height by orthogonal design method, which indicated that the optimization condition for ethanol was at vapor feed flow rate of 1.1g/min, bed temperature of 83 and bed height of 25cm. The packing density was 445kg/m^3. The residence time is determined by vapor feed flow rate and bed height. The residence time is more influenced than bed temperature. The longer the residence time, the more the output is. But it cannot be ignored that the adsorption selectivity may decreases. Low bed temperature is benefited to the adsorption process which agrees with the result of IGC. Five cycles experiments of adsorption-regeneration were carried out at the optimum adsorption and regeneration condition, the results showed that there is practically no change in the regeneration efficiency of CSA after the cyclic process adsorption–regeneration. It is due to the properly chosen regeneration conditions assuring complete desorption of the water adsorbed without changes in the CSA structure. This unchanged adsorption capacity after a number of cycles demonstrates reusability of the adsorbent.

To study the adsorption mechanism, experiment was conducted to study the dynamic saturated adsorbance of CSA. The water adsorption capacity of CSA has been found to be similar to some molecular sieves, which is about 0.15g/g. Besides, field emission scanning electron microscopy (FESEM) and mercury porosimetry (MP) were applied to investigate the change of surface morphology and microstructure of CSA before and after the adsorption, as well as after regeneration, as shown in Fig.1 and Fig.2. The FESEM images showed that compared with the particles before the adsorption, they became plump and the void spaces became smaller after adsorption, which indicated that the CSA was saturated with water. However, the CSA particles after regeneration became wizened like the nature particles, which demonstrated that almost all the water had been desorbed from CSA. The pore size distribution (PSD) indicated that compared with the PSD curve before adsorption, the peak at 6000nm disappeared and the peak at 17000nm became remarkably smaller after adsorption, which manifested that CSA is mainly composed of macropores which play a major role in the adsorption. However, the peaks at 6000 and 17000nm reappeared after regeneration, and the curve looked exactly the same with that before adsorption, which further demonstrated that CSA after regeneration can be totally regenerated and recovered to the original structure.

These results indicated that the starch-based adsorbent can be considered as an effective and low-cost bio-sorbent to dehydrate alcohol.

349781 Purification of 2-Amino-1-Phenylethanol Enantiomers By a New Technique Combining Distillation and Crystallization

Abstract:

Stripping crystallization (SC) is a new separation technique which combines distillation and crystallization. Basically, SC is operated at a triple-point condition, in which the liquid is simultaneously vaporized and crystallized due to the three-phase equilibrium. In essence, the process is continued until the liquid phase is completely eliminated and only the pure solid crystals remain in the feed. Thus, crystal washing is not required since only pure solid crystals remain in the feed and no impurities are adhered on the crystal surfaces at the end of SC. In the previous research, SC has been successfully applied to separate p-xylene from the mixed xylenes by Shiau et al. (2005, 2008).

Phenylethanolamine, also known as 2-amino-1-phenylethanol (2-APE), is an important basic skeleton in many adrenergic drugs and is often used as ingredients of cold medicines and anorectics. As the biological activity and pharmacokinetics of the enantiomers of a chiral drug can be different, the objective of this research is to apply SC to purify R-2-APE from the chiral 2-APE mixture. The experiments show that no crystallization occurs during SC without seeding due to subcooling of the liquid enantiomers. On the other hand, seeding can eliminate subcooling during SC, leading to the R-2-APE crystals gradually formed preferentially on the seeds. Thus, seeding with magnetic stirring can be successfully employed to purify R-2-APE from the 2-APE enantiomers.

1. Shiau, L. D., Wen, C. C. and Lin, B. S. Separation and purification of p-xylene from the mixture of m-xylene and p-xylene by distillative freezing, *Ind. Eng. Chem. Res*, **44(7)**, 2258-2265, 2005.

2. Shiau, L. D., Wen, C. C. and Lin, B. S. Separation of p-xylene from the multicomponent xylene system by stripping

350841 Fabrication of Low Cost Insulating Material from Kaolin Clay for Construction Purposes

Abstract:

A new low cost, thermal insulator, construction mineral polymeric material was synthesized from local Saudi Arabia resources using Inorganic Mineral Polymerization technique (MiP). In this technique, the mineral polymer is produced from the following components: clay mineral rich in koalinite, sand, sodium hydroxide solution, and water. Inorganic polymerization experiments are conducted by mixing known amounts of clay minerals and sand in a variable speed stainless-steel mixer for some period of times until a homogenous solid mixture is obtained. Sodium hydroxide solution is added to the mixture and the polymerization reaction takes place at different mixing speeds and times. The produced mixture is divided into a number of samples of known weights. Each sample is moulded in a stainless-steel cylinder at a medium pressure. Then the moulded samples are cured in an oven for 24 hours. The curing temperature was 80°C. The mineral polymerization conditions such as the ratio of clay to sand filler, particle size of the clay, particle size of the sand, concentration and amount of sodium hydroxide, speed and time of mixing were optimized to obtain stable and hard mineral polymer. Compressive strength for different specimens of fabricated mineral polymers was tested under dry and wet conditions. Values of compressive strength with values ranging from 11.5 MPa to 27.9 MPa were obtained.

351141 Thermodynamic Study of Binary PAH (Anthracene + Phenanthrene) Solid Mixtures

Abstract:

Polycyclic aromatic hydrocarbons (PAHs) are common components of many materials, such as petroleum and various types of tars. They are generally present in mixtures, occurring both naturally and as byproducts of fuel processing operations. It is important to understand the thermodynamic properties of such mixtures in order to better understand and predict their behavior (i.e., fate and transport) in the environment and in industrial operations. The Knudsen effusion technique was combined with gas chromatography and differential scanning calorimetry to measure the vapor pressure, composition, and phase behavior of binary anthracene and phenanthrene mixtures at near ambient temperatures. Mixtures of anthracene and phenanthrene exhibited non-ideal mixture behavior: In mixtures with less than 80% phenanthrene, this more volatile component preferentially sublimed and the vapor pressures decreased as the mixtures became enriched in anthracene. The vapor pressure data do not follow Raoult's law (that is, ideal mixture) predictions. Mixtures with high phenanthrene content (> 80%) showed azeotrope like behavior, in that anthracene and phenanthrene sublimed at nearly constant vapor pressure and molecular ratio. These results confirm that the behavior of few-component PAH mixtures does not follow ideal mixture models. Hence, as tars and petroleum (and potentially other organic compound-containing mixtures) weather to few component mixtures over time, complicated, non-ideal phase behavior might be observed.

351160 How Confident Are You That Your Major Accident Risks Are Under Control?

Abstract:

Major accidents in the process industries including those at Texas City , Buncefield UK and in the Gulf of Mexico have highlighted the critical role played by senior managers and executives. Effective leadership is essential if companies are to develop a positive safety culture that remains constantly vigilant towards the risk of a major process safety incident. In the UK the HSE has made leadership a key priority for the high hazard industries and there is a cross-industry initiative to develop and deliver training focussed towards board members responsible for process safety performance.

This paper explores the leadership failings that contributed to recent major accidents, and the cost to both the company reputation and the individuals involved. Key principles are discussed that can help improve leadership in the major hazard industries. These include; attitudes and decisions of senior manager affecting the investment strategy and the safety culture of the organisation, reinforcing the importance of safety by personal example, thorough understanding of major accident hazards and key risk control systems, investigating process safety incident and near misses to find the underlying causes, developing world class safety management systems, and identifying weaknesses in these systems using targeted performance indicators.

The author bases this paper on the development of a training standard for 'Process Safety Leadership' involving a cross-industry 'expert panel'. Training of senior executives in compliance with this standard has been carried out from late 2011 and feedback from these courses and the key requirements for effective leadership will be discussed.

351236 Sil Determination: Shortcomings with the Use of LOPA

Abstract:

Proposed Topic Area: Managing Risks

For a number of years, industry has been trying to apply the principles outlined in IEC 61508. In the process sector, this has resulted in a focus on the requirements of IEC 61511 (2003) and ANSI/ISA 84.00.01 (2004). These are risk based standards. They involve the identification of potential hazardous events, the causes that could lead to each event and the necessary risk reduction for each event. Where the risk reduction involves an instrumented function (e.g. a trip or an interlock), the required performance of the instrumented function can described in terms of the average probability of failure on demand. In the standards, the performance for such instrumented functions is assigned to one of four Safety Integrity Levels. The assessment of what performance level is needed in respect to a specific hazardous event is referred to as SIL Determination. In the field of SIL Determination, IEC 61511 provides examples of a number of different approaches – these are shown in IEC 61511 Part 3. These include Event Trees, Risk Graphs, Safety Matrices, Fault Tree Analysis and Layer of Protection Analysis (LOPA).

In the first few years, following the publication of IEC 61508 Edition 1 in 2000, industry generally embraced the use of Risk Graphs as the method of choice. However, over the subsequent few years, many companies using this method found that it was (a) giving higher safety integrity level targets than really needed and (b) was less flexible than other methods and consequently more difficult to use in many real situations. This has resulted in a general move to other methods and in particular to the use of Layer of Protection Analysis. Indeed, LOPA may now be the predominant "Method of Choice" for SIL Determination. As a method, LOPA gives the impression of simplicity and appears quite straightforward. It almost invites anyone with a minimum of understanding of the method to be a practitioner.

However, there are a number of shortcomings in the use of the LOPA method that have become evident through its use across the globe.

This paper will review the application of LOPA and provide insights into what may be seen as the proper use of the method and its core principles. It will also highlight various shortcomings in use that have been found during expert reviews carried out of a number of existing studies. The paper will be of benefit both to those who are new to the LOPA method, but also of value to those who have been using the method for some time.

Shale Oil and Gas – Wellhead to Fuel

Dr. Lawrence Kremer - Baker Hughes
AIChE - Hilton New Orleans Riverside
April 1, 2014

14 AIChE
Spring Meeting, New Orleans, LA

BAKER HUGHES

Outline Unconventional Shale Oil

- Properties of shale oils
- Problems
  - Transportation
  - Refining

14 AIChE
Spring Meeting, New Orleans, LA

BAKER HUGHES

General Characteristics of Shale Oils

- API gravity typically varies 30-55+
- Very paraffinic
- Low sulfur
- Can have hydrogen sulfide and mercaptans
- Low TAN
- Very low asphaltenes
- Light crude with very little residual material
- Many samples show great quality variation

Factor	Bakken	WCS
API	39.7	21.6
S%	0.25	3.44
Light ends	2.6%	
LSR	3.5%	3%
Light Nap	12.4	3.0
Heavy Nap	9.7	5.0
Jet	10.5	10.0
LFO	27.6	12.0
Gasoil	27.2	37.0
VacTar	6.7	29.8

14 AIChE Spring Meeting, New Orleans, LA

BAKER HUGHES

Shale Oil Property Comparison

Parameter	Eagle Ford	Bakken
API	52	40.8
TAN (g KOH/g)	< 0.05	0.09
Sulfur (wt %)	<0.2	0.304
Asphaltene (wt %)	0.1	0.41
Resin (wt %)	1.6	4.95
Filterable Solids (PTB)	225	76

14 AIChE Spring Meeting, New Orleans, LA

BAKER HUGHES

Utica Shale Crude - Three Phase

14 AIChE Spring Meeting, New Orleans, LA

BAKER HUGHES

Eagle Ford Crude Samples

14 AIChE Spring Meeting, New Orleans, LA

BAKER HUGHES

Analytical Breakdown

Value	Yellow	Red	Black
API	55.0	44.6	52.3
TAN	<0.05	0.07	<0.05
Sulfur	<0.20%	<0.2%	<0.2%
Na	1.0 ppm	1.6 ppm	1.6 ppm
K	0.3 ppm	0.4 ppm	0.5 ppm
Mg	3.4 ppm	2.9 ppm	3.0 ppm
Ca	2.6 ppm	2.8 ppm	3.8 ppm
Asphaltene	0%	0%	0.1%
Resin	0.5%	3.2%	1.6%
Filterable solids	176 PTB	295 PTB	225 PTB
P	5.3 ppm	4.6 ppm	5.0 ppm

14AIChE Spring Meeting, New Orleans, LA

BAKER HUGHES

Micrograph of Sludge

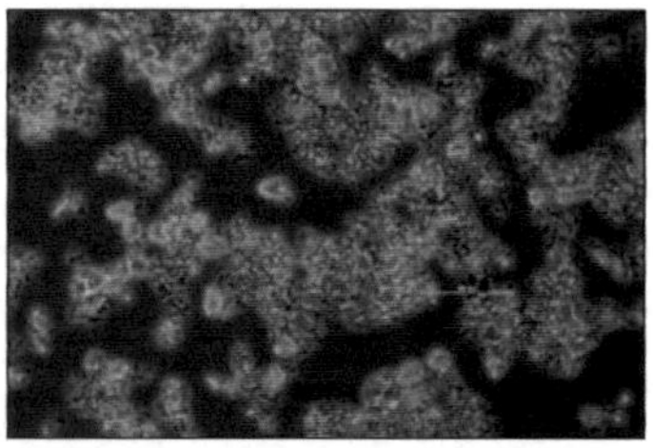

100 X

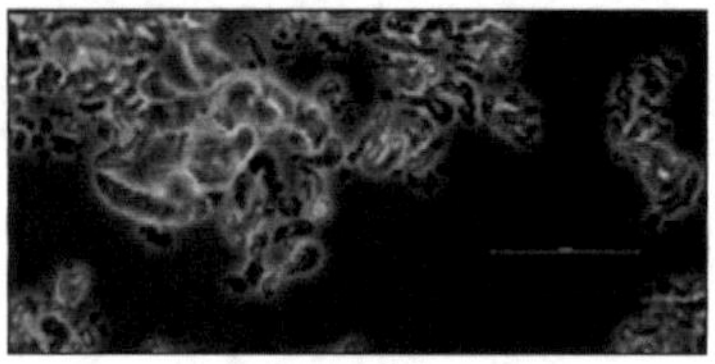

400 X

14AIChE Spring Meeting, New Orleans, LA

BAKER HUGHES

Paraffin Analysis of Sample

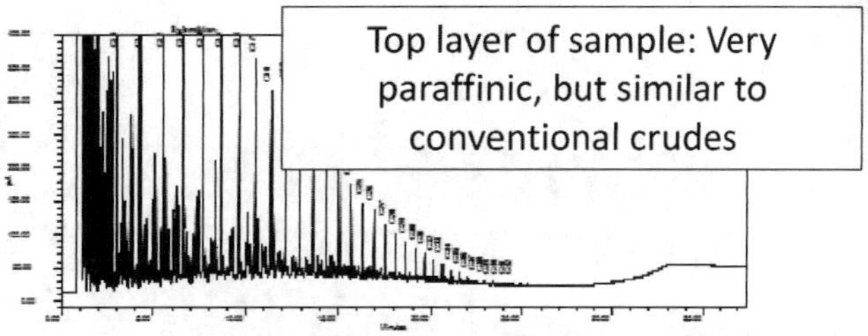

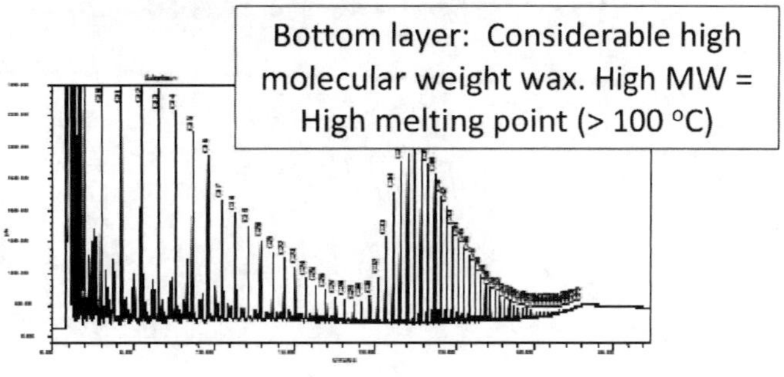

14 AIChE Spring Meeting, New Orleans, LA

BAKER HUGHES

Transport of Shale Production

- Infrastructure does not exist to meet needs
  - Most shale oil moved by rail or trucks
  - Limited pipelines expanding to meet transportation needs
- Unit trains to East Coast, West Coast, and Gulf Coast
- Problems:
  - Tankers and barges with heavy wax deposits ROB
  - Tight emulsions that look like oil or white condensate but high BS&W
  - Cars not uniformly treated for H_2S
  - Corrosion in rail cars

14 AIChE Spring Meeting, New Orleans, LA

BAKER HUGHES

Viscosity of ROB Before Treatment

ROB from untreated compartment

14AIChE Spring Meeting, New Orleans, LA

BAKER HUGHES

Eagle Ford Shale Oil Wax Dispersion Testing

Several products were effective in dispersing the paraffin wax in an Eagle Ford shale oil sample at ambient temperature for over a month.

14AIChE Spring Meeting, New Orleans, LA

BAKER HUGHES

Product 'A' Effect on Wax Deposition (Utica)

Blank 200 ppm 400 ppm

14 AIChE Spring Meeting, New Orleans, LA

BAKER HUGHES

Refinery Issues

- Crude Blending
- Fouling
- Corrosion
- Product quality

14 AIChE Spring Meeting, New Orleans, LA

BAKER HUGHES

Crude Unit Fouling

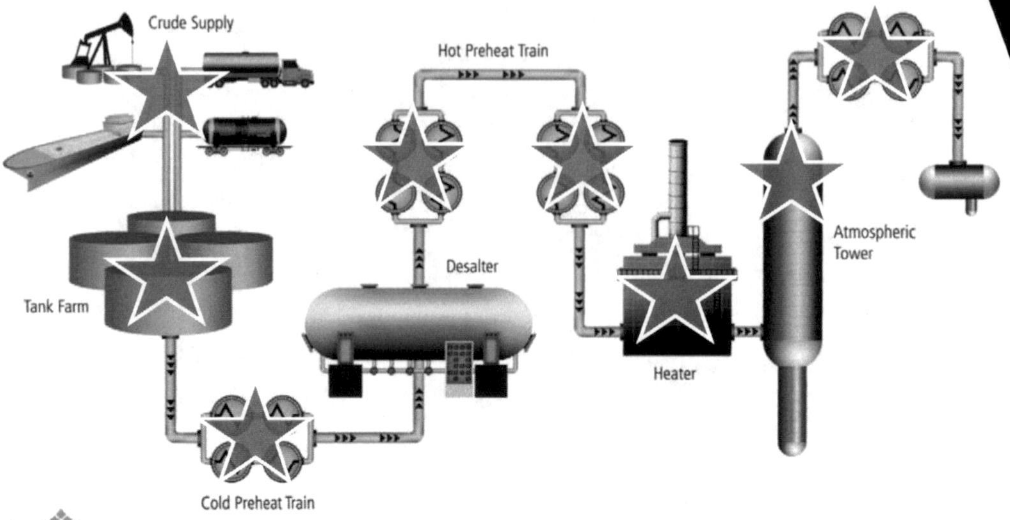

14AIChE®
Spring Meeting, New Orleans, LA

BAKER HUGHES

Asphaltene Precipitation

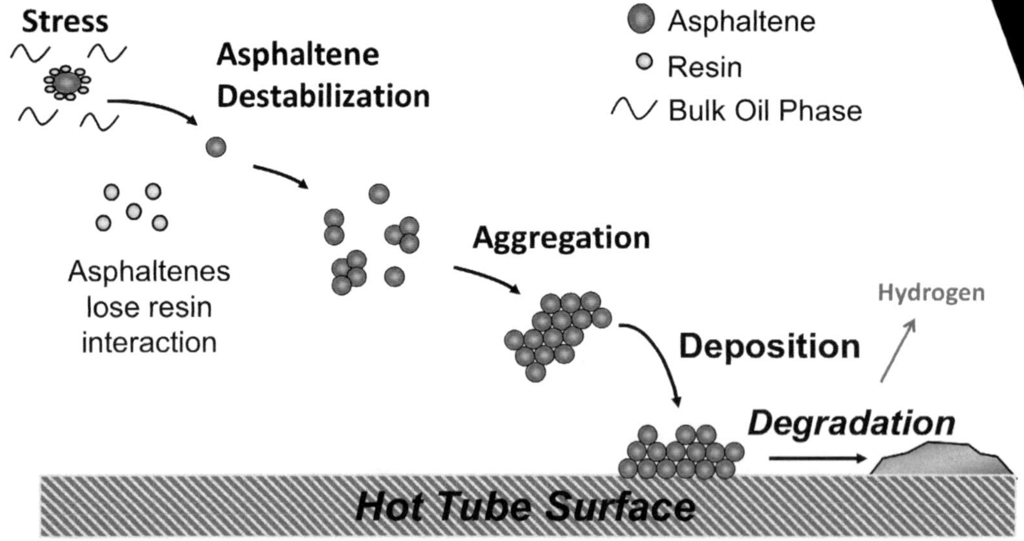

14AIChE®
Spring Meeting, New Orleans, LA

BAKER HUGHES

Asphaltene Stability Index Test

Field ASIT services™ technology measures the onset of the flocculation of the asphaltenes with high accuracy by inducing the asphaltene precipitation via titration with a paraffinic solvent.

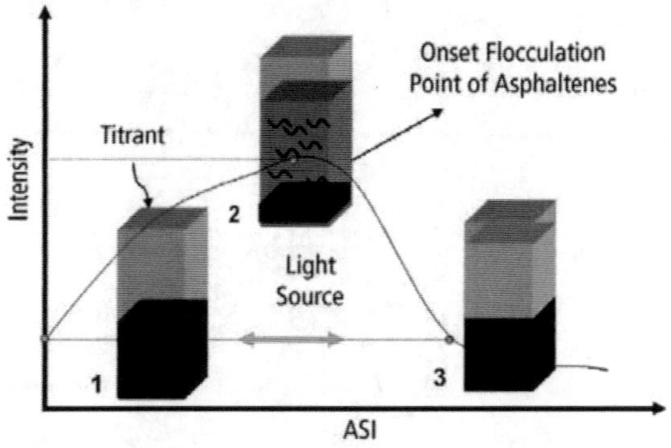

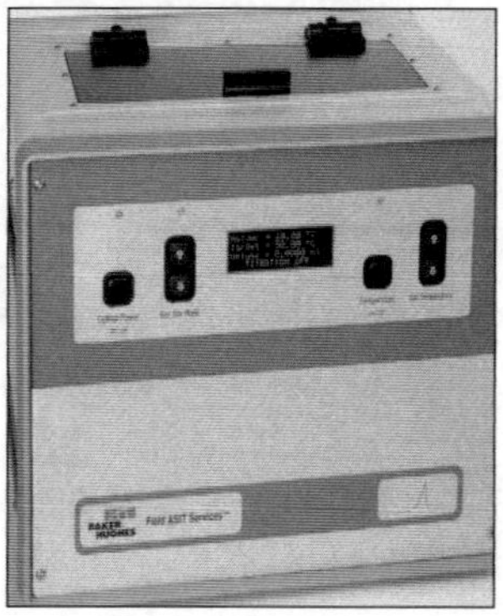

14AIChE
Spring Meeting, New Orleans, LA

Field ASIT services is a trademark of Baker Hughes Incorporated

BAKER HUGHES

ASI Shows Eagle Ford Can Destabilize Crude

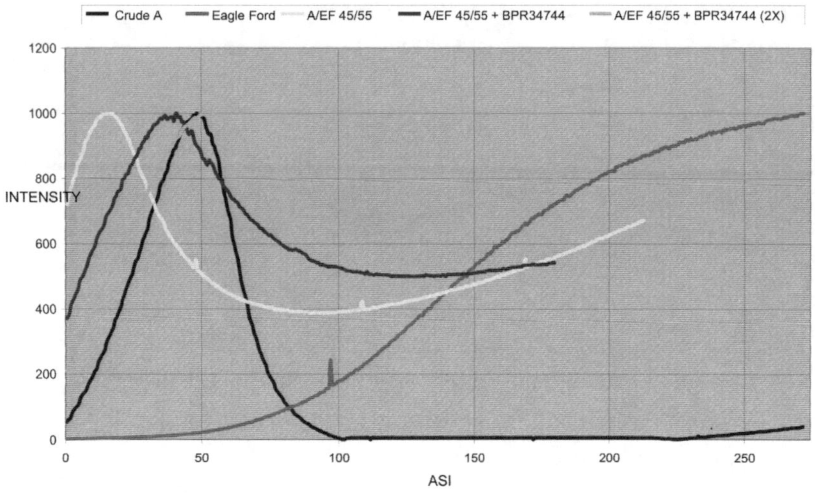

14AIChE
Spring Meeting, New Orleans, LA

BAKER HUGHES

Tower Tray Fouling - Phosphorus

- Seen by refineries running Bakken, Eagle Ford, and WTI
- Operationally manifested as increasing tower dP
  - Typically in the kero/jet zone
- Tower inspection shows deposits on trays in kero/jet zone
  - Some deposition may also show a few tray levels above and/or below kero/jet zone
  - Deposit scale typically tightly bonded to tray surface restricting flow and valve function

Above kero/jet zone

In kero/jet zone

Below kero/jet zone

14 AIChE Spring Meeting, New Orleans, LA

BAKER HUGHES

Shale Oil Impact – MEA Contamination

Corrosion and Fouling

- Shale oils transported by truck, rail, barge.
- Many shipments are treated with H_2S scavengers:

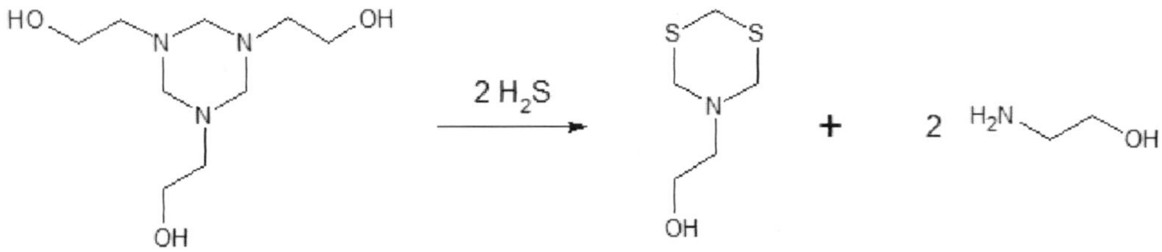

- Potential Impacts:
  - H_2S scavengers leave byproduct amine in crude oil (MEA, MA).
  - Tramp amines distill to OH system, create amine-HCl salts:
  - Corrosive to system metallurgy.

14 AIChE Spring Meeting, New Orleans, LA

BAKER HUGHES

Impact of Tramp MEA on OH Water

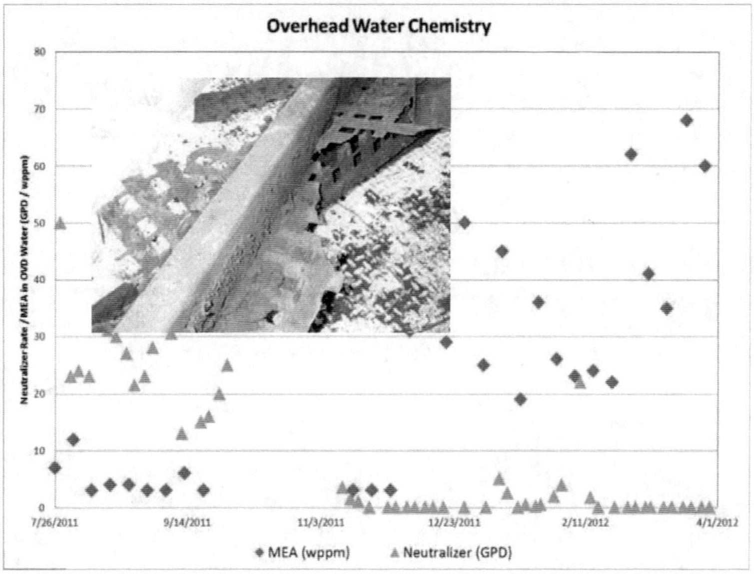

14 AIChE Spring Meeting, New Orleans, LA

BAKER HUGHES

Salt Risk w/MEA Concentration

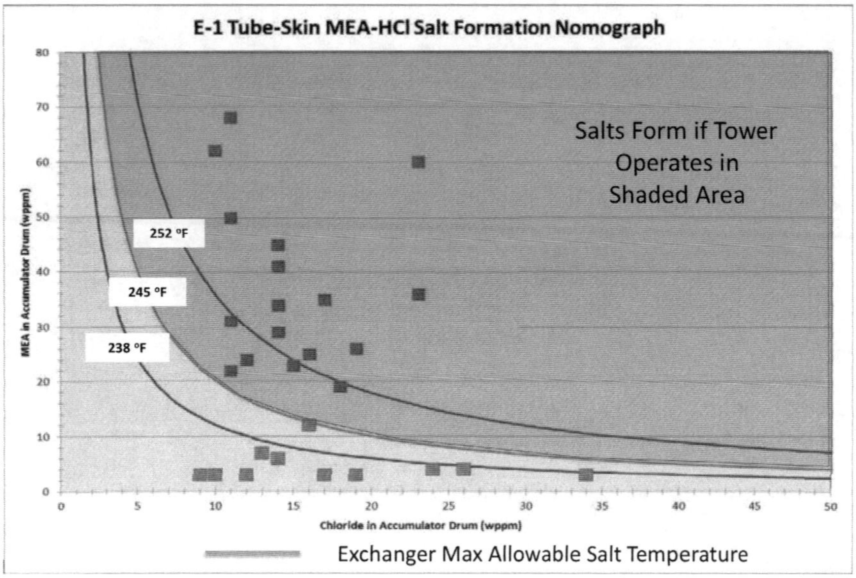

14 AIChE Spring Meeting, New Orleans, LA

BAKER HUGHES

Shale Oil – Refinery Process Concerns and Monitoring

BAKER HUGHES

CHARACTERISTIC OF SHALE OIL	CONCERN IN REFINERY	PROCESS DATA	SAMPLES AND ANALYSES Samples	 Analysis	OUTPUT DATA	FREQUENCY	BHI PROGRAMS
Crystalline Wax	Tank bottom sludge	• Tank Sludge Depth	• Tank bottom	• Visual sludge • GC analysis	• Tank Sludge Depth • Crystal Wax Presence	• Monthly	• XERIC™ Pretreatment
	Cold Preheat Fouling	• Desalter Temperature			• Temp vs. Time	• Update 2x/mo.	• LIFESPAN™ Wax Dispersants
	Desalter emulsions	• Emulsion layer growth	• Triline emulsions	• Visual wax • GC analysis	• Crystal Wax Presence	• After Shale Oil begins • When emulsion grows	• XERIC Demulsifiers
Asphaltene Stability in Blends	Desalter emulsions	• Emulsion layer growth	• Triline emulsions	• Precipitated Asphaltenes	• Precipitated Asphaltenes	• After Shale Oil beings • When emulsion grows	• XERIC Asphaltene Stabilizers • Field ASIT services™
	ColdPreheat Fouling	• Desalter temperature	• Crude blend	• ASIT	• Temp vs. Time • ASIT #	• Pre-Shale Oil • Update 2x/mo.	• LIFESPAN Antifoulants • Field ASIT services
	Hot Preheat Fouling	• Furnace inlet temp • Furnace tube temps • Furnace fuel usage • Preheat DP	• Crude blend	• ASIT	• Temp vs. Time • Fuel use vs. Time • ASIT #	• Pre-Shale Oil • Update 2x/mo.	• LIFESPAN Antifoulants • MILESTONE™ Furnace Fouling Control • Field ASIT services
Paraffinic	Furnace Coking	• Furnace tube temps • Furnace fuel usage	• Crude blend	• HLPS	• Coking tendency	• Pre-Shale Oil • W/Shale Oil	• MILESTONE Furnace Fouling Control
Tramp MEA	Tower/OH corrosion (Early Warning)	• OH Neutralizer rate • EXCAL usage rate	• OH water • Desalter brine	• OH pH • Brine pH	• Early Sign of Amine • Early sign of Amine	• Pre-Shale Oil • W/Shale Oil	• TOPGUARD™ Overhead Corrosion Control • EXCALIBUR™ Contaminant Removal Technology
	Tower/OH corrosion (Quantitative)	• Inputs for CRM	• OH water • Desalter ww • Desalter brine	• IC Amines	• Confirm MEA in Crude • OH MEA ppm • Salt Risk Tower • Salt Risk OH System	• Pre-Shale Oil • W/Shale Oil • Update 2x/mo.	• TOPGUARD Corrosion Control • EXCALIBUR Technology • TOPGUARD Corrosion Risk Monitor (CRM)
Phosphorus	Tower P Fouling	• Tower DP			• Tower DP Increase	• 2x/mo.	• LIFESPAN Antifoulants Dispersants
			• Crude	• ICP for P	• P in Crude	• 2x/mo.	
			• Kerosene	• Filt Solids • XRF Filter P • ICP for P	• Incr. Kero Solids • P in Solids • P in Kerosene	• 2x/mo.	
			• Kerosene • Diesel	• ASTM D86	• Incr. in Diesel IPB – Kero EP overlap	• 2x/mo.	

EXCALIBUR, FIELD ASIT SERVICES, LIFESPAN, MILESTONE, TOPGUARD and XERIC are trademarks of Baker Hughes Incorporated.

351307 The Shale Gale is Blowing: Plotting a Course that Avoids the Shoals and Rocks

Abstract:

Shale gas has revitalized the Chemical Industry in the U.S. The economic benefits have been widely described, but there is little discussion if the impacts of the great increase in ethane cracking. The shifting feedstock slate creates both challenges and opportunities for new technologies. The shoals and rocks caused by the shale gale will be detailed and a course described that can provide an even brighter future for the industry will be presented.

352096 Development of Shale Oil & Gas Programs - Critical Design Considerations for Economic Viability and Sustainability

Abstract:

The quantity and quality of crude oil & natural gas (from producing fields) change with time. These changes are large enough to require design and operational flexibility of the processing plants. Establishing trends for these changes are also very challenging. In addition, dynamics of the oil, gas, and gas liquids pricing bring additional demands in operational flexibility requirements. The paper will present these issues and discuss approaches to address the challenges.

352681 Future Energy and the Shale Gas Revolution: Implications and Opportunities

Abstract:

"Future Energy" will entail an unprecedented growth in demand, together with increasing stakeholder concerns for the impact of energy use on climate and the environment, and ensuring security of supply. These factors will drive diversification of energy sources as well as delivery systems and technology, to address both global and regional needs. Recent developments in shale gas and oil production via hydraulic fracturing together with increased use of bio-based and other renewable energy sources, will drive new opportunities for chemicals in delivering energy system infrastructure, improving energy efficiency, and re-optimizing process routes to minimize cost, as well as carbon and environmental footprints. In particular, low cost of gas relative to oil on an energy content basis will drive the development of new process technologies for commodity chemicals derived from C1 chemistry. Low energy costs will drive a new era of "smart manufacturing".

This seminar will examine trends and opportunities in responsible energy and chemicals resulting from the shale gas "revolution", and end with an open discussion of what emerging chemical sciences and engineering innovations are most likely to be impactful in addressing future energy and sustainability needs.

352789 A Summary of the Solids Founds at the Modular Caustic-Side Solvent Extraction Unit Process

Abstract:

Since March 2008, the Actinide Removal Process-Modular Caustic-Side Solvent Extraction Unit (ARP-MCU) facility has successfully processed more than three million gallons of supernate. The process has experienced a few outages arising from plugged coalescers. Recently, the facility began an outage to perform maintenance activities aimed at extending the original life cycle for the facility in light of the future deployment of the Next Generation Solvent (NGS).

The outage included cleaning and maintenance of the centrifugal contactors. Inspection of the extraction centrifuges revealed the presence of accumulated solids in the bottom channels of Extraction Contactor SEP-401. SEP-401, the first centrifuge the supernate enters the solvent extraction process, is where radioactive supernate from ARP meets solvent scrubbed with 0.05 M nitric acid. The aqueous carry over from the solvent mixes with the supernate possibly resulting in a lowering of its pH to the point of solid precipitation.

Any large variance in the concentration of the aluminum and silicon concentrations, elements known to readily precipitate throughout the tank farm, may indicate precipitation driven by a mild pH swing. A similar deposit discovery (sodium aluminosilicate) was also found on September 2010 at the Salt Solution Feed Tank.

This poster presents a summary of the solids found throughout the different components that are part of the MCU at the Savannah River Site (SRS)

352853 Fighting Scales in Oil and Gas Production Systems – a Challenging Flow Assurance Case Study

Abstract:

Maturing oilfields has not just caused problems of large water handling but also demands novel flow assurance solutions which are practical and cost-effective. One of the major problems experienced in the oil production industry is the formation of mineral scale. Scales can form from the formation water as it undergoes changes in pressure and temperature, and tend to deposit on surfaces. Its formation causes flow reduction or even blocking of pipes, valves and other process equipment.

The main implications of the problem are:

- Economic; due to downtime equipment replacement, reduced or lost production
- Safety; Equipment failures due to under-deposit corrosion can cost lives.

Summarized, the impact of scales in Oil & Gas surface equipment can be serious and challenges in removal, prediction and prevention are numerous.

An approach was developed for utilizing prediction-based knowledge on the scaling problem to formulate a workflow for flow assurance monitoring. The workflow is also intended to aid in arriving at practical and cost-effective mitigation strategy for problems related to scaling in surface equipment. The following are the brief steps involved in the workflow:

- Identify & define the key Issues (such as production shortfalls, pressure anomalies, etc.)
- Define objectives and specify boundaries
- Model process (multiphase network model)
- Analyze Results and employ a variety of techniques to pin down the causes
- Conduct a structured assessment to determine/evaluate solutions
- Develop an action plan / process redesign for implementing the solutions

The root causes were confirmed as calcite scale deposition in flow lines. A detailed assessment was also performed to ascertain thermodynamic envelopes for solids deposition and identify specific hot spots related to pressure, temperature in the network to ensure trouble-free production.

The workflow suggested was valuable for troubleshooting bottlenecks of existing facilities & addressing production shortfalls. The lessons learned in this endeavor were also used to make better decisions for designing new projects.

352615 Computational Study of Oil-Water Wetting Regimes Established in Horizontal Flows: Wettability Maps and Corrosion

Abstract:

Wettability effects in pipelines carrying oil-water phases at varying proportions are crucial in undermining the net effects of corrosion in the system. Prediction of phase distribution, especially the information whether oil or water wets the pipe wall is an important factor, which can affect the corrosion mitigation strategy and increase confidence in the integrity of the pipeline. There are large knowledge gaps in this area of research and consequently only empirical criteria are used in the industry.

In the present work, detailed computational study involving, a turbulent oil-water multiphase flow system in a horizontal carbon steel pipeline, is carried out to predict different water wetting regimes as well as corrosion kinetics of emanating from the fluid-metal interaction. Computational Fluid Dynamics simulations include Eulerian multiphase models with appropriate sub-models for interfacial forces, turbulent interaction and population balance equations. Population balance techniques identify and follow the dispersed (water) phase dynamics such as droplet breakage and agglomeration in a detailed fashion. In addition, concentration equation for ferrous ion ($Fe2+$) is modeled to capture corrosion kinetics. The current study focuses on flow pattern computations for water cut ranging from 0 - 20%.

Our simulations captured various water wetting regimes inclusive of intermittent, fully wetting, dispersed and semi-dispersed water-in-oil behavior. Results were in good agreement with the experimental observation. Simulations clearly predicted that for given superficial water velocity, elevating the superficial oil velocity leads to transition of regimes from stable water wetting to intermittent water wetting and subsequently stable oil wetting scenario. Phasic fraction and velocity distribution at different flow cross-section reveal the flow modulations as a function of the turbulence inherent to the two-phase system. Corrosion computations clearly indicate a substantial generation of Fe^{2+} ion under stable water wetting regime while the concentration and corrosion levels deteriorate with intermittent water wetting, an effect well characterized in flow loop tests.

352338 Modeling of Conductive Antifouling Paints for Marine Applications

Abstract:

Protection from fouling organisms (biofouling) is a key function of coatings on marine hydrokinetic (MHK) devices. Accumulation of biomass on devices risks performance degradation and possible failure. At the present there are no proven reliable options for prevention of fouling in the marine environment with applications to match device lifetimes expected to be on the order of 10-20 years. Removal, cleaning, and re-application of short-lived antifouling paints is an expensive prospect. Long term solutions for antifouling protection need to be identified and tested for reliability. Electrified conductive paints have been shown to be effective in the prevention of marine biofouling and may be a viable solution for long term antifouling performance[1] . The present work seeks to further the understanding of the mechanism of action and to develop the engineering knowledge needed to enable application.

The mechanism of action has been proposed variously to be direct electron transport[1], electromagnetic field repulsion[2], and electrochemical reaction[3, 4]. Evidence suggests that electrochemical oxidation of chloride and bromide present in seawater produces strongly oxidizing agents which sanitize the surface being protected. These oxidizing agents are short lived in the environment with end states being the original chloride and bromide. Comparison of the experimentally determined minimum potential needed for biofouling protection with Tafel plots in sterilized and newly acquired seawater support this hypothesis, as does cyclic voltammetry.

Chloride oxidation at graphitic anodes has been studied in highly saline water at high over-potential for the chloralkali process. The salinity, current density, product concentration, and pH conditions of the chloralkali process are vastly different from the conditions at a protected surface in the natural marine environment with current density on the order of 20μA/cm^2. Determination of reaction products and reaction selectivity under the applicable conditions is necessary to allow design of electrochemical antifouling systems. The present work utilizes a well characterized millimeter scale laminar flow electrochemical reactor and amperometric methods to determine reaction products, which yields reaction selectivity.

In order to evaluate the longevity of conductive antifouling coatings in marine applications accelerated aging methods are required. Standard procedures for rapid aging of marine paints involve exposure to ultraviolet light, moisture, temperature swings, and salt fog[5]. The two primary concerns for MHK applications are adhesion and electrochemical performance. The existing test regimens do not address electrochemical performance; therefore in addition to the standard tests accelerated aging methods using high over-potentials and current density have been developed. The electrode selectivity is characterized throughout the aging process, which enables prediction of performance losses.

Aged coatings have also been deployed in the marine environment to validate performance after aging. A standard procedure for washing test coupons to remove loosely adhered debris using consistent shear stress followed by visual and microscopic inspection has been developed as a means to standardize determination of antifouling efficacy.

[1] T. Matsunaga, T. Nakayama, H. Wake, M. Takahashi, M. Okochi, and N. Nakamura, "Prevention of marine biofouling using a conductive paint electrode," *Biotechnol. Bioeng.,* vol. 59, pp. 374-378, // 1998.

[2] T. Nakayama, H. Wake, K. Ozawa, N. Nakamura, and T. Matsunaga, "Electrochemical prevention of marine biofouling on a novel titanium-nitride-coated plate formed by radio-frequency arc spraying," *Appl. Microbiol. Biotechnol.,* vol. 50, pp. 502-508, // 1998.

[3] R. E. Perez-Roa, M. A. Anderson, D. Rittschof, C. G. Hunt, and D. R. Noguera, "Involvement of reactive oxygen species in the electrochemical inhibition of barnacle (Amphibalanus amphitrite) settlement," *Biofouling,* vol. 25, pp. 563-71, // 2009.

[4] J.-R. Huang, W.-T. Lin, R. Huang, C.-Y. Lin, and J.-K. Wu, "Marine biofouling inhibition by polyurethane conductive coatings used for fishing net," *J. Coat. Technol. Res.,* vol. 7, pp. 111-117, // 2010.

[5] *Paint and Coating Testing Manual : 15th edition of the Gardner-Sward handbook*: ASTM International, 2012.

2014 AICHE Spring Meeting
March 30 to Apr 03, 2014

Burns & McDonnell
1700 West Loop South, Suite 300
Houston, TX 77027

Authors/Presenters:
Mark W. Lockhart, P.E., Process Technology Manager, Chemicals and Petrochemicals
Burns & McDonnell, Houston, TX
832-214-1975

Elliott Robertson, P.E., Senior Process Engineer, Chemicals and Petrochemicals
Burns & McDonnell, Houston, TX
832-214-2916

Expanded Abstract:

"Shale Gas Boom – A New Era of Technology for Ethylene and Ethylene Derivatives"

There is a new era in the U.S. with the rise of the Shale Gas Boom that has emerged in this country. Large production increases are occurring from shale gas plays ranging from C1's – C4's, C5's/condensate and "tight oil". The plentiful supply of C1's – C4's versus current demand is driving prices for these molecules down. The recent result is directionally lower prices for methane, ethane, propane and butane as well as heavier products.

The plentiful supply and low prices for these building block molecules is having a marked effect on downstream industries that use the molecules either as fuel for energy or for raw materials as feeds. Steam crackers are moving to maximally change feedstock from naphtha's and gas oils to ethane to make ethylene and recently experiencing very attractive margins as a result.

Ethylene yield is very high when using ethane as feed which results in very favorable economics given the low price of ethane that has been prevalent over recent years. Ethylene producers have already made moves to feed as much ethane as possible within the limits of their existing production facilities. Ethylene producers are also moving to debottleneck existing plants including various debottlenecks that have already been announced. Additionally, as many as seven (7) new grass-roots facilities have been announced or are under consideration primarily along the gulf coast. Although, at many as two new ethylene plant projects are under review to be located in the North East to enable taking advantage of ethane production from the Marcellus Shale Play located in the states of Ohio, Pennsylvania and West Virginia.

Ethylene capacity totaling over 7 mm MTons/yr have been announced or are under study between debottlenecking of existing plants and new grass-roots ethane crackers. The debottlenecking projects can be accomplished in a shorter timeframe with the capacity increases continuing through 2013 to 2015. The new grass-roots capacities are slated to come on-line through 2016 and 2017. Several of the new ethane crackers will also install additional polyethylene capacity as part of the new facilities.

Historically, naphtha and gas oils have been used as feed to steam crackers. These feeds typically come from processing crude oils in refineries and, as such, the pricing for naphtha and consequently ethylene is directly related to the price of crude. Ethane production from shale gas plays provides crackers a new source of feed that is not related to the refining of crude oil or world-wide oil markets. This represents a significant paradigm shift in the traditional economics for ethylene production.

Ethylene is a major building block for numerous chemicals and petrochemicals. Consequently, the shale gas renaissance is affecting a broad spectrum of the chemical industry.

Shifting ethylene production from naphtha and gas oil to ethane as feed has a dramatic effect on the production of other byproducts that traditionally come from steam crackers. The yields of propylene, C4's, C5's and heavier products that are produced when feeding ethane are very low.

The reduction in propylene production is affecting the economics of producing propylene through the "on purpose" route. Numerous propane dehydrogenation projects are already in design with others being announced. This is just one example of the changing markets and the necessary transition that is taking place in the U.S. as the markets are in transition.

The total spending for the chemical/petrochemical industry for the period 2013 to 2030 is now estimated at approximately $120 billion. There is an anticipated sharp increase in spending starting in 2014 to peak in 2016/2017 at over $13 billion. The total spending for the period 2013 to 2017 totals nearly $48 billion.

With the advent of the announced ethylene plants slated to come on line by 2017, ethylene will be in greater supply. The economics of supply and demand suggest an oversupply of ethylene produced in the U.S. which will result in lower ethylene prices. A lower ethylene price has the effect of swinging economics in favor of producing more derivatives made from ethylene which will have to be exported from the U.S. to the global market place in order to maintain a supply and demand balance.

It is estimated that along with ethylene production various ethylene derivatives including polyethylene, ethylene oxide, ethylene glycol, EDC, VCM, PVC, among others will also increase. This has the effect of increasing the capital spending through 2020 - and beyond.

Where will all this production go? It is forecasted that the U.S. will be increasing net exports of the ethylene derivatives.

Lower energy cost along with cheap ethane has caused a marked shift in where U.S. stands on the global ethylene cost of production curves. An impressive result of the Shale Gas Boom and consequent lower cost of energy and cheap ethane has been the effect on U.S. cost of production for ethylene on the global cost curve. Comparing the 2007 and 2012 cost of production curves it is clear that the shift has now brought the U.S. much closer to the lowest cost producers including Canada and the Middle East.

Ethylene Dichloride (EDC) is one of the derivatives of ethylene that is forecasted to experience capacity increases in the U.S. EDC is used to produce Vinyl Chloride Monomer (VCM) which in turn is used to produce PVC. The U.S. is likely to increase export of PVC for use in home building and plumbing products world-wide. Burns & McDonnell has chosen to highlight the EDC -> VCM value chain in this presentation.

The process to produce chlorine is very energy intensive at 20,200 Btu/lb. Reduced U.S. energy costs represent a significant advantage in lowering the cost of production for chlorine which is another advantage in the EDC, VCM and PVC value chain. There are two widely accepted methods for producing EDC: direct chlorination and oxychlorination.

In direct chlorination, chlorine and ethylene are reacted to form the EDC molecule such that two chlorine molecules are used for each molecule of ethylene. The typical reaction is between the dissolved gases in liquid over an iron chloride catalyst at 15 to 75 psig pressure and 200 to 300 deg F. The use of high purity reactants is typical to avoid the formation of a broad range of impurities which may further complicate EDC purification. An excess of chlorine is typical to ensure complete ethylene conversion.

In oxychlorination, the reaction occurs over a copper chloride impregnated alumina catalyst in the liquid phase at a temperature of approximately 300 deg F and a pressure of up to 75 psig. The typical fluid bed technique offers heat transfer, prevents the occurrence of hot spots and allows more efficient catalyst regeneration. Ethylene conversion of 93 to 97% can be achieved with selectivity in EDC of 91 to 96%.

The EDC purification section removes trace amounts of impurities in the furnace feed which can inhibit the cracking reactions and cause fouling. In addition, EDC must be dry to prevent excessive corrosion in equipment and piping downstream of the reactor. Therefore the combined EDC from three different sources must be purified to greater than 99% EDC from direct chlorination, EDC from oxychlorination and EDC recovered and recycled from the cracking step. This combined EDC is washed with water and scrubbed with caustic to remove chlorine, HCl and other water soluble impurities including chloral. The crude EDC and unconverted EDC from the VCM column are combined and dried and purified in two distillation columns. In the first "Lights Column", water and lower boiling point impurities are separated as the overhead product. The bottoms product is fed to a "Heavies Column". Pure EDC of approximately 99% purity is taken as the overhead product while the heavy impurities leave the column as the bottoms product.

The pure EDC is forwarded to a furnace for thermal cracking of EDC to vinyl chloride. A typical furnace consists of a radiant section, a convection section, and a stack. The furnace fire box operates at 950 deg F. The convection utilizes combustion gases to preheat the feed before entering the firebox.

The vinyl purification section consists of two distillation columns in series which are used to separate vinyl chloride from EDC, HCl and remaining by-products. The first column ("HCL Column") distills the HCL mixture to a pure overhead product which is recycled to the oxychlorination reactor. The HCL Column is a trayed column and operates at a pressure of 135 psig. The bottoms product of the HCL Column is fed to the second column ("VCM Column") which operates at 65 psig. A VCM product of greater than 99 wt% is produced as the overhead product of the VCM Column. The bottom of the VCM Column is recycled to the Lights Column for re-purification.

A brief overview of the economic analysis for EDC/VCM plant investment indicates that the total installed cost (TIC) for a 300,000 metric ton/yr facility is in the range of $450MM to $500MM. With current EDC pricing of $245-275/ton and assuming an ethylene and chlorine cost of $1320-1362/ton and $275-310/ton, respectively and assuming the energy, labor, overhead and tax contributions are in the range of $70-85/ton results in a payback of less than five years. These rough economics do not include cash costs for land, start-up nor the capital investment required for utilities and infrastructure.

Constructing a new EDC plant at an existing facility where existing infrastructure and utilities are present will make the investment more attractive than a pure green field site. Debottlenecking an existing EDC facility is likely yield an ROR that represents another attractive investment.

Furthermore, installing or debottlenecking existing VCM and PVC facilities make this value chain one of keen interest in the coming low cost ethylene environment that is predicted.

In conclusion:

- The Shale Gas Boom is resulting in high production for C1's – C4's, which has resulted in reduced prices for these chemical and petrochemical building blocks.
- Several new grass-roots ethane crackers and ethylene plant debottlenecks to increase ethane feed capacity have been announced.
- Increased ethylene production in the U.S. is forecasted to lower ethylene prices in the long run.
- Ethylene is a central building block in producing various chemicals and petrochemicals.
- Therefore, a large sector of the U.S. chemical industry will be affected.
- Ethylene derivatives benefit from reduced ethylene prices.
- U.S. exports of ethylene derivatives are forecasted to increase to supply the world-wide market.
- The U.S. can compete well in the world chemical and petrochemical market place with the advent of the Shale Gas Plays.
- Chlorine cost of production is and will continue to benefit from the low energy prices in the U.S.
- EDC and VCM are examples chemicals that are forecasted to benefit as more ethylene production comes on line along with consequent lower ethylene pricing.
- The economics of producing EDC and VCM in a low cost ethylene environment is compelling.
- The EDC->VCM->PVC value chain will benefit. Export of these chemical products to the world-wide market is forecasted to increase in the future.

The Shale Gas Boom has shifted the U.S. energy picture as well as the outlook for chemical and petrochemical production. The U.S. is once again a key actor and leader in the world chemicals stage.

This truly is in a new era of Technology!

347676 Opportunities in Chemical Engineering Being Created by New US Shale Gas Supplies

Abstract:

Recent technical breakthroughs in extraction techniques have allowed the combination of hydraulic fracturing with long horizontal wells to enable the release of unconventional hydrocarbon sources. The ability to tap into these previously unreachable shale gas resources has completely transformed the US natural gas and crude oil supply and demand outlook. These recent technology breakthroughs have resulted in natural gas production in the United States increasing by 25% over the last 5 years, largely due to shale gas production. New US natural gas supplies are playing a key role in "turbocharging" US manufacturing over the next 10 to 20 years, creating an industrial renaissance. This paper summarizes potential opportunities in chemical engineering that could arise as a result of new shale gas production.

(97c) Shale Gas Boom and Tight Oil - a New Era for the Refining Industry

Lockhart, M. W., Burns & McDonnell

Varraveto, D.,

Originally presented on: 4/1/2014 15:30:00 - 16:00:00

Production of Tight Oil from the Shale Gas Plays in the U.S. is having a profound effect on the Refining Industry. Refiners are maximizing the amount of advantaged, "tight oil", they can feed their crude units some with geographic locations that allow them to capitalize the most on the sometimes attractive spreads provided by what is mostly light and sweet crude. New topping columns, crude units and condensate splitters are being constructed in various refineries in the U.S. This paper provides an overview of some of the studies and projects that are taking place and provides an overview across the spectrum of possibilities that refiners are facing. Burns & McDonnell has much experience with actual projects that will be highlighted to provide an exciting overview of this activity.

347649 Economically Treating Sour Gas from Tight Oil Formations

Abstract:

GT-SPOC – Economically treating sour-gas in the shale plays

In the shale plays, there are pockets of sour gas, or sour associated gas. Treating this gas to pipeline specifications with amine units gets the gas to pipeline specifications. However, operators have to deal with the resulting acid gas stream that cannot generally be vented due to permit limits on sulfur emissions. If the gas is not very sour, sulfur scavengers are the obvious solution, however, as the gas contains more than 0.1 ton/day of sulfur, the treating method of choice is not as clear.

GT-SPOC offers an alternative to the conventional furnace based "modified Claus" method for sulfur recovery. The technology is very suitable in the 0.5 – 20 ton/day sulfur production range, which is typical of the sulfur rates we see in the shale plays.

The approach improves upon the Claus process which is used to produce >95% of sulfur around the world, by using a patented catalyst at the front of the plant for SO2 production. The catalyst utilizes constituents that address typical operator issues relating to heavy hydrocarbon contamination of the gas, start-up and shut-down issues, sour-gas composition variations and turndown. The presentation will discuss the capital and operating economics, as well as specific technical aspects of the technology that allow modularization, reduced footprint, and a tremendous reduction in capital cost.

352260 Monetising Gas Hydrates: New Mode for Natural Gas Storage and Transportation Against LNG Storage and Transportation

Abstract:

Gas Hydrates are considered in the petroleum industry e.g. natural gas transportation through pipelines, during natural gas production from the crude oil wells. But this paper focuses on converting the nuisance of gas hydrates to an attractive means of transportation. New Technology development has been focusing on using gas hydrates as a way of converting gas to solids.

The Natural Gas Hydrate (NGH) is a clatherate compound in icy state and stabilised around -20 degree Celsius under atmosphere. This paper summarises about the cost effectiveness of the NGH value chain against the LNG (Liquefied Natural Gas) value chain and also in this paper entire processes and operations to create NGH pellets are designed and discussed. The calculations show that this process has a high initial plant cost but on a long term this process of NGH production is cost effective compared to LNG liquefaction process. Moreover, it is easy to handle the NGH at -20 to -30 degree Celsius compared to LNG at around -160 degree Celsius.

Thus, this unique method of transporting the natural gas in bulk solid form leads to reduction of the total investment cost for the entire value chain. Moreover, NGH is expected to be a plausible solution in monetising small to medium gas fields which are considered to be difficult to be developed economically using LNG.

352647 Towards the Development of a Control-Relevant Model of the Hydraulic Fracturing Process to Investigate Various Control Strategies

Abstract:

Hydraulic fracturing or 'fracking' is an increasingly important technique for developing energy resources across both the U.S. and the world, in which high pressure fluid is injected into the reservoir to crack the rock formation and provide highly conductive channels for oil and gas transport back to the wellbore. During this process, the underground environment is characterized by complex fluid dynamics, geo-mechanics, and multiphase flow phenomena. Surface equipment exerts control over the process from between hundreds of feet to several miles from the moving front of the fracture(s) below ground. The fluid pumped underground is a slurry consisting of water, proppant (often a special type of sand) to keep the created fractures open after the process finishes, and chemical additives such as cross-linkers, acids, and surfactants to maintain fluid viscosity, reduce corrosion, and perform other duties. During the fracturing process, pumping rate and fluid properties (viscosity and the concentrations of proppant and other additives) can be varied at the surface. The main objectives are to keep the fracture growth within a predefined region of the reservoir open by distributing the proppant evenly throughout the fracture so that oil and gas can flow after the fracturing fluids have leaked into the formation. Complications can include excessive growth of the fracture external to the target region that results in the proppant being spread so thinly that it is ineffective to keep the fracture open. Conversely, if a high concentration of proppant is added to the fracturing fluid mix too early in the process the condition called 'screen-out' can occur wherein the proppant enters the advancing fracture front and bridges the tip preventing further fracture growth and causing rapid pressure buildup in the system. A related issue is proppant settling, that is a large sand bank or dune forms near the entrance of the fracture at the wellbore resulting in only a small open region around the wellbore after the fluid leaks off. These three potential proppant-related challenges may have a large negative impact on oil well production.

Applying process control techniques to regulate the hydraulic fracturing process, via setting optimal pumping schedules, is a challenging problem not only due to the remoteness of the control action, but also due to the limited feedback information available because of the sensing system associated with the bottom-hole situation. To better understand these challenges, a mathematical model of the physics of the fracturing system is developed to determine the effect of different pumping schedules on the final fracture conductivity, the most widely used calculated performance metric. The physics-based model incorporates equations governing fracture propagation, fluid dynamics, proppant transport, and bank formation. The resulting system of nonlinear, partial differential equations is solved numerically using a moving grid method to track the fracture growth and particle tracking to predict the proppant movement and settling. Additionally, the 'shut-in' process, during which pumping is stopped and the rock formation walls are allowed to close as the fluid leaks off is a part of this model development. The final geometry of the channel available to oil and gas flow is predicted. Using this model, improved pumping schedules can be developed and tested to achieve near-optimal fracture conductivity.

352284 Shale Gas as a Potential Source of Unconventional Reserve for Crude Oil

Abstract:

With a decrease in the conventional reserves available for the recovery of oil, the alternative fossil fuel extraction industry is on a growth. As the development of unconventional oil and gas resources has been progressed, the heterogeneity and complexity of shale gas and tight oil reservoirs have become apparent. In the U.S. primarily shale gas and tight oil is expected to continue growing dramatically in coming decades. According to the U.S. Energy Information Administration, an assessment of 137 shale formations are identified in 41 countries other than those identified in U.S. At the local level, though the environmental impacts of shale gas and tight oil development may turn out to be significant.

Shale gas and tight oil are unconventional sources of petroleum that are found within the reservoirs with a very low permeability. The oil and gas contained within these reservoir rocks typically will not flow to the wellbore at the economic rates without assistance from technologically advanced drilling and completion processes. Usually, horizontal drilling coupled with multi stage fracturing is used to access these difficult to produce reservoirs. Unconventional natural gas and oil deposits are difficult to characterize overall, but in general are often lower in resource concentration, more dispersed over large area, and require well stimulation or some other extraction or conversion technologies.

Hydraulic Fracturing

Water Reuse Challenges & Solutions

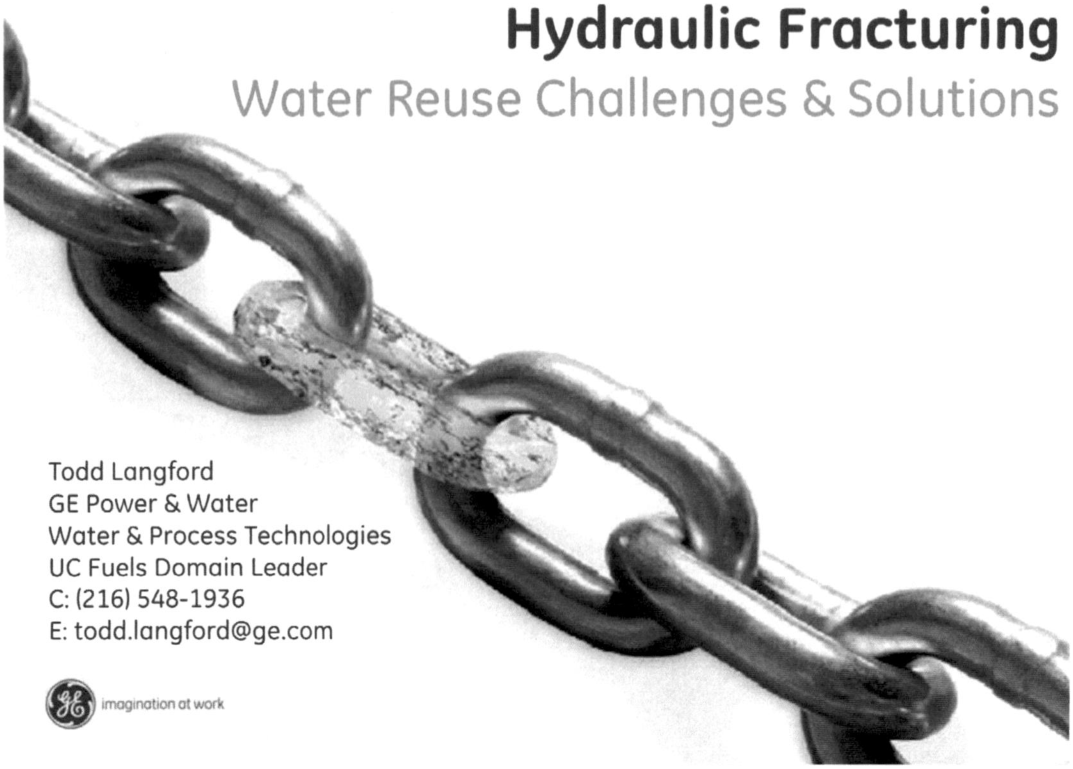

Todd Langford
GE Power & Water
Water & Process Technologies
UC Fuels Domain Leader
C: (216) 548-1936
E: todd.langford@ge.com

imagination at work

Hydraulic fracturing

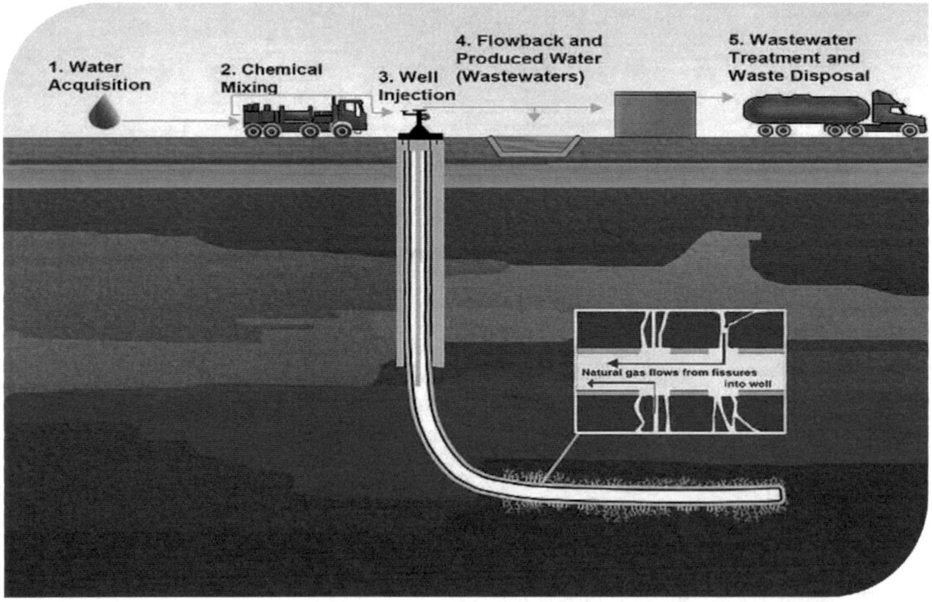

imagination at work

Source: US EPA

Shale gas *Evolution*

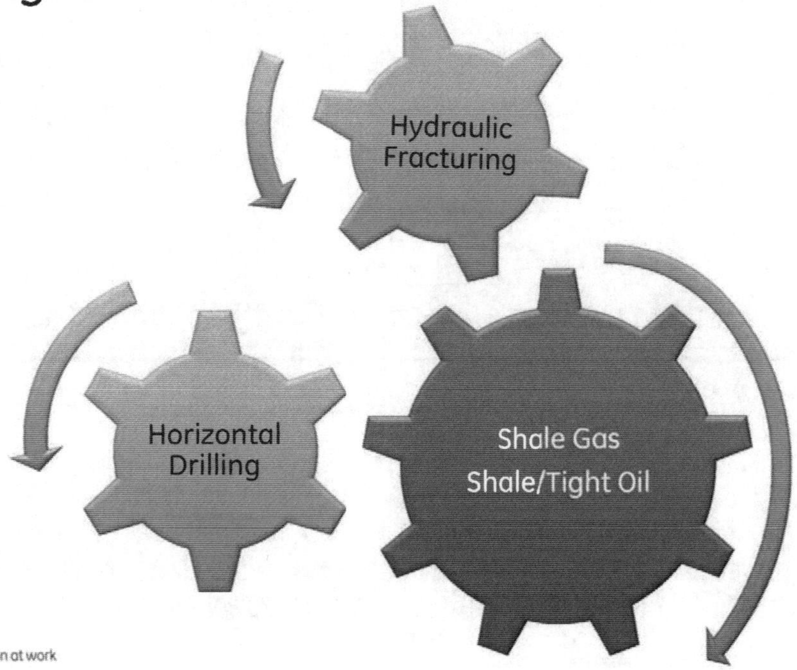

imagination at work

3
Langford
March 2014

Active shale and tight oil plays

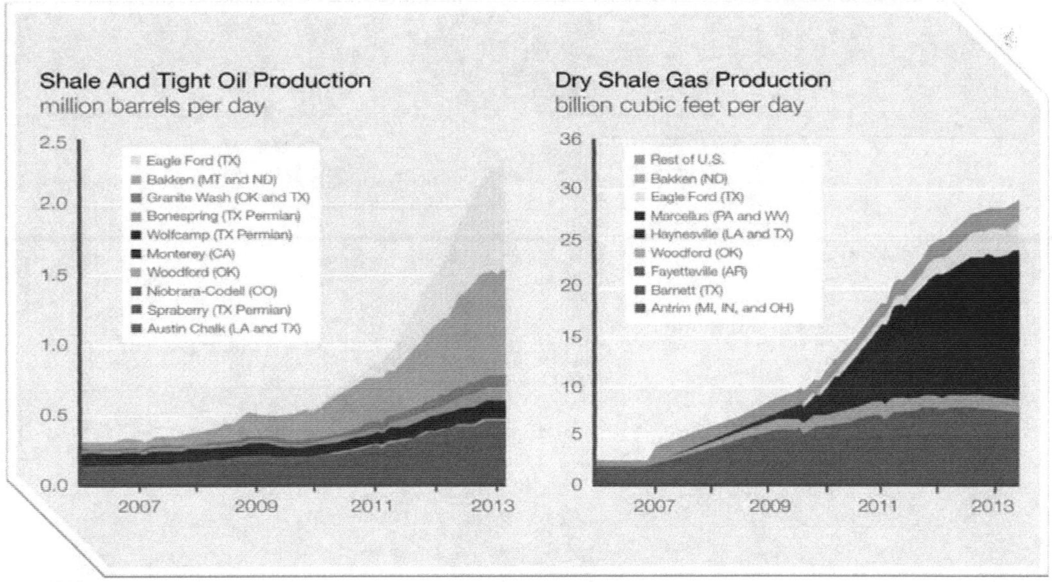

imagination at work

Source: American Petroleum Institute

4
Langford
March 2014

Energy demands on water resources

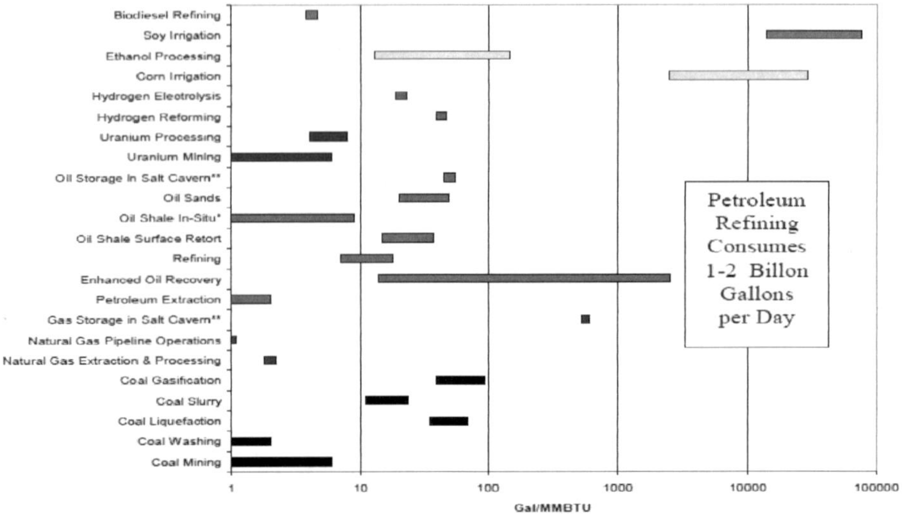

Figure V-4. Water Consumption Per-Unit-Energy and Current Water Use for Fuel Extraction and Processing
See Appendix B for Data References

imagination at work

Source: U.S. Department of Energy, 2006
Report to Congress: Interdependency of Energy and Water

5
Langford
March 2014

What's driving recycle/reuse?

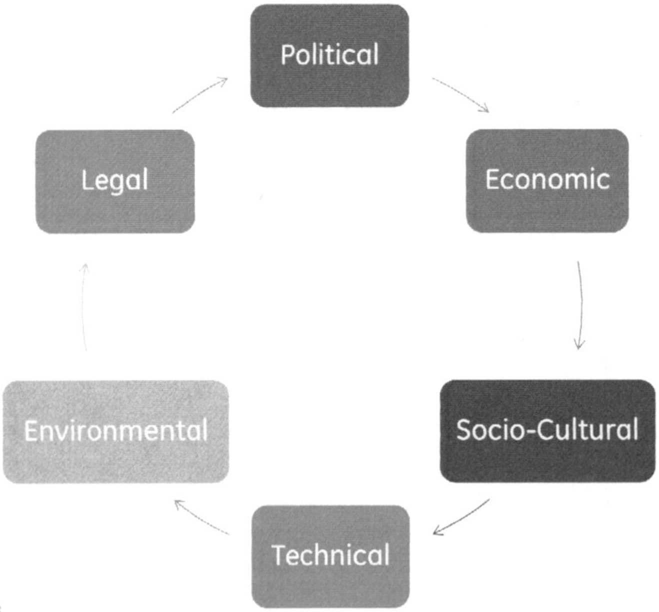

imagination at work

6
Langford
March 2014

Enviro concerns in unconventional O&G

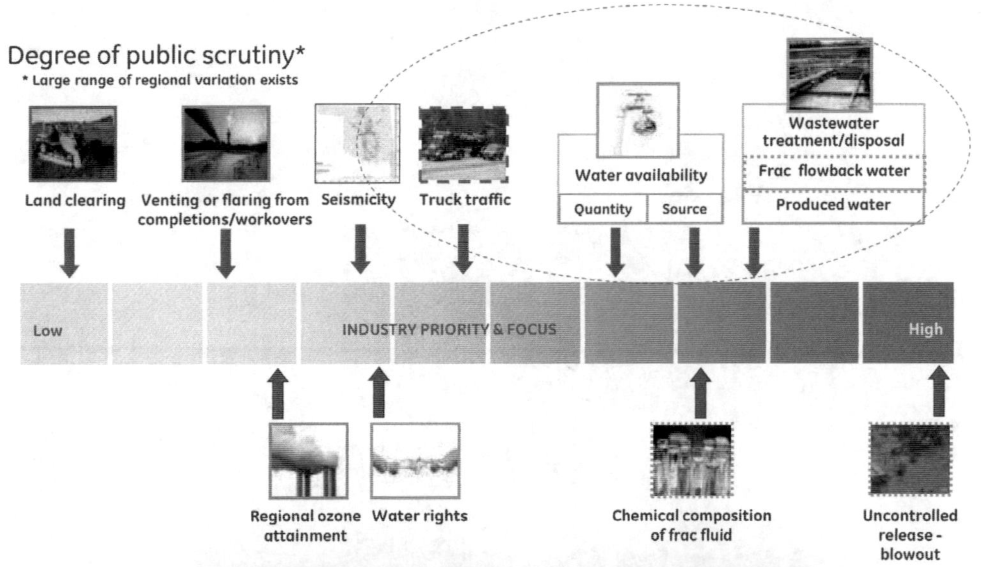

Water issues elevated in national debate
Water issues are LOCAL

imagination at work

Source: GE Energy, Global Strategy & Planning; based on customer discussion not a statistical survey

7
Langford
March 2014

Key drivers for projects to move forward

Regulatory *...limited disposal + discharge standards*

Economic *...expensive trucking to disposal*

Sustainable *...multi-generational planning*

imagination at work

8
Langford
March 2014

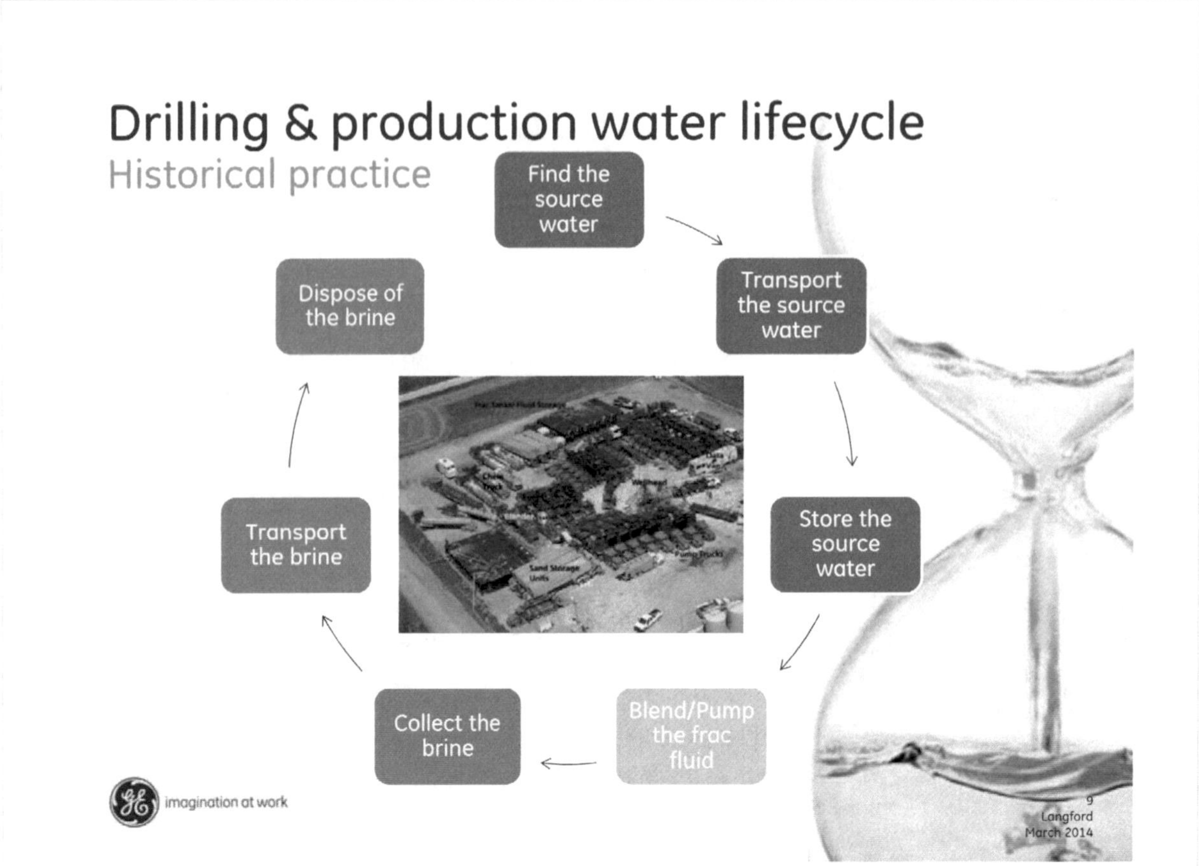
Drilling & production water lifecycle
Historical practice
Find the source water
Transport the source water
Store the source water
Blend/Pump the frac fluid
Collect the brine
Transport the brine
Dispose of the brine
imagination at work
9
Langford
March 2014

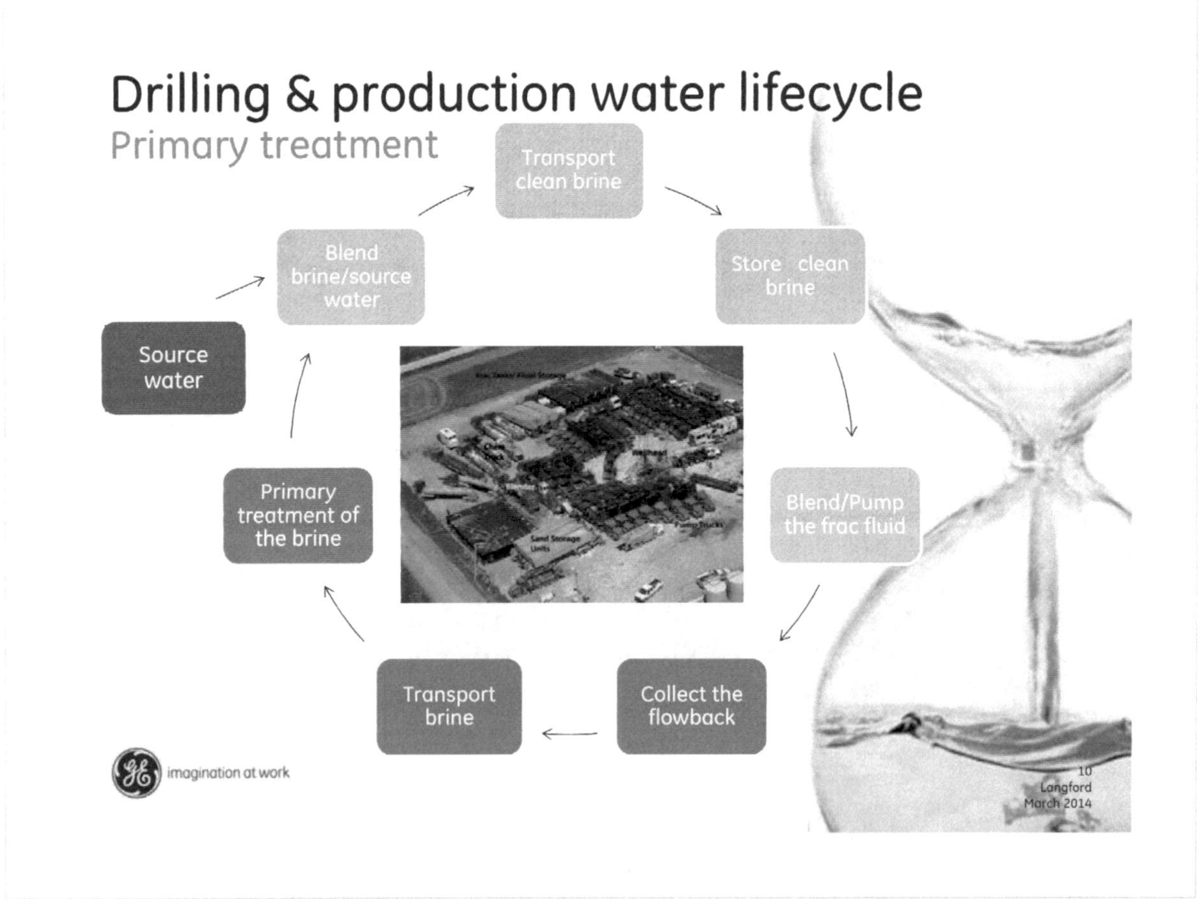
Drilling & production water lifecycle
Primary treatment
Transport clean brine
Store clean brine
Blend/Pump the frac fluid
Collect the flowback
Transport brine
Primary treatment of the brine
Source water
Blend brine/source water
imagination at work
10
Langford
March 2014

Drilling & production water lifecycle

Advanced treatment

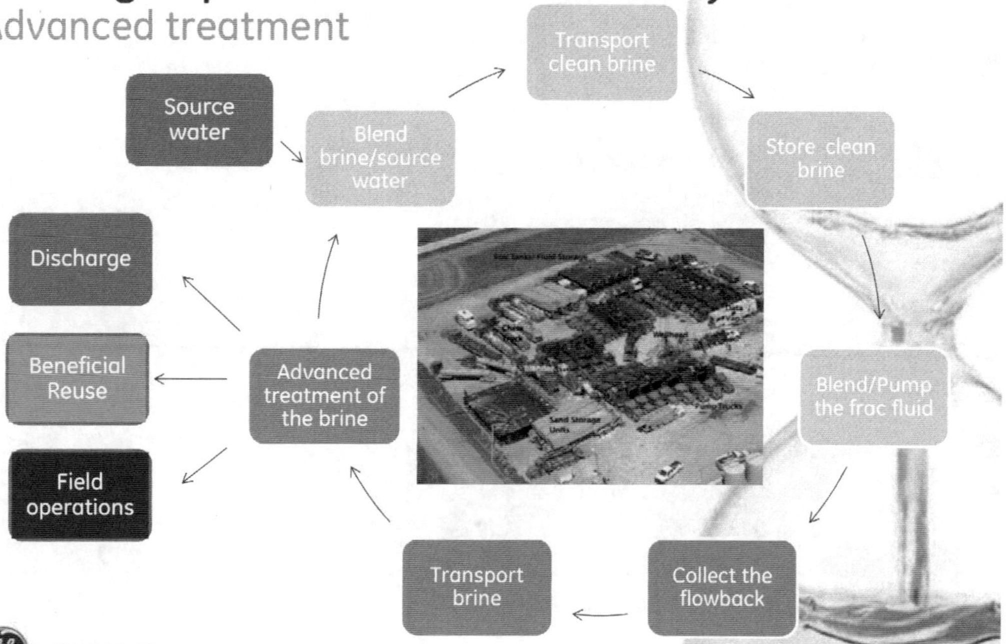

imagination at work

11
Langford
March 2014

imagination at work

12
Langford
March 2014

Source: FracFocus.org

Flowback/produced water economics

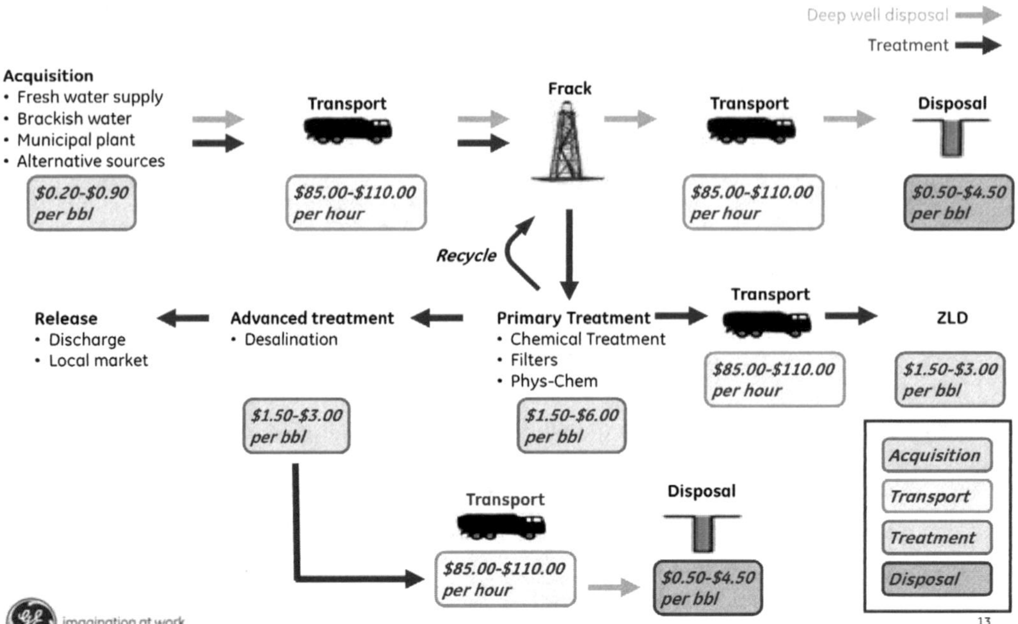

imagination at work

Water spend

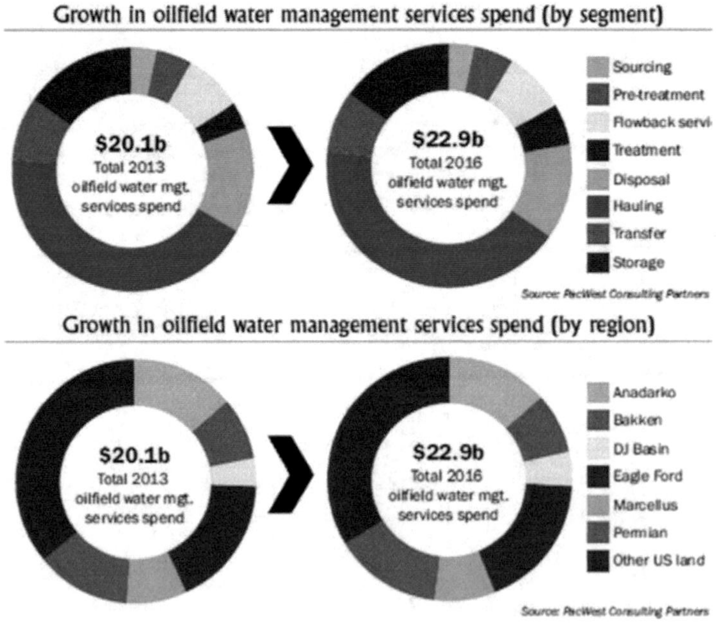

imagination at work

Source: PacWest Consulting Partners/GWI March 2014

Example frac flowback & produced water

(SINGLE WELL)

Day	Percent of Fracture	Barrels
1	7.56%	6124.0
2	7.56%	6119.0
3	7.50%	6071.4
4	7.44%	6023.8
5	7.26%	5881.0
6	4.56%	3690.5
7	2.88%	2333.3
8	1.21%	976.2
9	0.88%	714.3
10	0.65%	523.8
11	0.44%	357.1
12	0.35%	285.7
13	0.26%	214.3
14	0.18%	142.9
15	0.12%	95.2
16	0.07%	57.1
17	0.04%	28.6
18	0.04%	33.3
19	0.03%	23.8
20	0.03%	26.2
21	0.04%	28.6
22	0.04%	29.8
23	0.03%	27.4
24	0.03%	27.1
25	0.04%	30.5
26	0.04%	28.6
27	0.04%	33.3
28	0.03%	23.8
29	0.08%	61.9
30	0.04%	28.6
31	0.04%	29.8

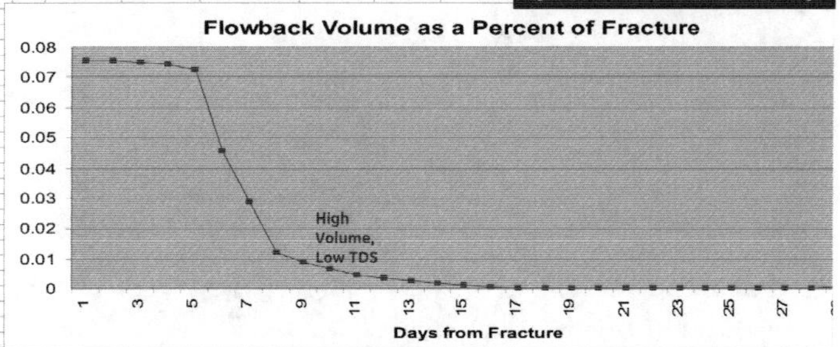

Day	Specific Gravity	Ionic Strength	TDS	pH	Ca (mg/L)	Mg (mg/L)
1	1.002	0.0083	559.09		40.00	3.72
1.5	1.005	0.1025	6437.60		379.75	31.26
2	1.009	0.2087	12773.03		935.18	77.34
3	1.017	0.3727	23818.18		1312.51	101.69
4	1.028	0.6434	40803.17		2447.30	190.57
5	1.031	0.7192	45758.81		2679.66	209.47
12	1.040	0.9371	59633.81	5.9	4395.06	381.2
15	1.052	1.2300	78681.52		4462.10	372.03
21	1.102	2.4501	156570.39		10528.19	872.28
25	1.111	2.6452	170014.09		12084.61	1013.11
31	1.121	2.9331	185728.6		14575.36	1141.79

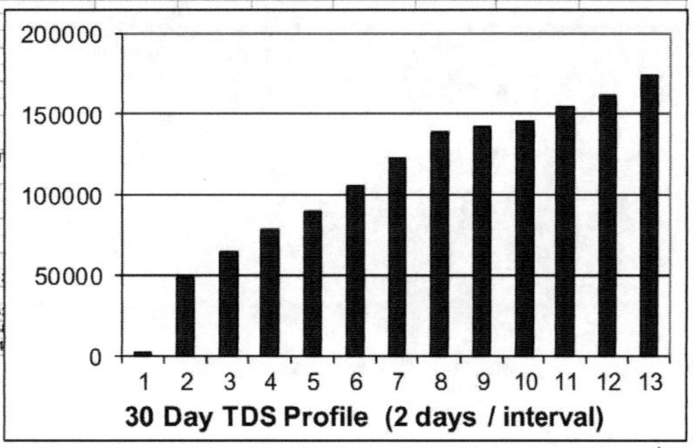

imagination at work

Source: GE Water & Process Technologies

Multiple wells...

Assumptions	
# wells	250
Days drilled apart	15
Volume per frack (gal)	4,000,000
Volume per frack (bbl)	95,238
% returned on days 1-5	5.00%
% returned on day 6-7	2.50%
% returned on day 8-9	1.00%
% returned on day 10-30	0.25%
% returned on day 30-5475	0.005%
% that returns over time	64%

No reuse + fresh water

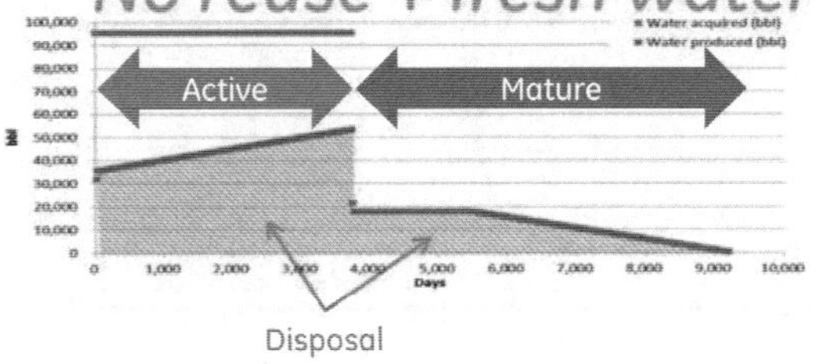

Reuse + clean brine

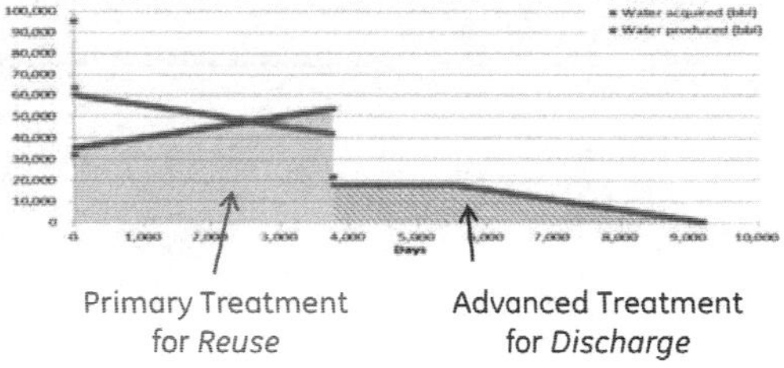

Reuse + fresh water

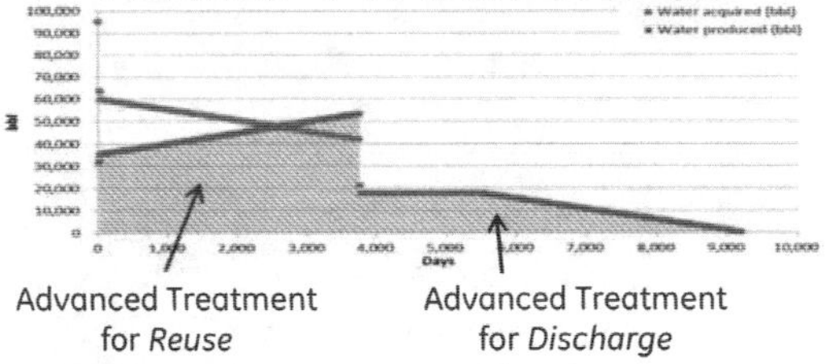

imagination at work

Recycle/reuse items to consider

Target quality

- Fracture fluid
- Discharge

Source quality

- Fresh water
- Flowback water
- Produced water
- Blended quality

Process issues

- Varying feed quality
- Varying feed quantity
- System mass balance
- System mobility
- System footprint
- Power supply
- Operations
- EHS
- Permitting

Economics

- Storage
- Transport
- Processing charge
- Process waste disposal
- Beneficial reuse

imagination at work

17
Langford
March 2014

For illustration only

Customer economics to determine value

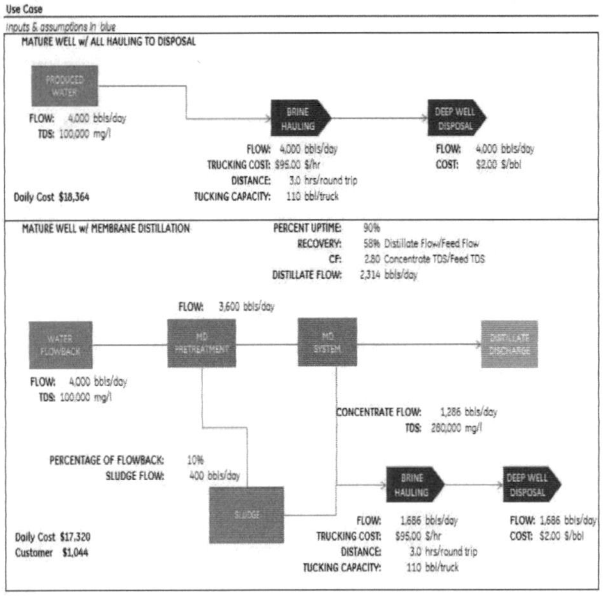

Summary

Annual customer savings			**$381,065**

Disposal Pathway

Hauling & disposal			
Brine hauling cost	$/bbl	$2.59	
+ Deep well disposal cost	$/bbl	$2.00	
= Total hauling & disposal cost	$/bbl	$4.59	
x BBL sent to disposal	bbl/day	4,000	
= Daily disposal costs			**$18,364**

VS.

Treatment Pathway

Treatment			
Volume to treat	bbl/day	4,000	
x Cost to treat	$/bbl	$2.26	
= Daily treatment cost			$9,039
Distillate (revenue)/cost			
Volume of distillate	$/bbl	2,083	
x Distillate (value)/cost	$/bbl	($0.25)	
= Total distillate (revenue)/cost	$/bbl		($520.71)
Hauling & disposal cost			
Pretreatment brine	bbl	360	
+ Concentrate flow	bbl	1,157	
+ Average daily volume caused by downtime	bbl	400	
= Average daily hauling & disposal volume	bbl	1,917	
x Total hauling & disposal cost	$/bbl	$4.59	
= Average annual hauling & disposal costs			$8,801
= Average daily cost-to-treat (w/ disposal)			**$17,320**

imagination at work

18
Langford
March 2014

Technology

	Primary Treatment			Advanced Treatment
	Preliminary	**Secondary**	**Tertiary**	**Desalination**
Application	• **TSS** • **Free/suspended oil** • **Bacteria**	• **TSS** • **Emulsified oil** • **Metals**	• **Dissolved organics**	• **Dissolved solids**
Technology	• Collection/Decant • API Separator • Hydrocyclone • Clarifier/Thickener	• Phys-Chem • MMF • Ultrafiltration • Walnut shell • Electrocoagulation • DAF, EAF • Ceramic membranes	• MBR • GAC • Optipore (DOW) • Biological Treatment • Advanced Oxidation	• EDR • BWRO • SWRO • HERO • Membrane Distillation* • Evaporation • Crystallization • Forward osmosis • Humidification-dehumidification

* Commercial scale in development

imagination at work

19
Langford
March 2014

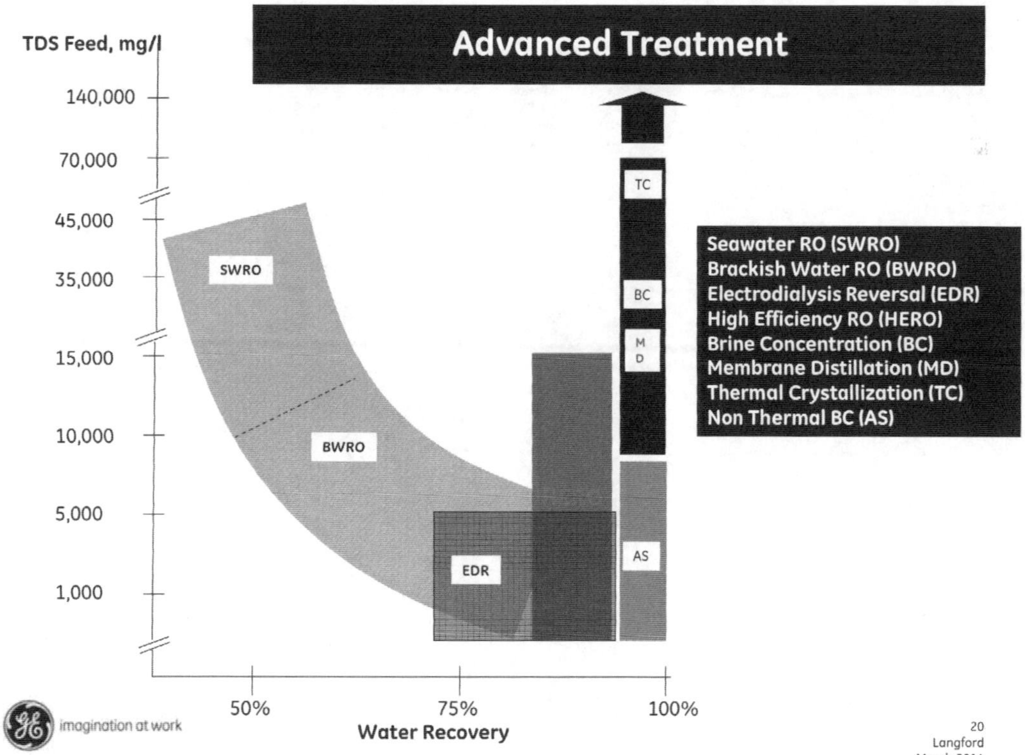

imagination at work

20
Langford
March 2014

Summary

- Evolving market
- Challenging water
- No well like its neighbor
- No silver bullet
- Evolving solutions

Need solutions tailored to today and tomorrow

imagination at work

21
Langford
February, 2014

351667 An Integrated Approach to Water Treatment and Recycle in Oil and Gas Production and Processing Via Thermal Membrane Distillation

Abstract:

Thermal membrane distillation (TMD) is an emerging technology in the field of water treatment and recycle. It is driven primarily by heat which creates a vapor-pressure difference across a porous hydrophobic membrane. It gained interest because of many advantages; it can work properly with feed temperature below boiling point and can provide high level of separation of the treated water. Because TMD relies on heating the wastewater feed stream, one of the key benefits is that it allows thermal integration with low-quality sources of heat. This thermal coupling can significantly enhance the treatment and recycle costs. The paper presents an integrated approach to the design of TMD networks. First, TMD is modeled based on key design and operating variables. Next, a framework is developed for cost-effective thermal coupling. Several applications in upstream production and downstream processing of oil and gas are presented. The results are compared with conventional separation techniques and with systems lacking heat integration.

SIMULATION AND SUSTAINABILITY ANALYSIS FOR PRODUCED WATER TREATMENT DRIVEN BY ON-SITE WASTE HEAT

Liwen Chen[1], Parag Saxena[1], Helen H. Lou[1], Lily Xu[2] and Danny Reible[3]
[1] Dan F. Smith Department of Chemical Engineering, Lamar University
[2] Environmental and Water Resources Engineering, University of Texas at Austin
[3] Department of Civil & Environmental Engineering, Texas Tech University

Topics

- Produced Water Management
- On-site energy source
- Simulation of several thermal evaporation technologies (MSF, MED, MD) in Aspen Plus
- Sustainability analysis in terms of performance, economics and environmental criteria

What's Produced Water?

3

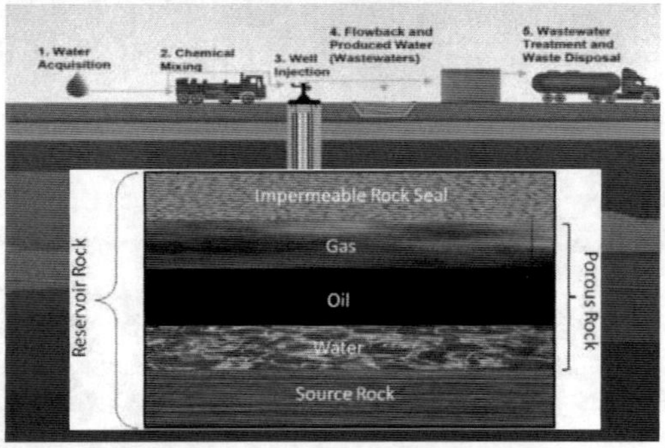

- Water that comes to the surface with oil and gas;
- Rich components (Salt, Oil and grease, Chemical additives, NORMs, HCs, etc);
- Largest volume waste stream from oil and gas production;
- Ratio of water-to-oil in the production
  - ➢World-wide estimate – 2:1 to 3:1
  - ➢U.S. estimate – 7:1
  - ➢Many older U.S. wells have ratios > 50:1!!!

source: http://www2.epa.gov/hfstudy/hydraulic-fracturing-water-cycle; E.T. Igunnu and G.Z. Chen Produced Water Treatment Technologies, International Journal of Low-Carbon Technologies 2012.

PW Composition

4

√ Salts, Medals,
√ Oil, Grease, HCs
√ Radioactive materials
...

Total Dissolved Solids (TDS): 8000 ~ 360,000 ppm

Parameter	Minimum value	Maximum value		Heavy metal	Minimum value (mg/l)	Maximum value (mg/l)	
Density (kg/m^3)	1014	1140		Calcium	13	25 800	Cations
Conductivity (μS/cm)	4200	58 600		Sodium	132	97 000	
Surface tension (dyn/cm)	43	78		Potassium	24	4300	
pH	4.3	10		Magnesium	8	6000	
TOC (mg/l)	0	1500		Iron	<0.1	100	
TSS (mg/l)	1.2	1000		Aluminium	310	410	
Total oil (IR; mg/l)	2	565		Boron	5	95	
Volatile (BTX; mg/l)	0.39	35		Barium	1.3	650	
Base/neutrals (mg/l)	—	<140		Cadmium	<0.005	0.2	
Chloride (mg/l)	80	200 000	Anions	Copper	<0.02	1.5	
Bicarbonate (mg/l)	77	3990		Chromium	0.02	1.1	
Sulphate (mg/l)	<2	1650		Lithium	3	50	
Ammoniacal nitrogen (mg/l)	10	300		Manganese	<0.004	175	
Sulphite (mg/l)	—	10		Lead	0.002	8.8	
Total polar (mg/L)	9.7	600		Strontium	0.02	1000	
Higher acids (mg/l)	<1	63		Titanium	<0.01	0.7	
Phenol (mg/l)	0.009	23		Zinc	0.01	35	
Volatile fatty acids (mg/l)	2	4900		Arsenic	<0.005	0.3	
				Mercury	<0.005	0.3	
				Silver	<0.001	0.15	
				Beryllium	<0.001	0.004	

Rich components, Wide concentration range, High Salinity!

1% = 10,000 ppm (mg/L)

Source: Igunnu et al. 2012 Produced water treatment technologies

Federal & State Regulations

Federal:

- U.S. Environmental Protection Agency (EPA);
- Department of the Interior's Bureau of Land Management (BLM);
- Bureau of Ocean Energy Management, Regulation and Enforcement (BOEMRE, formerly MMS);

EPA requirement for oil and grease in PW (NPDES):

Offshore [40CFR§435.13]:

- Daily Maximum Discharge (EPA): **42** mg/L
- Monthly average limit: **29** mg/L

Onshore [40CFR§435.32]:

- Zero discharge

State:

- Regulations varied widely;

For example, Texas Commission on Environmental Quality (TCEQ)

Rule 3.8 (Water Protection);Rule 3.9 (Disposal Wells)

Require permits to discharge into surface/subsurface water

source: http://www.netl.doe.gov/research/coal/crosscutting/pwmis/fed-state-regulations

PW Management

Three hierarchy:

Minimization

- **Prevent PW entering well** (Mechanical blocking, Water shut-off chemicals)
- **Prevent PW reaching the surface** (Downhole separation, Sea floor separation)

Recycle/reuse

- **Injection for enhanced oil recovery** (TX: ~38540 wells (5.3 billion bbl/yr))
- **Injection for disposal** (TX: ~12000 wells (1.2 billion bbl/year), PA: 8 wells!)

Treatment

- **Dilution** (blending with freshwater) → ~$1.5-2/bbl
- **Settling** (allow enough residence time for solids to settle) → ~$2-2.5/bbl
- **Filtration** (sand filters, bag filters, disk filters) → ~$2-3/bbl
- **Chemical Precipitation** (pH adjustment, flocculants) → ~$2.5-4/bbl
- **Evaporation** (Highest effluent quality with highest energy demand. Energy cost can be mitigated by waste heat) → ~$5.5-8/bbl

Permeate Standard (WHO): TDS ≤ 500mg/L

source: Veil J.A. 2009 Produced Water Management Options –One Size Does Not Fit All

7

Gas Flaring

Why Flare
- Safety concerns:
  - ➢ Equipment malfunction;
  - ➢ Off-spec production;
  - ➢ Depressurizing equipment;
  - ➢ Start ups and emergency shut-downs;
- Lack of infrastructure (pipeline)

1/3 flared in Bakken Shale, $100 million/month!

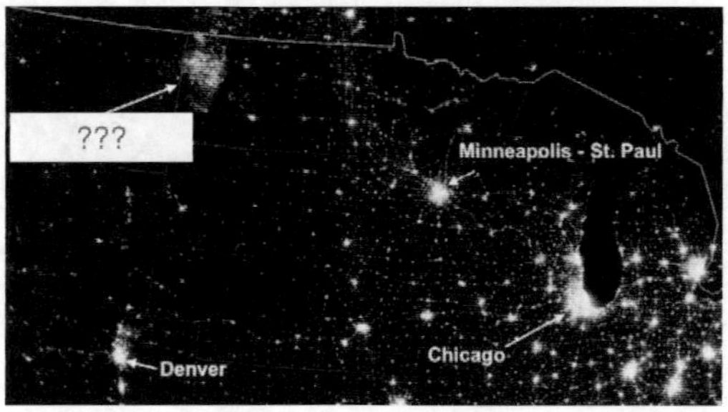

source: http://www.pennenergy.com/articles/pennenergy/2013/07/bakken-shale-natural-gas-flaring-tops-100-million-each-month.html

8

Energy Savings

- 5% of global gas production (150 Billion m³/yr)
- 23% US NG use, 30% EU NG use
- 160 MMCFD NG in Bakken Shale flared out of 480 MMCFD production

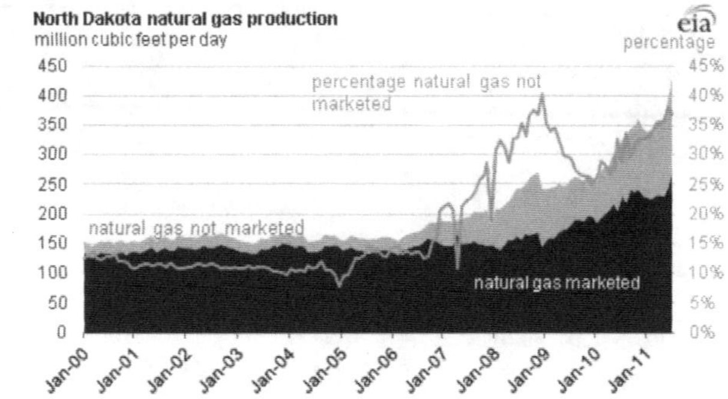

Source: U.S. Energy Information Administration, based on the North Dakota Department of Mineral Resources.

9

Flare Minimization Strategies

Legislation
- Kyoto Protocol
- Global Gas Flaring Reduction (GGFR);
- Economic Incentives

Reduction
- Reinjection EOR
- Compress to liquid;
- Dynamic Control System

Utilization
- Heat source for co-generation

10

Energy-Water Nexus Scheme

Pros:
Cost becomes profit;
Reduce environmental impacts;
On-site water treatment supply;
…

Cons:
Unsteady supply;
Safety issue;
Infrastructure;
Capital cost;
…

11

PW Management Desalination Technologies

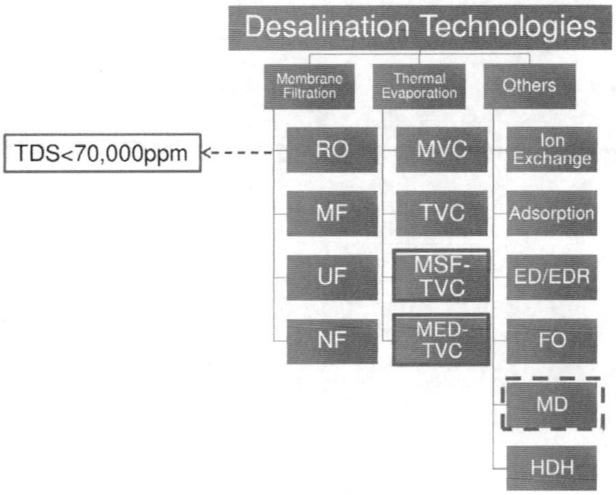

source: Shaffer et al. Environ. Sci. Technol. 47 (2013)

12

Sustainability Analysis

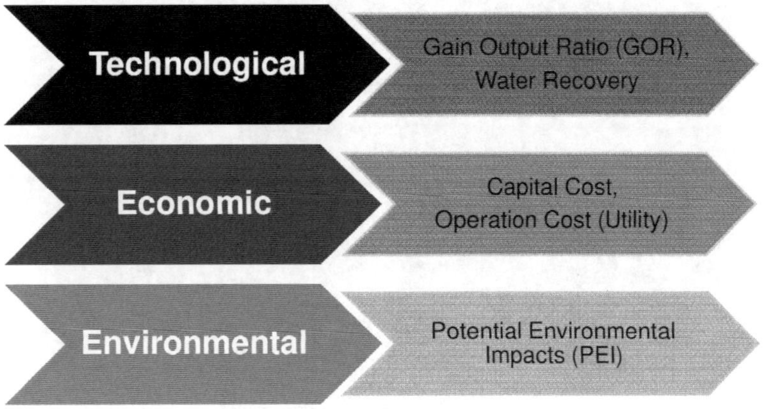

13

Environmental Impact Analysis

- WAR (Waste Reduction) Algorithm
  - Developed by Young & Cabezas (EPA), 1999
  - Database on environmental impacts of chemicals
  - 4 general categories, 8 sub-categories
  - http://www.epa.gov/nrmrl/std/war/sim_war.htm
- PEI (Potential Environmental Impact) index to represent environmental friendliness

  "A low PEI score indicates a lower environmental impact."
- Green house gas emission by CO_2 is the major potential environmental impact

14

WAR Algorithm: Impact Categories

General Impact Category	Impact Sub-Category	Measure of Impact Category
Human toxicity	Ingestion	HTPI
	Inhalation/Dermal	HTPE
Ecological toxicity	Aquatic toxicity	ATP
	Terrestrial toxicity	TTP
Global atmospheric impacts	Global warming potential	GWP
	Ozone depletion potential	ODP
Regional atmospheric impacts	Acidification potential	AP
	Photochemical oxidation potential	PCOP

Potential Environmental Impact (PEI)

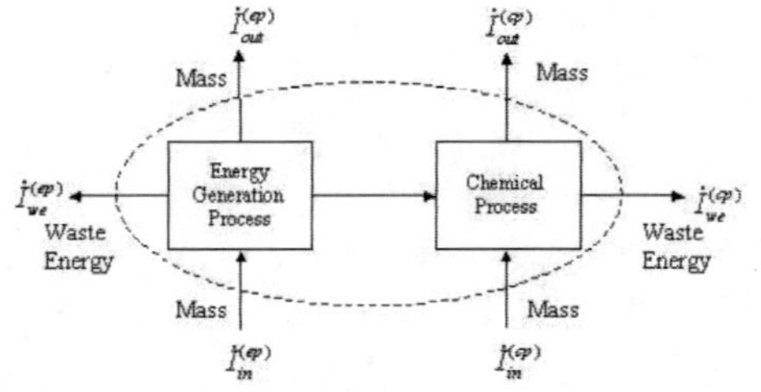

The potential environmental impact balance is as follows:

$$\cancel{\frac{\partial I}{\partial t}} = I_{in}^{(cp)} + \cancel{I_{in}^{(ep)}} - I_{out}^{(cp)} - I_{out}^{(ep)} - \cancel{I_{we}^{(cp)}} - \cancel{I_{we}^{(ep)}} + I_{gen}^{(t)}$$

$$\boxed{0 = \dot{I}_{in}^{(t)} - \dot{I}_{out}^{(t)} + \dot{I}_{gen}^{(t)}}$$

The total input rate of PEI:

$$\dot{I}_{in}^{(t)} = \sum_{i}^{EnvCat} \alpha_i \dot{I}_{i,in}^{(t)} = \sum_{i}^{EnvCat} \alpha_i \sum_{j}^{Streams} \dot{M}_{j,in} \sum_{k}^{Comps} x_{kj} \psi_{ki}^{s}$$

The total output rate of PEI:

$$\dot{I}_{out}^{(t)} = \sum_{i}^{EnvCat} \alpha_i \dot{I}_{i,out}^{(t)} = \sum_{i}^{EnvCat} \alpha_i \sum_{j}^{Streams} \dot{M}_{j,out} \sum_{k}^{Comps} x_{kj} \psi_{ki}^{s}$$

To evaluate the PEI output index on a production basis,

$$\boxed{\hat{I}_{out}^{(t)} = \frac{\dot{I}_{out}^{(t)}}{\sum_{p}^{ProdStreams} \dot{P}_p}}$$

17

Multi-Stage Flash (MSF) Process

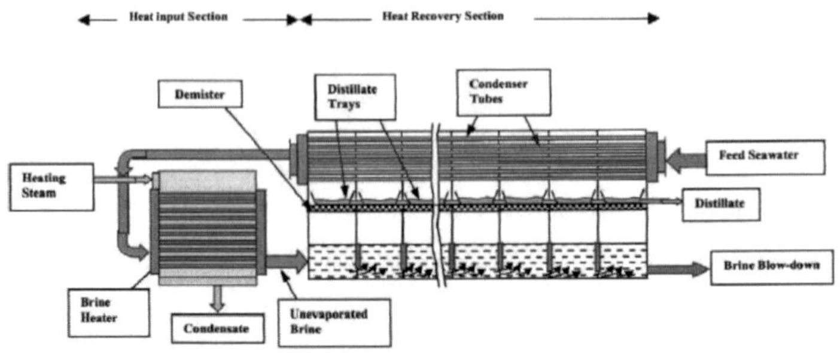

MSF-OT (once-through w/o brine recirculation)

Source: El-Dessouky et al./ Chemical Engineering Journal 1999

18

Specification

Feed Components: Na^+, K^+, Ca^{2+}, Mg^{2+}, Cl^-, HCO_3^-, SO_4^{2-} ($CaSO_4$ ↓)
Feed Flowrate: 10,000 ton/day

Three Cases: (Low TDS, Medium TDS, and High TDS)

Location	Powder River (MT)	Barnett Basin (TX)	Bakken Basin (ND)
Compnents (mg/L)	Low	Medium	High
Na+	6022	20935	71466
K+	75	61	512
Ca++	327	5441	8993
Mg++	93	1734	1467
Cl-	9074	45989	129461
SO4--	1529	1394	1885
HCO3-	1329	408	322
TDS	18447	75962	214106

Low TDS (Powder River) (<35,000ppm)

Medium TDS (Barnett) (35,000~200,000ppm)

High TDS (Bakken) (200,000~400,000ppm)

Source: US Geological Survey Produced Water Database

19

Multi-Stage Flash w/ Thermal Vapor Compression (MSF-TVC) Simulation in Aspen Plus

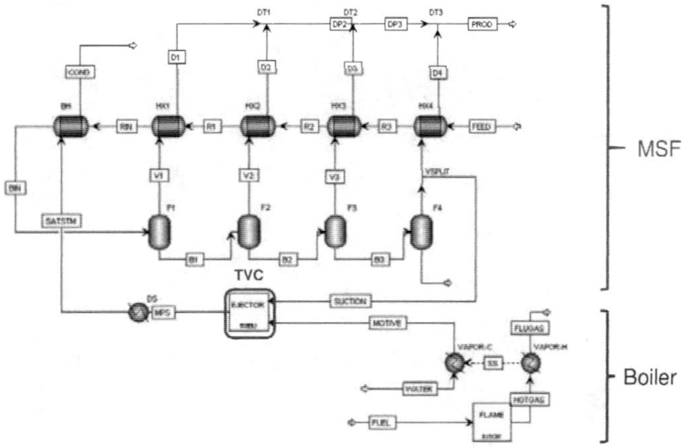

20

Simulation Results of MSF-TVC

MSF-TVC stages		F1	F2	F3	F4
Temperature (°C)		90	88	86	84
Pressure (bar)		0.671	0.618	0.568	0.521
Vapor (ton/day)		970	873	786	707
Prod (ton/day)		2840			
Steam Supply	(°C)	133.6			
	(bar)	3			
	(ton/day)	440			
GOR (Prod/Steam)		6.5			
Product Recovery (Prod/Feed)		28.4%			
Energy Consumption (MMBTU/day)		1099			
Utility Cost ($/day)		4130			
CO_2 Emission (ton/day)		62			

5-year (2009-2013) average Henry Hub NG Price = $3.76/MMBTU (http://www.eia.gov)

Case Results

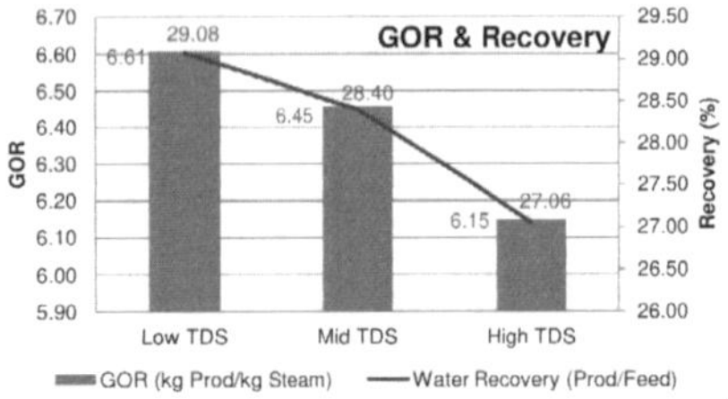

	Utility Cost
$/ton	1.42-1.53
cents/gal	0.54-0.58
cents/bbl	16.94-18.21

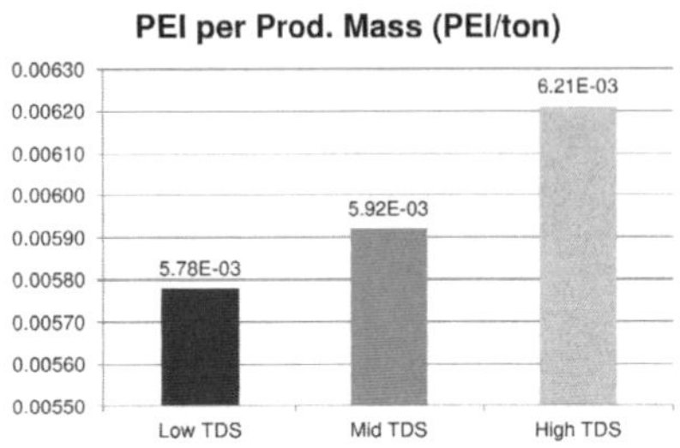

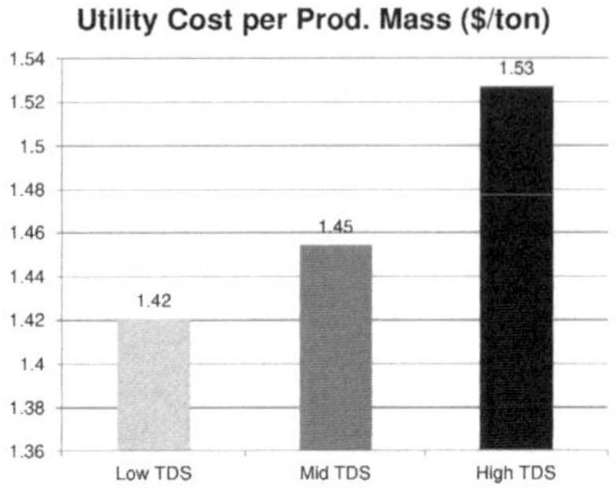

PEI in each categories (MSF-TVC)

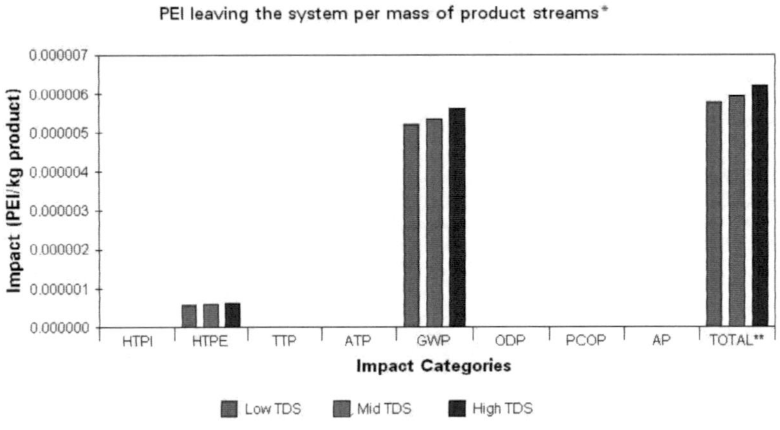

HPTE: Human Toxicity Potential by Inhalation or Dermal Exposure
GWP: Global Warming Potential

Multi-Effect Distillation (MED)

Features:

- Higher heat transfer coefficient;
- Max Brine Temperature 70°C (avoid scalants);
- Economic and durable material (aluminum alloy vs. copper);
- Possibility of using low grade heat;

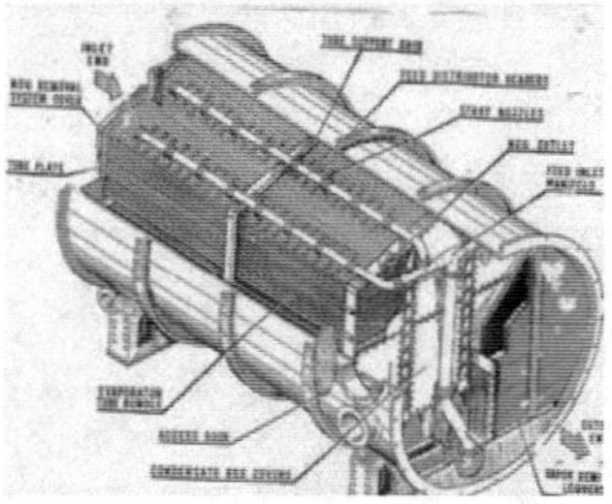

Horizontal tube assembly
A. Ophir, F. Lokiec et.al. Desalination 182 (2005)

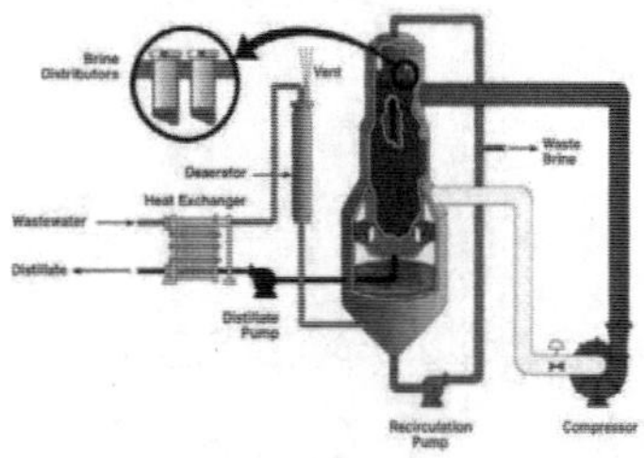

Falling Film Evaporator
(GE Power & Water 2012)

MED-TVC Schematic Diagram

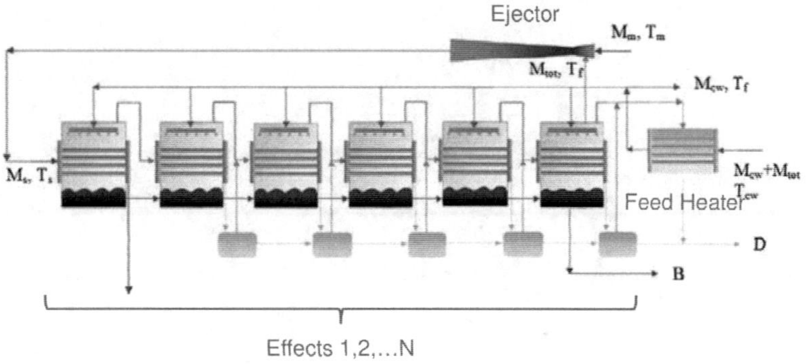

source: Manesh et al. Desalination 316 (2013)

MED-TVC Simulation in Aspen Plus

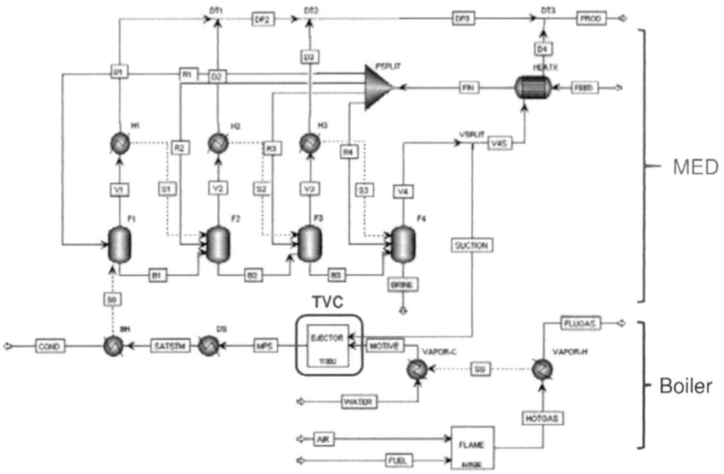

Results for MED-TVC

MED-TVC stages		F1	F2	F3	F4
Temperature (°C)		60	56	52	48
Pressure (bar)		0.182	0.152	0.125	0.103
Vapor (ton/day)		1156	1099	1067	1059
Prod (ton/day)		3640			
Steam Supply	(°C)	133.6			
	(bar)	3			
	(ton/day)	500			
GOR (Prod/Steam)		7.3			
Recovery (Prod/Feed)		36.4%			
Energy Consumption (MMBTU/day)		1249			
Utility Cost ($/day)		4693			
CO_2 Emission (ton/day)		71			

5-year (2009-2013) average Henry Hub NG Price = $3.76/MMBTU (http://www.eia.gov)

27

MED-TVC Results Summary

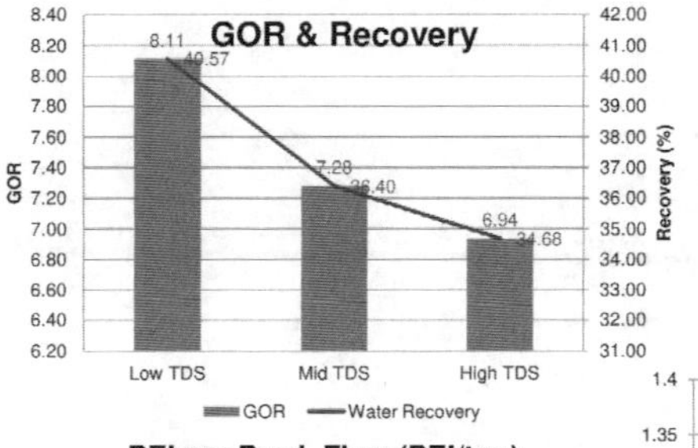

	Utility Cost
$/ton	1.16-1.35
cents/gal	0.44-0.51
cents/bbl	13.80-16.14

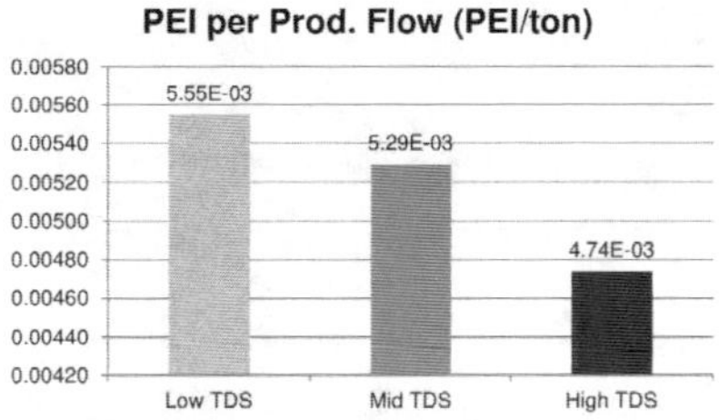

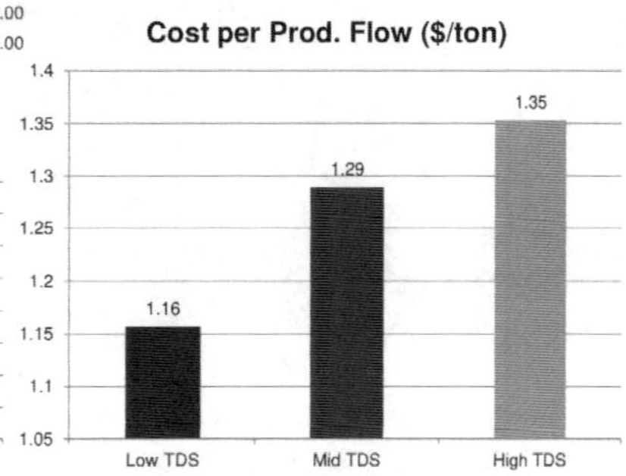

28

Membrane Distillation (MD)

Advantages:

- Lower operating temperatures and vapor space than MSF & MED;
- Lower operating pressure than RO;
- 100% rejection of non-volatile solute;
- Performance not limited by high osmotic pressure or concentration polarization;

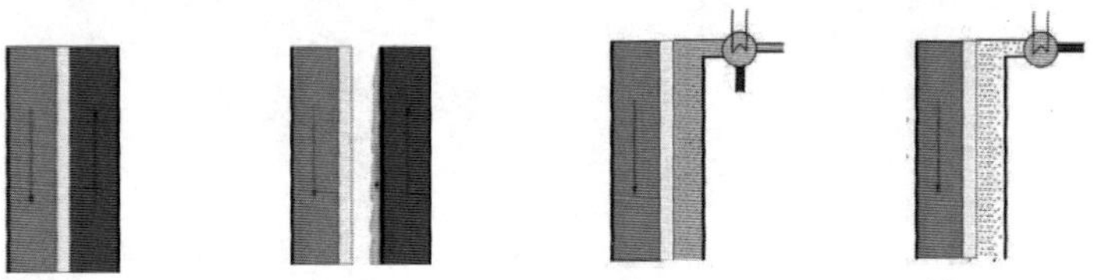

Direct Contact | Air-gap | Sweeping Gas | Vacuum

Source: Wikipedia

29

DCMD Concept

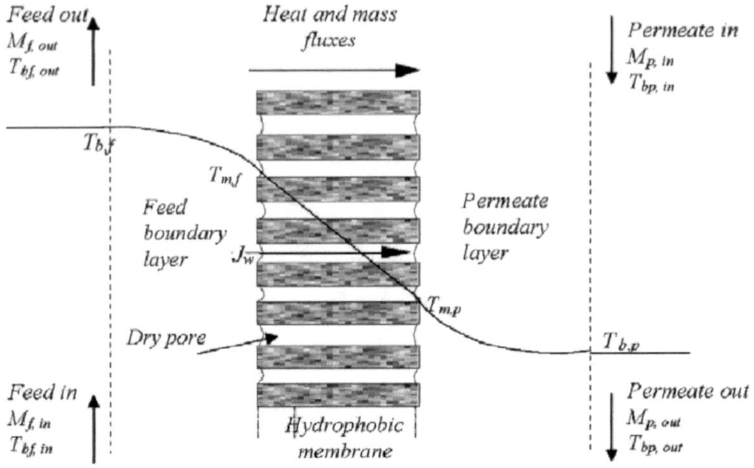

Source: M. Qtaishat et al. / Desalination 219 (2008) 272–292

30

Simulation of DCMD in Aspen Plus

15-20 effects for good approximation

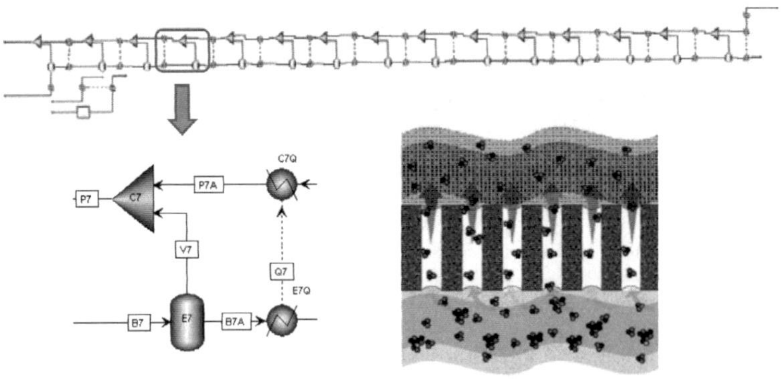

Source: http://www.nap.edu/openbook.php?record_id=12027&page=128

31

Hot Feed (E1, E2)

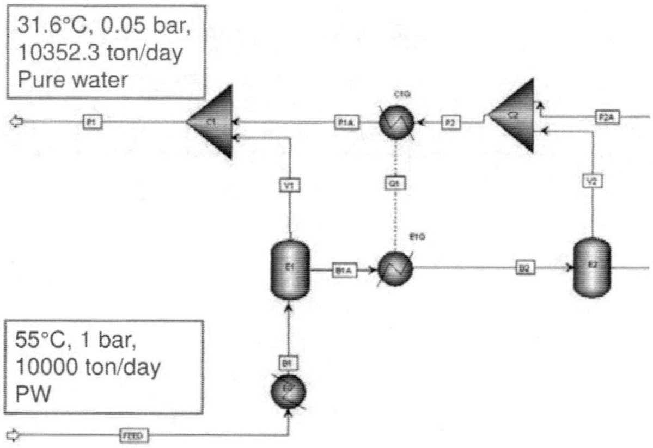

32

Cold Feed (E14, E15)

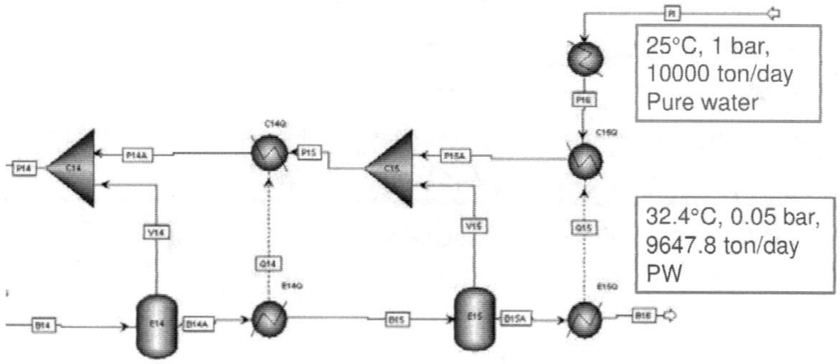

33

Conclusion Remarks

- All 3 desalination technologies (MSF, MED-TVC and MD) have potentials to treat PW with pros and cons;
- Coupling flared gas as heating source is beneficial but challenging;

34

Acknowledgements

- Professor Chau-Chyun Chen
  (Department of Chemical Engineering, Texas Tech University)
- Professor John L. Gossage
  (Department of Chemical Engineering, Lamar University)
- Texas Hazardous Waste Research Center
- Lamar University Graduate Scholarship

35
감사합니다
Natick
Grazie
Danke
Ευχαριστίες
Dalu
Obrigado
Thank You
Köszönöm
Tack
Спасибо
Dank
Gracias
谢谢
Merci
Seé
ありがとう

* DESMAR SOLAR

NATALY AGUDELO
YURANI GUTIÉRREZ
ANDRES OLAYA
FELIPE SASTOQUE

Energetic and exergetic analisys from a seawater solar desalinator

DESMAR SOLAR

*SCHEDULE

- Introduction
- Justification
- Objectives
- Resources available at the university.
- Main processes of water desalination.
- Material and energy balances.
- Preliminary test.
- Environmental considerations.
- Results
- Bibliography.

DESMAR SOLAR

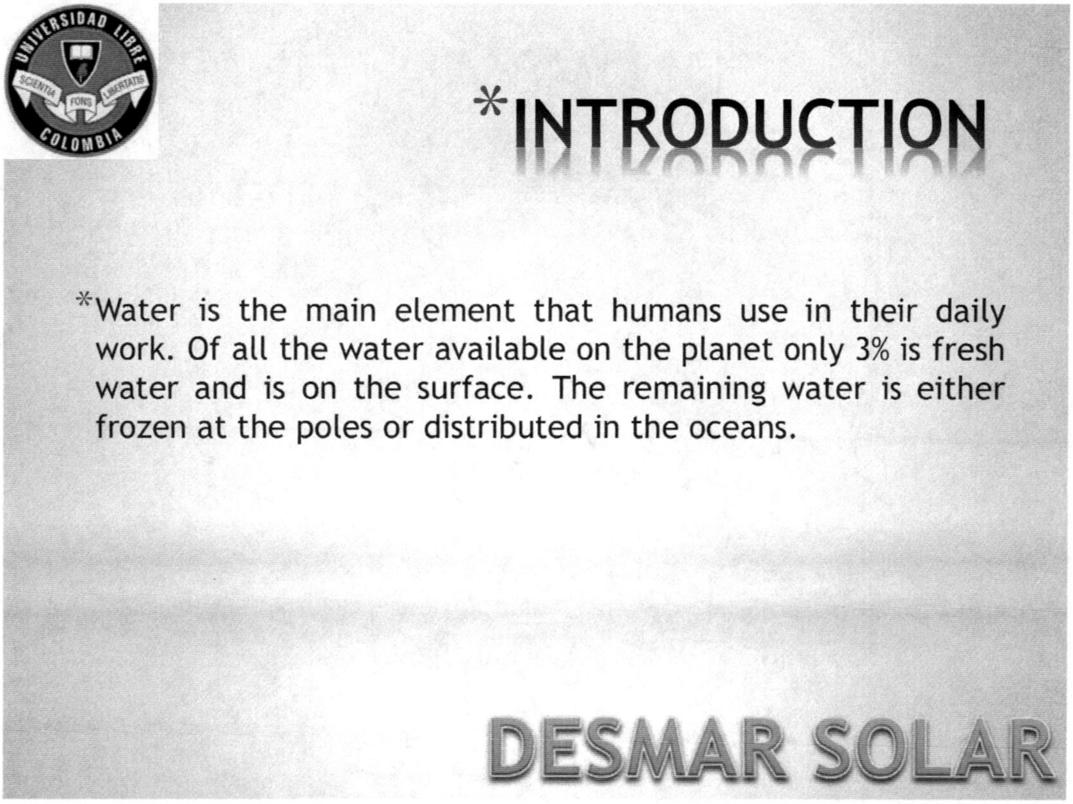
UNIVERSIDAD LIBRE
SCIENTIA FONS LIBERTATIS
COLOMBIA
*INTRODUCTION
*Water is the main element that humans use in their daily work. Of all the water available on the planet only 3% is fresh water and is on the surface. The remaining water is either frozen at the poles or distributed in the oceans.
DESMAR SOLAR

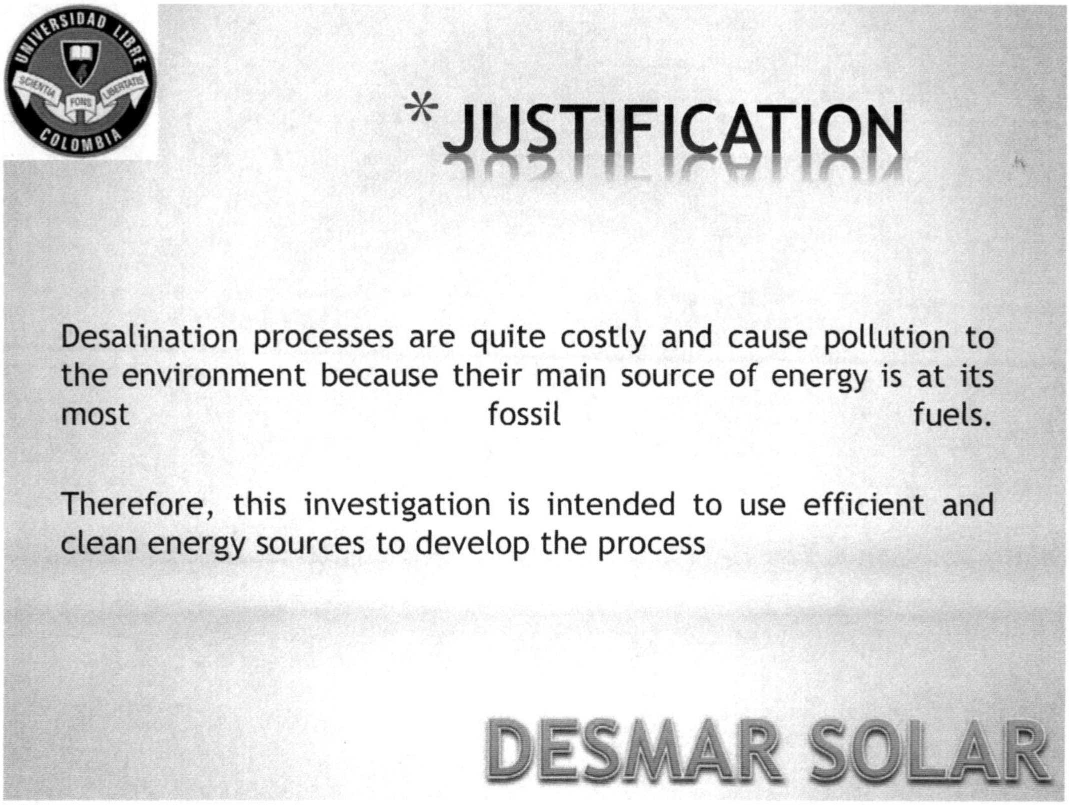
UNIVERSIDAD LIBRE
SCIENTIA FONS LIBERTATIS
COLOMBIA
*JUSTIFICATION
Desalination processes are quite costly and cause pollution to the environment because their main source of energy is at its most fossil fuels.
Therefore, this investigation is intended to use efficient and clean energy sources to develop the process
DESMAR SOLAR

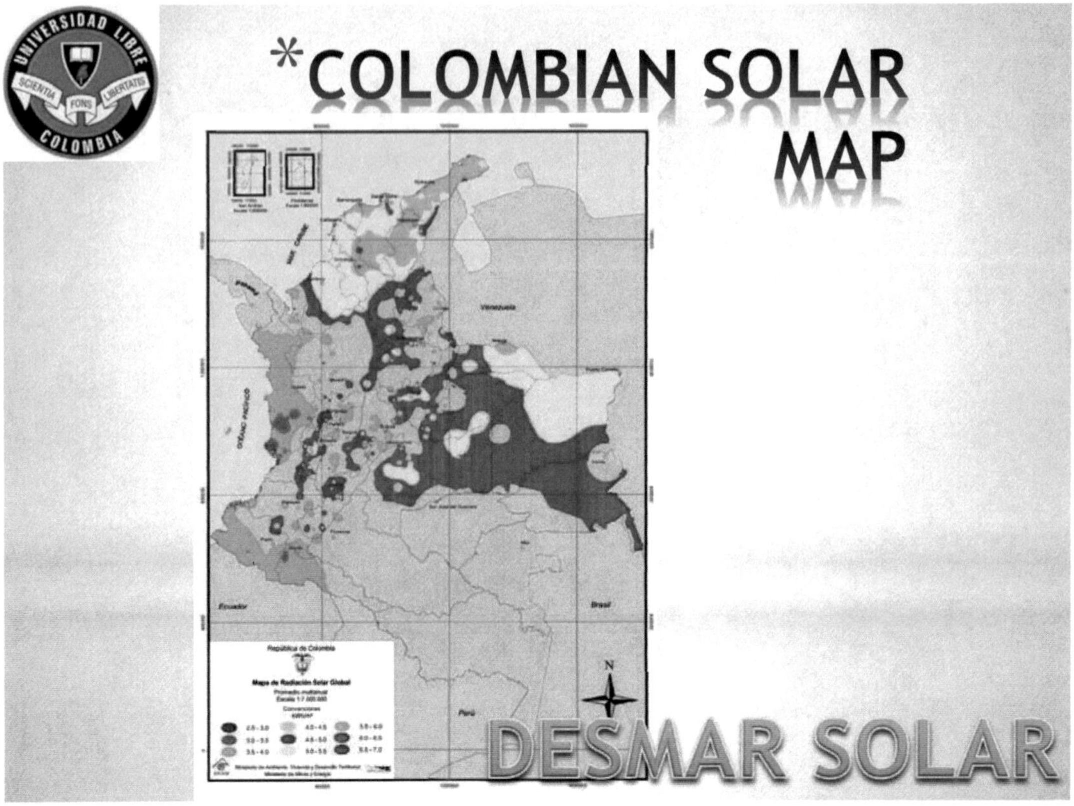
*COLOMBIAN SOLAR MAP
República de Colombia
Mapa de Radiación Solar Global
DESMAR SOLAR

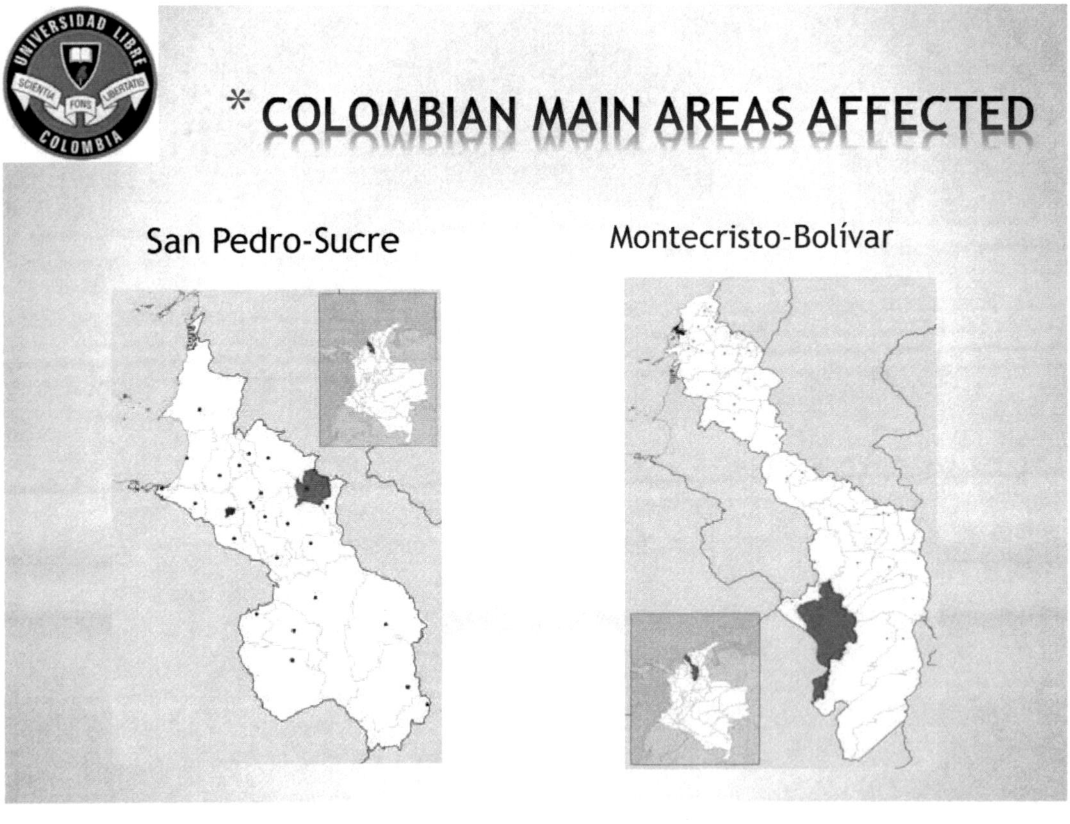
* COLOMBIAN MAIN AREAS AFFECTED
San Pedro-Sucre
Montecristo-Bolívar

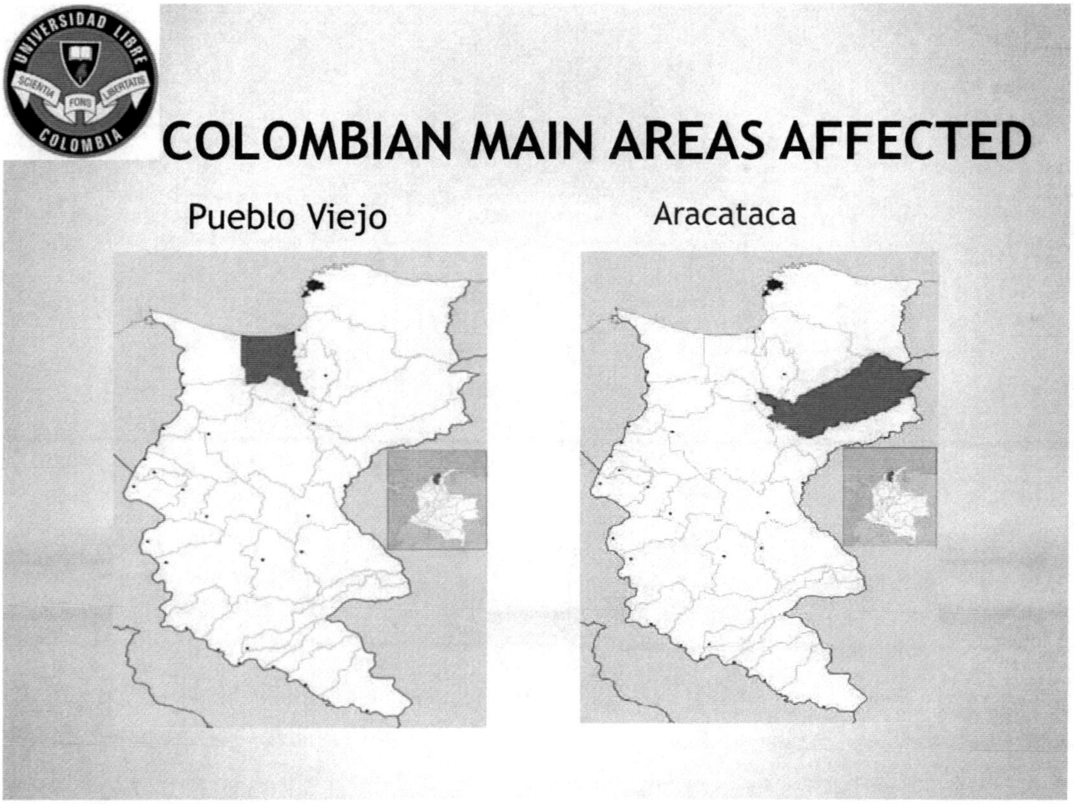
COLOMBIAN MAIN AREAS AFFECTED
Pueblo Viejo
Aracataca

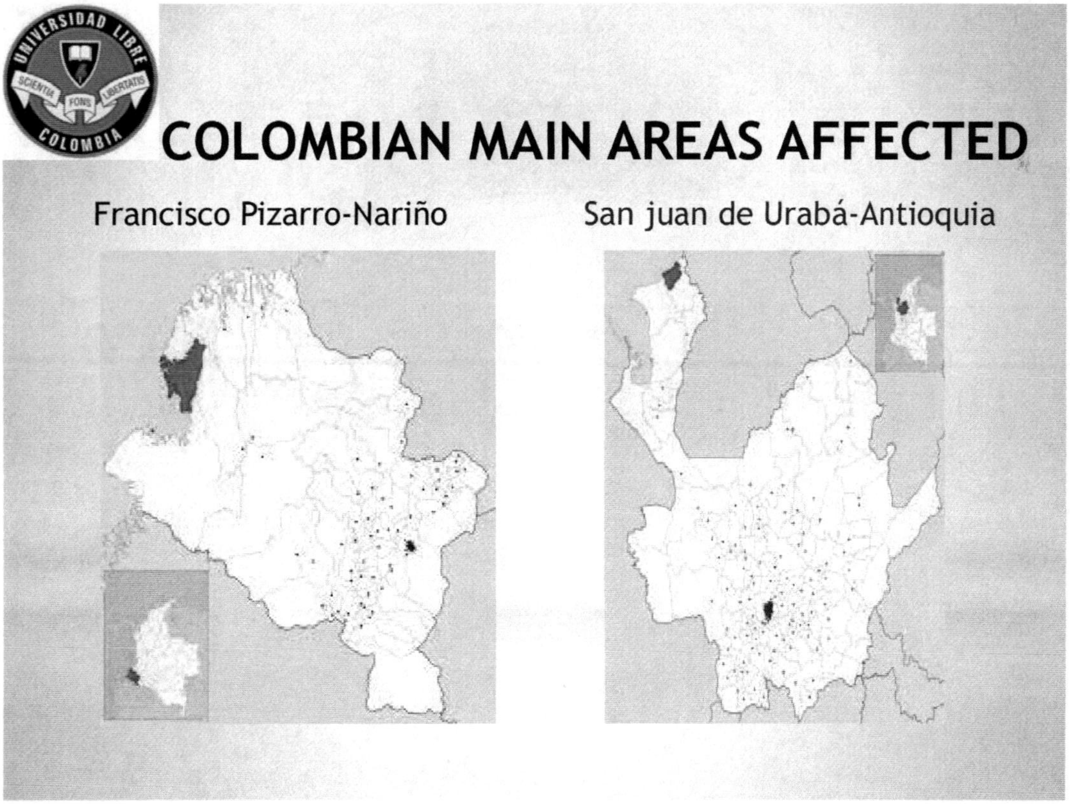
COLOMBIAN MAIN AREAS AFFECTED
Francisco Pizarro-Nariño
San juan de Urabá-Antioquia

COLOMBIAN MAIN AREAS AFFECTED

Timbiquí

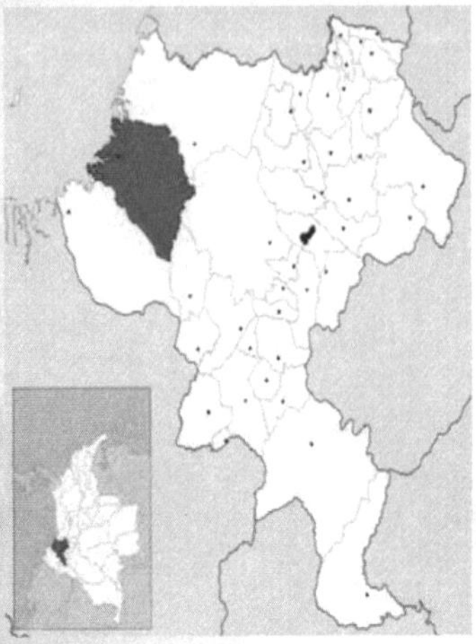

Villa rica

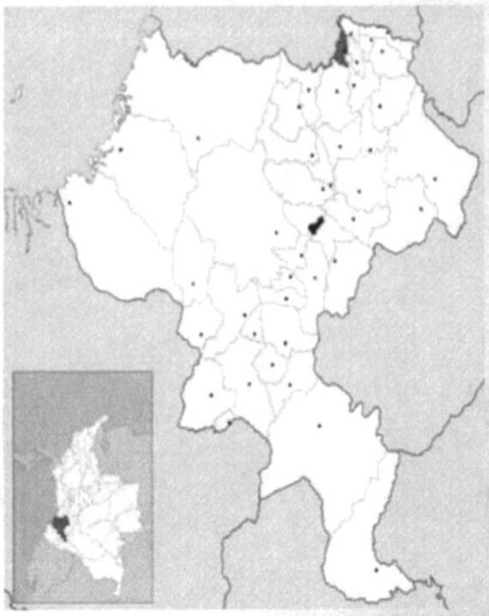

* GENERAL PURPOSE

Establish the energy and exergy analysis of a desalination system using solar energy.

DESMAR SOLAR

*SPECIFIC PURPOSE

* Determine the energy requirements of each of the stages of a solar desalination process.
* Establish the efficiency of different models of solar energy collection system to be used in the desalination process.
* Evaluate the exergy of the process and establish improvements in the design selecting the most suitable model system for the process
* Determine physical properties of the materials.
* Quantify the effect of temperature and the properties of the materials for their selection and use in a solar desalination.
* Establish the lifetime of the material chosen from the tests already performed.

DESALINATION PROCESS

* Desalination by reverse osmosis

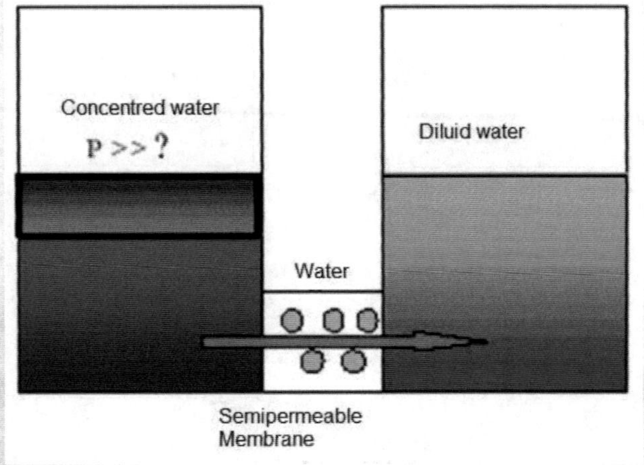

DESMAR SOLAR

*Desalination by Freezing

*Desalination by hydrate formation

*Flash point desalination

DESMAR SOLAR

*Desalination by distillation

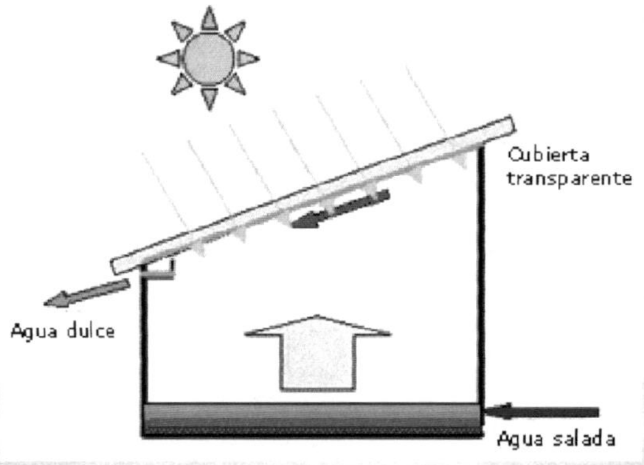

DESMAR SOLAR

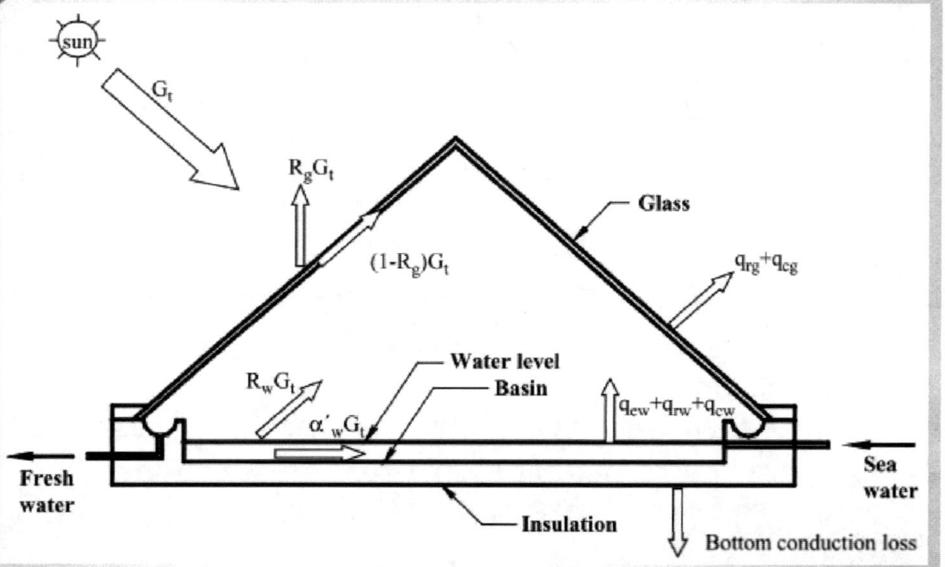

DESMAR SOLAR

*Energetic and exergetic models

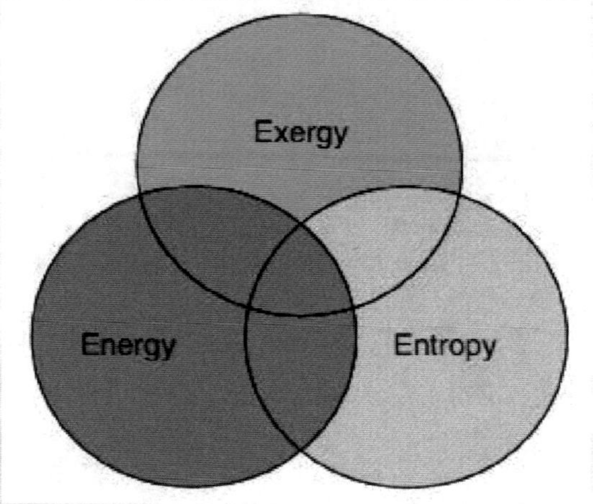

DESMAR SOLAR

ENERGY AND EXERGY DEFINITION?

* Energy: The ability that have the matter to produce work in the form of motion, light, heat, etc..

* Exergy: A thermodynamic property that determines the useful work potential of a given amount of energy that can be achieved by the spontaneous interaction between a system and its environment. Reports the system's potential usefulness as a source of labor.

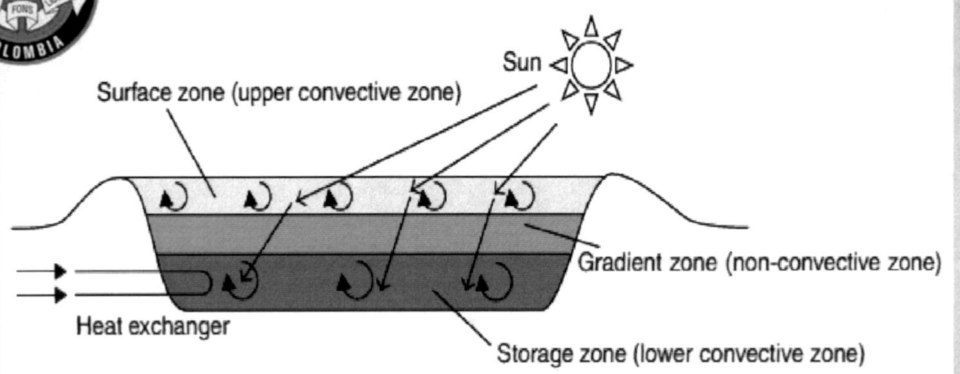

$$n = \frac{m_s h_s + m_w h_w}{G_t}$$

*EXERGY

$$Q_{net} = Q_{HSZ,solar} - Q_{bottom} - Q_{up} - Q_{side}$$

$$n_{HSZ} = 1 - \frac{(Q_{bottom} + Q_{up} + Q_{side})}{Q_{HSZ,solar}}$$

DESMAR SOLAR

*ENTROPY

$$\begin{aligned} s &= x_s s_s + x_w s_w \\ &= x_s [s_{s,\text{pure}}(T,P) - R\ln(x_s)] + x_w [s_{w,\text{pure}}(T,P) - R\ln(x_{\mathbf{w}})] \\ &= x_x s_{s,\text{pure}}(T,P) - R[x_s \ln(x_s) + x_w \ln(x_w)] \end{aligned}$$

DESMAR SOLAR

*EXPERIMENTS

*EXPERIMENT 1.

Distillation through heating plate

DESMAR SOLAR

*HEATING DATA

	TEMPERATURE (°C)	TIME (min)
Evaporated	67	02:46
Finished	91	43:33

FLOW COOLING

Measure	Volume (mL)	Time(s)	Cooling flow(mL/seg)	Cooling flow(L/min)
1	1000	25,57	39,1	2,35
2	1000	25,86	38,7	2,32
3	1000	25,26	39,6	2,38
	PROMEDIO	25,56	39,1	2,35

DESMAR SOLAR

RESULTS

Solution	Temperature (°C)	Volume (mL)	ppm
Initial	20	200	$90x10^3$
Final	20	128	$148x10^3$
Desalinated	20,4	68	0

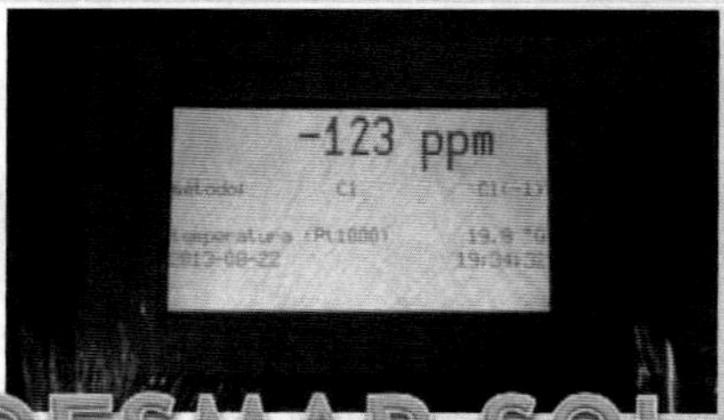

DESMAR SOLAR

BALANCE RESULTS

1

Matter balance		
Input	Output	
200 ml	128 mL	68 mL
90×10^3	148×10^3	0

me = m1+ m2	$90x10^{3} * 0{,}200 = 0 * 0{,}068 + 0{,}128 * ppm$ output
ppm output=	140×10^{3}

PERCENTAGE OF ERROR

$$\frac{ppm_{plancha} ppm_{salida}}{ppm_{plancha}} * 100 \qquad \frac{148000 - 140000}{148000} * 100 = 5.40\%$$

DESMAR SOLAR

2

Energy Balance		
	w*t	
Input	$2{,}8 \times 10^7$	W * seg
	7,8	kWh

$$\rho v C_{H2O} \Delta T = m \Delta H + m C_{H2O} \Delta T$$

ρ	1037,5	kg/m^3
v	0,0002	m^3
C H2O	4184	kJ/(seg * K)
ΔT	70,68	K
m	0,2075	kg
ΔH	173	kJ/(kg * K)
61362	=	61398
	36 J	

DESMAR SOLAR

EXPERIMENT 2

Distillation through incandescent bulbs

DESMAR SOLAR

HEATING DATA SAMPLE

	TEMPERATURE (°C)	TIME(min)
Evaporated	38	46
Finished	90	380

DESMAR SOLAR

BALANCE RESULT

Matter balance		
Input	Output	
200 ml	148mL	52 mL
90×10^3	115×10^3	0

me = m1+ m2	$90\times10^{3} * 0{,}200 = 0 * 0{,}052 + 0{,}148$ *ppm output
ppm output=	$121{,}62\times 10^{3}$

PERCENTAGE OF ERROR

$$\frac{ppm_{plancha} ppm_{\ salida}}{ppm_{plancha}} * 100 \qquad \left|\frac{115000 - 121620}{115000}\right| * 100 = 5{,}76\%$$

DESMAR SOLAR

RESULTS

Solution	Temperature (°C)	Volume (mL)	ppm
Initial	20	200	$90x10^3$
Final	20	148	$121{,}62x10^3$
Desalinated	54	52	0

DESMAR SOLAR

*MATERIALS FOR THE DESALINATOR

DESMAR SOLAR

*TYPES OF MATERIALS

*Metallic Materials

*Ceramic Materials

*Plastics Materials

DESMAR SOLAR

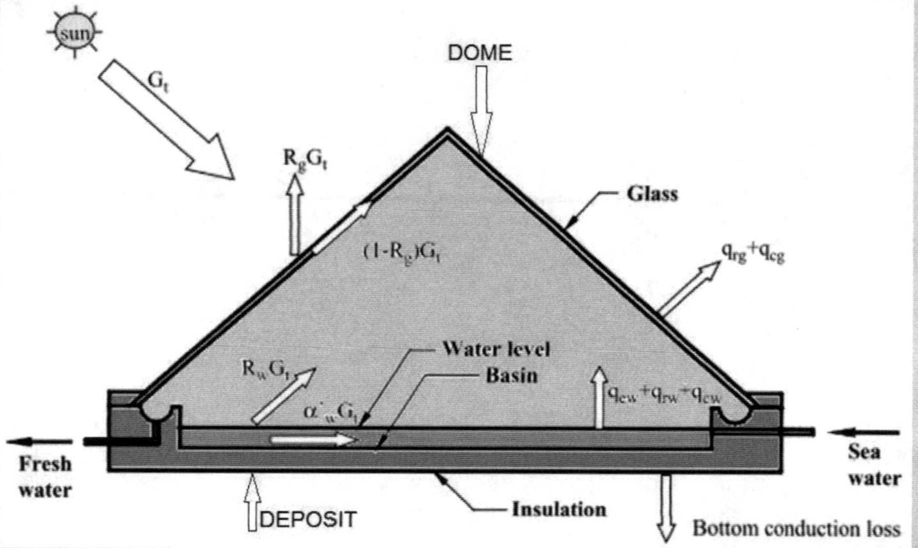

DESMAR SOLAR

*MATERIALS FOR THE DOME

DESMAR SOLAR

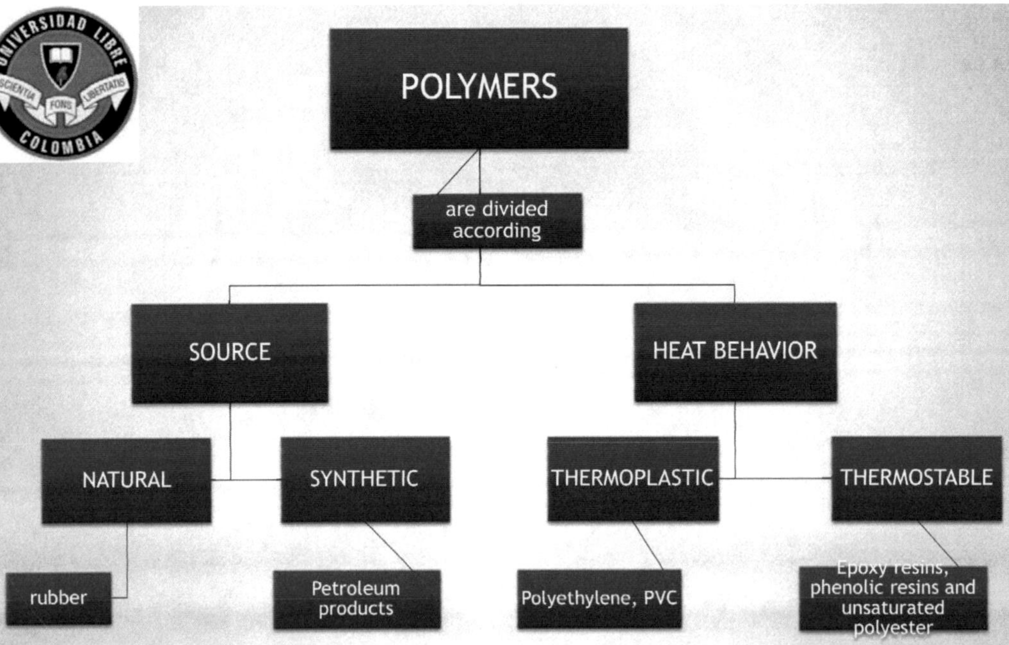

DESMAR SOLAR

*PROPERTIES

*Elongation:** Quantity that measures the increase in length of a material

***Thermal conductivity:** Ability of a material to conduct heat

***Specific heat:** The heat to be supplied to the unit mass of a substance to raise its temperature by one unit

***Coefficient of thermal expansion:** Coefficient that express the length change when a body undergoes a temperature change

DESMAR SOLAR

	•Polietileno de baja densidad (LDPE)	•Polietileno de alta densidad (HDPE)	•Polipropileno (PP)	•Cloruro de polivinilo rigido(pvc)	•Cloruro de polivinilo plastico(pvc)
Elongation [%]	90-800	50-1000	200-700	002-40	200-450
Thermal conductivity [kw/mk]	3,35	4,6-5,2	1,17	1,3-2,9	1,3-1,7
Specific heat [kj/kgk]	2,3	2,3	1,9	0,84-1,17	1,26-2,09
Coefficient of linear thermal expansion [K^-1]	1e-4 - 2e-4	1,1e-4 - 1,3e-4	5,8-10,2	5-18,5	7-25
Continuous use temperature [K]	355,37-373,15	394,26	394,26-433,15	338,71-352,59	338,71-352,59
Deflection temperature [K]	100-121	140-190	200-230	135-180	....
Solvents	N/a	N/a	N/a	Cetonas, ésteres,aromaticos,hidrocarburos	Cetonas, ésteres,aromaticos,hidrocarburos
Elasticity in flexure limit [pa]	....	6894,76	41368,56-55158,08	68947,6-110316,16	N/a

*APPROPRIATE MATERIALS

* Low density polyethylene (LDPE)
* High density polyethylene (HDPE)
* Polypropylene (PP)
* Polyvinyl chloride (PVC)

DESMAR SOLAR

DEPOSIT MATERIALS

DESMAR SOLAR

CERAMIC MATERIALS

* Nonmetallic inorganic solids produced by heat treatment.
* Compared with metals and plastics are hard, non-combustible and non-rusting.
* They can be used in environments with high temperature tribological and corrosive.

DESMAR SOLAR

CERAMIC´S PROPERTIES

* High melting temperature.
* Low density.
* Strength, hardness, structural strength.
* Corrosion resistance.
* Good thermal and electrical insulators.

DESMAR SOLAR

ENVIRONMENTAL CONSIDERATIONS

- By using solar energy production CO2 is removed.?
- The use and advancement of renewable energy is implemented. (In our case solar energy is used).
- Drinkable water in coastal areas.

DESMAR SOLAR

*BIBLIOGRAPHY

- Robert Cook (1986). Computer Simulation of Stress-Strain Behavior in Polymeric Materials. MRS Proceedings, 79, 409 doi:10.1557/PROC-79-409.
- David L. Edwards and Mircea Chipara (2004). Polymeric materials for Solar Sail: The combined effects of polymer thickness, radiation, and temperature. MRS Proceedings, 851, NN1.6 doi:10.1557/PROC-851-NN1.6.
- Alice B. Paterakis (1992). The Consolidation and Desalination of Ceramic Impregnated with Sodium Sulphate. MRS Proceedings, 267, 1071 doi:10.1557/PROC-267-1071.
- T. Y. Wu and D. L. Questad (1992). Mechanical Characterization and Constitutive Description of Thin-Film Polymeric Materials. MRS Proceedings, 264, 143 doi:10.1557/PROC-264-143.(Jr., n.d.)
- Jr., W. D. C. (n.d.). *Materials Science Engineering, An Introduction.* (third edit., p. 785).

DESMAR SOLAR

- Suzana P. Nunes, Ali Reza Behzad and Klaus-Viktor Peinemann (2013). Self-assembled block copolymer membranes: From basic research to large-scale manufacturing. Journal of Materials Research, 28, pp 2661-2665. doi:10.1557/jmr.2013.253.
- Paterakis, A. B. (1992). The Consolidation and Desalination of Ceramic Impregnated with Sodium Sulphate. *MRS Proceedings, 267.*
- Paul C. Painter, M. M. C. (1996). *Fundamentos de ciencia de polimeros: un texto introductorio* (p. 467). Cambridge University Press.
- Zalba, B. (2003). Review on thermal energy storage with phase change: materials, heat transfer analysis and applications. *Applied Thermal Engineering, 23*(3), 251–283. doi:10.1016/S1359-4311(02)00192-8

DESMAR SOLAR

352037 Models Describing the Thermal Recovery as a Productive Option for the Heavy Oil Reservoirs

Abstract:

The crude oil requirement is increasing relatively at a higher rate in comparison to the amount of new crude reserves being identified. To meet the requirement, the recovery from the available crude reserves should be maximized. After the primary recovery, there is a still a good proportion of crude is still left in the reserve. There is a need for the secondary and tertiary methods for the recovery of the left out crude.

Thermal methods are one of the successful techniques which can be implemented to enhance the recovery. I have been tested since 1950's, and they are the most advanced among all the enhanced oil recovery methods available, as far as field experience and technology are concerned. The best utilization of thermal methods can be in the reserve left with heavy oils (of 10-20° API) and tar sands (≤10° API). Thermal methods provide a source of energy in the reservoir resulting in the large reduction in the viscosity of the heavy oil left. Decrease in the viscosity will result in a higher mobility ratio.

Thermal methods have been highly successful in Canada, USA, Venezuela, Indonesia and other countries. The general classification of Thermal enhanced oil recovery can be broadly classified in terms of hot water, steam, in-situ gasification or electrical heating. Among these methods the steam based methods are more successful commercially than others. In the present study, we will be discussing about the various models used to understand the behaviours of the reservoir in response to the thermal enhanced oil recovery methods. We will be discussing about what are the type of simulators being used to identify the behaviour.

Fire flooding as a budding source of enhanced oil recovery

- by KHUSHBOO SINGH (f2011409@pilani.bits-pilani.ac.in)

Fire flooding is a thermal method of oil recovery in which compressed air is injected into an injection well and the petroleum in the reservoir surrounding the injection well is ignited to establish a combustion zone or fire front .This fire front propagates in a radially expanding manner away from the injection well. The heat reduces the viscosity of oil, cracks some higher molecular weight hydrocarbons into smaller ones, vaporizes some lighter hydrocarbons to help miscibly displace oil and also produces steam that may steam distill the trapped oil .Thus, along with decreasing the mobility ratio, it also produces improved oil. Also, the gases produced due to combustion and nitrogen (in air) displace the oil towards reservoir away from fire front again enhancing production .Other methods of EOR cannot be used for highly viscous oils whereas thermal recovery or say, fire flooding can be used for both high and less viscous oils. Fire flooding also offers many advantages over the other method of thermal oil recovery i.e. steam flooding. Lot of fresh water and capital is used to produce steam for steam flooding whereas in this there is just the need to push the compressed air and ignite the oil thus, lowering the cost of production .Also, it has the potential to operate in wide variety of reservoirs that other methods. For example, lower in pressure (as it lowers the viscosity (improved mobility)as well as produces gases that supply pressure), containing more shale , lower in quality and thinner (as it requires less volume of air for producing the same amount of oil than other methods which employ gases like carbon dioxide method of oil recovery)and deeper (heat losses occur in steam flooding).But it also has some limitations like it can be applied only to reservoirs of high permeability , high porosity and oil saturation and has poor sweep efficiency in thick formations .Also, if excessive coke is deposited the rate of advancement of combustion zone will be slow and more air will be required. It also produces flue gases which cause environmental problems. Along with this it also faces operational problems like corrosion due to low pH hot water, serious oil water emulsions, and increased sand productions, deposition of carbon or wax and pipe failures to high temperatures.

Now however, oxygen enriched air or pure oxygen is being used as the combustion supporting medium for which about one-fifth as much volume is required for the injection process as compared to compressed air since nitrogen which accounts for about 80% of the volume of the compressed air is eliminated. Thus, the use of the pure oxygen in the process instead of air not only drastically reduces the volume of the combustion supporting medium required but also provides a decrease in the fluid volume following combustion which generally increases the injectivity of the reservoir matrix in the combustion zone .Also, it produces carbon dioxide which dissolves in oil and reduces its viscosity. Thus, one of the major limitations of fire flooding i.e. it requirement of high permeability in reservoirs is eliminated in this way .After an initial level of in situ combustion the production well is partially choked or shut in until the bottom hole pressure increases to a substantial fraction of the injection pressure .The production well can then be opened to a lower back pressure level which results in an immediate acceleration of fluid flow under the increases oil pressure gradient and leads to enhanced recovery. Thus, pressurization and producing cycles can be repeated using intervals found to be most effective for the particular system. Also, water or steam or mixtures of steam and carbon dioxide (or low molecular weight C3-C8 hydrocarbons) can be injected intermittently or subsequent to injection of combustion supporting medium to enhance production. Further, relatively heavy viscous crude oil from a subterranean stratum can be produced by the use of in situ combustion. In conventional direct drive or forward underground in-situ combustion, when applied to certain oil-bearing strata, particularly those containing a heavy viscous crude oil, it becomes impossible to continue to support an expanding combustion zone because of the formation of a viscous liquid bank of oil in advance of the combustion zone. This viscous liquid bank, substantially completely plugs the stratum and prevents continued injection of air to feed the combustion zone .So, now the oil is ignited in the stratum to be produced around a single well penetrating same and air is injected through the well to expand and move the ignited area (combustion zone) into the stratum away from the well whereby a liquid bank of viscous oil builds up in the stratum beyond the combustion zone so as to substantially plug the stratum and build up substantial pressure above the stratum pressure at the time of initiation of the process, and thereafter terminating the injection of air and opening the well to production so that the fluid pressure in the stratum beyond the combustion zone forces liquid oil from the liquid bank back into the combustion zone and the hot burned-over area behind the combustion zone so as to force the upgraded oil into the well from which it is recovered by conventional methods. After a substantial period of production during which the pressure-build up in the stratum during the injection period forces oil in upgraded condition into the well, the stratum is again ignited adjacent the well to establish a combustion zone which feeds on the oil filling the pores of the stratum at the end of the first production phase. In the event the stratum is below combustion supporting temperature, the ignition is effected in conventional manner as by heating with an electric igniter, an ice charcoal pack, a gas fired heater, or similar device and injecting air into the hot stratum. Also, recovery can be improved using water containing surfactant and alternate slugs of air during forward combustion. In an underground in situ combustion process for the recovery of oil from formations having high permeabilities, after combustion has been initiated, water containing a surfactant and air can be injected under conditions such that the water upon contact with the hot formation vaporizes, creating slugs of condensable foam in an amount sufficient to generate a significant resistance factor of at least 5, which improves volumetric sweep efficiency by retarding the mobility of the steam and subsequent slugs of air. Air and water, including a surfactant, can be injected simultaneously as a mixture or as alternate slugs of water containing a surfactant and air. It also provide a method for temporarily or partially plugging the more permeable strata and a process for producing a well by selectively blocking and plugging highly permeable sections of subterranean oil-bearing formations with a self-destructive foam generated in situ exhibiting a significant resistance factor. Thus, as we saw in situ combustion can be applied to a wide array of reservoirs whether it be thin, low pressure or deep reservoirs. In fact it is the only thermal method that can presently be applied to deep reservoirs, though deep down steam generation is being tested. It can be used at any stage of reservoir depletion. It can be used in special situations such as offshore or in arctic regions. Because of the lack of heat losses at the surface and in the injection wells, it is the most thermally efficient thermal recovery method. The inject ant (air) is readily available. Combustion allows wider well spacing than steam. Economic results are comparable to those of steam injection.

Several aspects of operating ISC projects are important. First is the large compression ratios and associated costs required to inject air into the formation. Second is planning and design requirements for a combustion project that are more difficult than steam. Third is extensive laboratory work to assess fuel availability, air requirements and burning characteristics of the crude that are required before designing in situ combustion projects. Fourth is the high degree of technical sophistication and careful monitoring needed to ensure proper operation of project Fifth is the limitation of numerical simulation and other techniques that make predictions of recovery more difficult than most other enhanced oil recovery methods .As mentioned before, considerable improvements in the application of in situ combustion have been made since the early projects .New developments like using oxygen enriched air, injecting water containing surfactant and alternate slugs of air during forward combustion etc improve the efficiency of fire flooding as well as eliminate the problems faced during its application.

352165 Design of Surfactant-Only Chemical Flooding in a Low Perm, High Tds Reservoir: Single Well Test

Abstract:

Low permeable oil reservoirs are often characterized by slow waterflood injectivity and low fluid productivity. For low perm formations, the permeability appears to be the limiting factor to deploy Surfactant-Polymer (SP) and Alkaline Surfactant Polymer (ASP) flooding for improved oil recovery. Injection of carbon dioxide has shown promising results in low perm formations and is commonly used in these fields when sufficient CO_2 can be obtained. However, even when sufficient CO_2 can be sourced, initial capital investments for CO_2 injection may be quite high and may limit its application for some mature fields. In this paper we report a single-well test of an improved surfactant-only formulation in a low permeability, high TDS reservoir. The target single well, SEH1, is located in south central of Oklahoma. The perforated zone is located 6,000 feet below the surface with an on-going waterflood. The average permeability of the formation is 14 mD, ranging from 0.2 to 44 mD. The produced brine contains total dissolved solids (TDS) at 102,300 mg/L and the reservoir temperature is 50 °C (123 °F). The reservoir crude exhibits a viscosity of 23 cp at reservoir conditions, presenting further challenges for an EOR process without mobility control using polymer. For the test, the hydrophilic-lipophilic difference (HLD) equation was used to predict possible salt-tolerant surfactant formulations, based on site-specific fluid properties and reservoir conditions. Among the binary and ternary surfactant mixtures developed, the surfactant formulations were further optimized by introducing low concentrations (between 400 to 600 mg/L) of hydrophilic linkers to achieve a stable surfactant solution in the reservoir brine and ultra-low interfacial tensions (< 0.008 mN/m) between site crude and produced brine. In laboratory testing injection of three pore volumes of a surfactant-only formulation in sand packs and core flood devices achieved tertiary oil recovery ranging from 45% to 70% of the residual oil (Sor) after water flooding. The levels of surfactant loading were varied from 0.23

wt% to 0.92 wt% to assess the impact of surfactant concentration along with slug size. The field single well test used a ternary surfactant formulation (i.e., a mixture of alkoxylated alcohol sulfates/ethoxylated alcohol sulfates/blends of monoalkyl and dialkyl diphenyloxide disulfonates) and was conducted in April-May of 2013 to further confirm laboratory results at the SEH1 site. The 3-week single-well test was a technical success. A surfactant only formulation was injected at 0.69% followed by injection of reservoir brine without mobility control. Based on the pre- and post-chemical tracer test results (i.e., use of ethyl formate as partitioning tracer plus other conservative tracers), approximately 90% of the residual oil at the target zone was mobilized. This field experience will be built into a design protocol using single-well tests to mitigate the risks of pilot-scale chemical flooding targeted at mature fields in the U.S. mid-continent region.

352753 Foam Mobility Control: An Enhanced Oil Recovery Technique

Abstract:

In Enhanced Oil recovery techniques a significant fraction of the oil-in-place is left in the ground after primary or secondary recovery due to various reasons. Gas injection is one method used to extract these oils. There are basically four types of gases used in gas injection: natural hydrocarbon gas, inert gas, carbon dioxide, and steam; but in all cases there is the problem of gravity override.

Gravity override meaning the preferential movement of the injected gas through the upper part of the reservoir due to the lighter density of gas compared with oil. Gravity override causes early gas breakthrough in production wells and hence, reduced oil recovery. The steam is injected into the ground to reduce the viscosity of the oil and to make the oil more mobile. Unfortunately, due to gravity over ride, as steam rises to the top of the reservoir. Together with reservoir heterogeneity, gravity override results in early steam production and reduced volumetric sweep efficiency. Hence recovery is not improved to the degree desired. The injection of a permeability blocking agent would do the required. Various scientist have reviewed the selective blockage of fluids in thermal recovery projects and concluded that foam is best suited for this purpose. The foam is found to be promising to work in alkaline/surfactant processes to replace or supplement polymer as mobility control agent. The present study deals with the advancements in the foam mobility control for the recovery of unrecovered crude oil from the reserve.

352632 A Qualitative in Field Test to Characterize Oil-Shale Reservoirs

Abstract:

Shale oil recovery, despite its extraction costs and low yield per shale matrix is becoming a more feasible solution to the rapidly emerging global fuel demands. However, many oil-shale reserves are located several thousand feet below the earth's surface, which makes it extremely difficult and expensive to extract. In addition, the quality of the shale matrix may have a broad range such that the potential yield through and other methods may vary significantly across neighboring reservoirs. And determining the quality of the oil-shale matrix and its potential yield of synthetic/tight oil using laboratory analytical methods can take several weeks. This work presents a rapid and simple qualitative "in-field" test-tube test method that categorizes and prioritizes a given shale matrix based upon its composite. This test method can be a significant cost benefit and a proof positive indicator of where to focus exploration and recovery interest.

Pressure Regulation and Kick Attenuation with Wired Drill Pipe Technology in Managed Pressure Drilling

Hector Perez
Reza Asgharzadeh Shishavan
John Hedengren
Brigham Young University

Outline

- Overview
- Motivation for Pressure Control
- Method and Structure of the Control System
- Results
- Results of Automated BHA Pressure

Motivation for Pressure Control

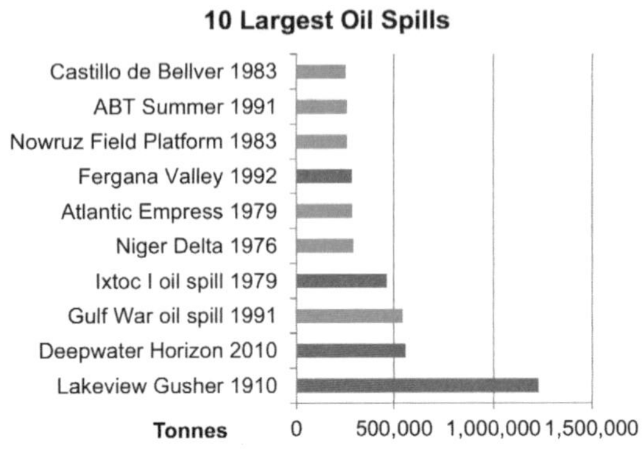

Apr 2, 2014 Pressure Regulation and Kick Attenuation with Wired Drillpipe in MPD

Motivation for Drilling Control

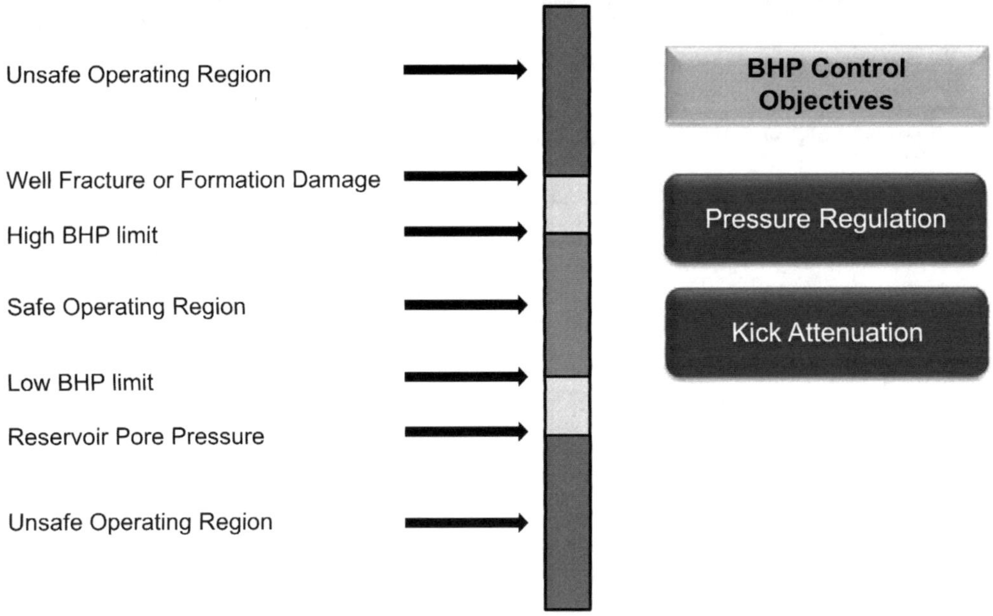

Apr 2, 2014 Pressure Regulation and Kick Attenuation with Wired Drillpipe in MPD

Managed Pressure Drilling System

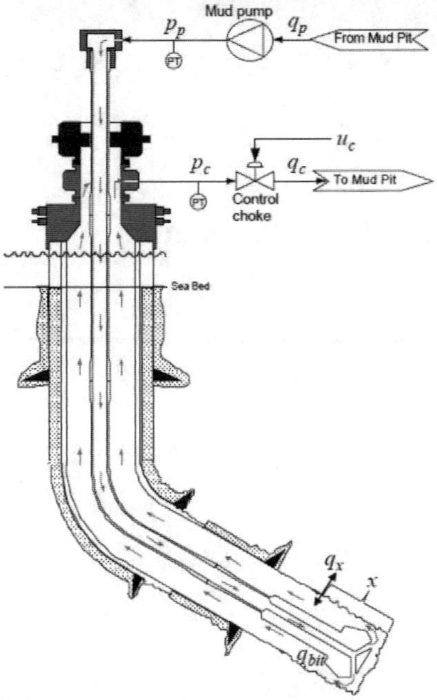

- ❑ Overbalanced Drilling
- ❑ Managed Pressure Drilling (MPD)
- ❑ Underbalanced Drilling

MPD offers:

- ❑ More precisely control of borehole pressure
- ❑ Decrease NPD and mitigate drilling hazards

Manual Control versus Automated Control

Challenges

- Large Time Delays
- Complex Behavior of Multiphase flow
- Sudden Change in Formation Properties

Benefits of BHP control Automation

- Improve Reliability of Pressure control
- Reduce the possibility of human errors

Objective

Designing a comprehensive controller which Simultaneously

- Regulates the BHA Pressure within specified range
- Attenuates Unwanted gas influx when it happens
- Uses latest automation technologies
- Has improved performance in comparison with manual and conventional controllers

Wired Pipe Technology to be used in Control Scheme

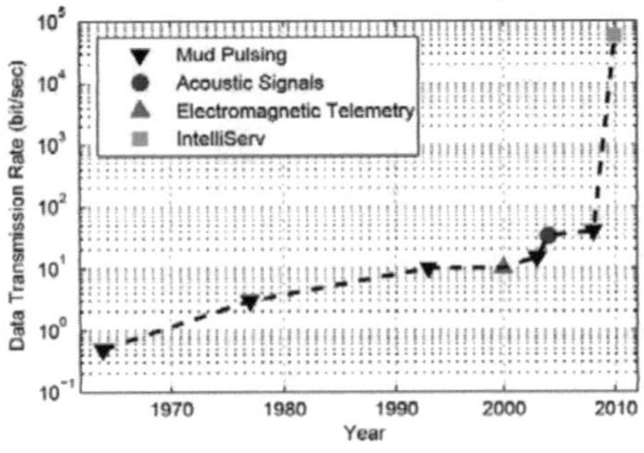

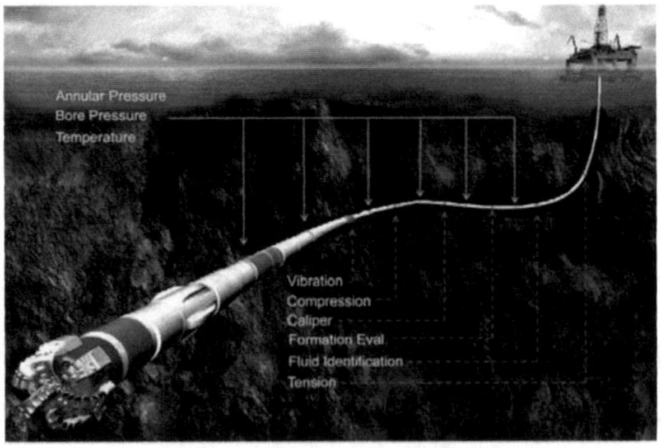

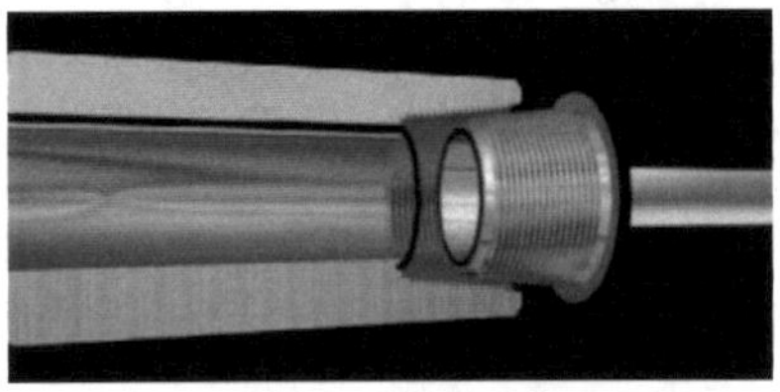

Challenges in the designed controllers

Mud Pulse Telemetry Based	Wired Pipe Telemetry Based
❑ Indirect control of BHP	❑ Directly control of BHP and annular pressures ✔
❑ High uncertainty in estimated variables values	❑ Increased precision in estimating the unmeasured process variables ✔
❑ Still Large time delay in kick detection	❑ Improved kick detection ✔

Schematic of the MPD Controller

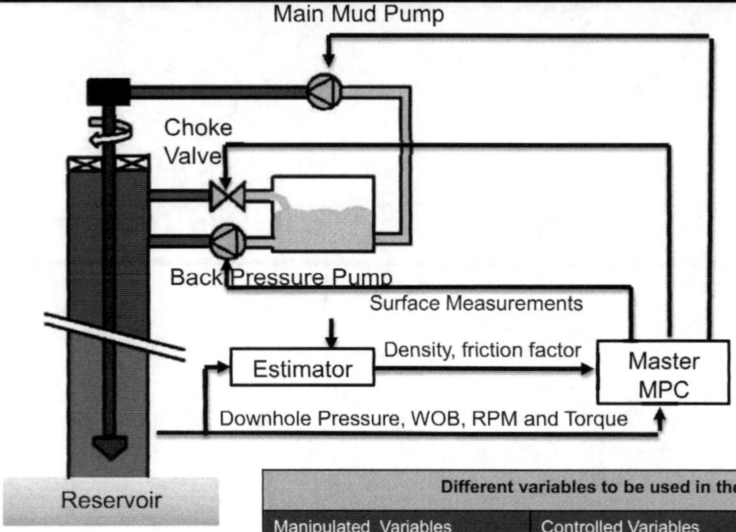

Different variables to be used in the current research		
Manipulated Variables	Controlled Variables	Estimated Variables
Rig Pump Flow Rate	BHP	Annulus Drill Fluid Density
Choke Valve Opening	Annular Pressures	Annulus Friction Factor
		Reservoir Pore Pressure

Controller Structure

Multivariate Nonlinear Model Predictive Controller (MPC)

What are advantages over conventional controllers?

Nonlinear Feature of the Controller

Multivariate Nonlinear Model Predictive Controller (MPC)

- Drilling is a non linear process
- Current controllers use linear (simplified) models
- A nonlinear controller can capture the dynamics better
- Designing nonlinear controllers are more challenging

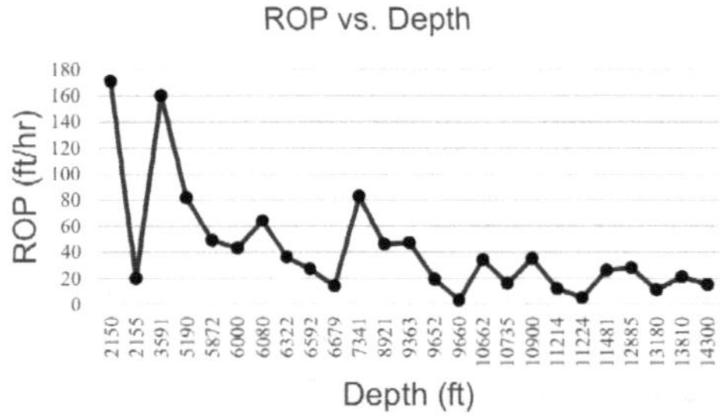

Figure 1. Change of ROP with drilling depth. Field data from (Kinabalu East-1, 1990)

Model Predictive Control and Moving Horizon Estimator

Controller

Multivariate Nonlinear Model Predictive Controller (MPC)

- Future prediction capability – Improved control performance
- MPC vs. Conventional Controllers

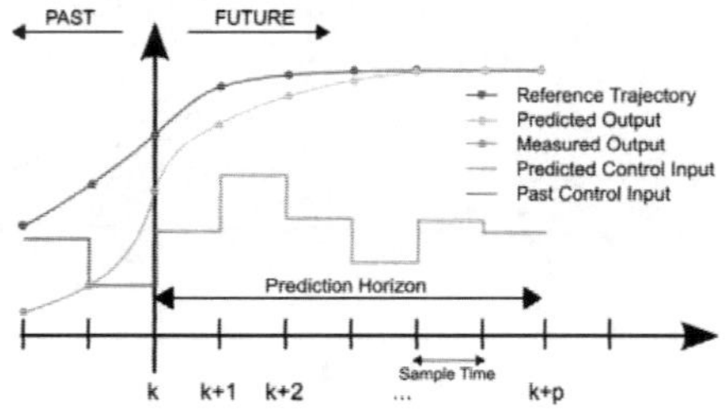

Estimator

- Moving Horizon Estimator
- Uses both downhole and surface measurements
- More precise and faster parameter and state estimation

Software

- APMonitor Optimization Suite
- Designed Especially for Dynamic Nonlinear Processes
- Using the Most Efficient Solvers Available
- Solution Time: Fraction of a Second

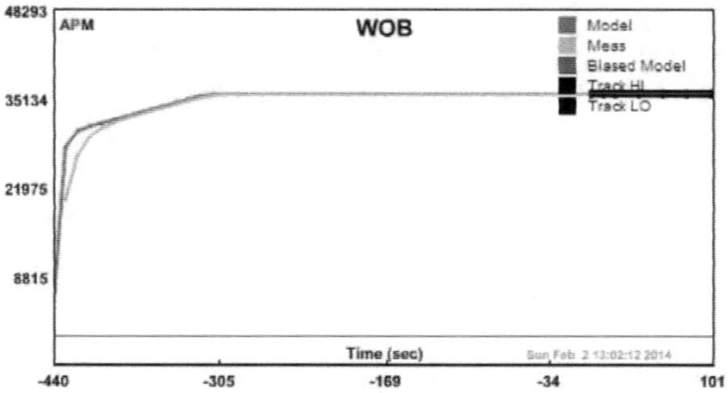

Lower Order Model

Balance Equations

$$\dot{p}_p = \frac{\beta_d}{V_d}(q_{pump} - q_{bit})$$

$$\dot{p}_c = \frac{\beta_a}{V_a}(q_{bit} + q_{back} - q_{choke} + q_{res})$$

$$\dot{q}_{bit} = \frac{1}{M}(p_p - F_d|q_{bit}|q_{bit} + \rho_d g h_{bit} - p_{bit})$$

$$\dot{\mathbf{h}}_{bit} = \boldsymbol{ROP}$$

Closure Relations

$$M_a = \rho_a \int_0^{l_w} \frac{1}{A_a(x)} dx$$

$$M_d = \rho_d \int_0^{L_{dN}} \frac{1}{A_d(x)} dx$$

$$M = M_a + M_d$$

$$p_{bit} = p_c + \rho_a \times F_a|q_{bit} + q_{res}|(q_{bit} + q_{res})h + \rho_a g(h_{bit} - (n-1)h)$$

$$q_{choke} = K_c z_{choke}\sqrt{\rho_{an}(p_c - p_0)}$$

Zhou, Kaasa, et al. (2008)

Apr 2, 2014 | Pressure Regulation and Kick Attenuation with Wired Drillpipe in MPD

Kick Detection and Attenuation

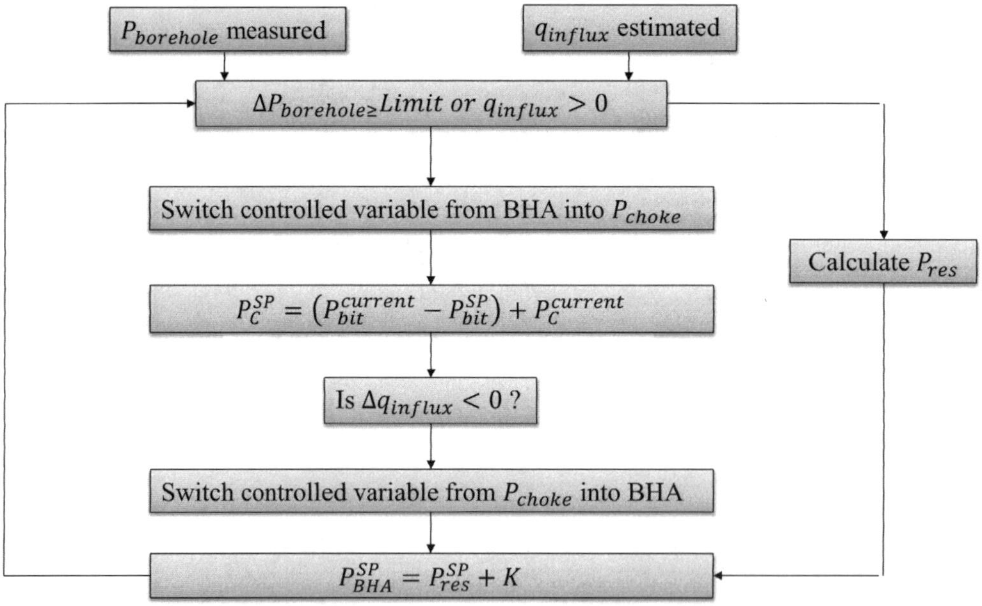

Apr 2, 2014 | Pressure Regulation and Kick Attenuation with Wired Drillpipe in MPD

Results

Apr 2, 2014 Pressure Regulation and Kick Attenuation with Wired Drillpipe in MPD

Automated BHA pressure control

Table 1. Initial well conditions used for simulations

Variable	Value (SI units)	Value (US units)
Distance between two successive pressure sensors h_a	150 m	492.1 ft
ROP	0.01 m/s	0.033 ft/s
Fluid density in the drill string - ρ_d	0.012 x 10^{-5} kg/m^3	74.9 lb/ft^3
Fluid density in the annulus - ρ_a	0.014[a] x 10^{-5} kg/m^3	87.4 lb/ft^3
Bulk modulus in the annulus β_d	14000 bar	2.03 x 10^5 psi
Bulk modulus in the annulus β_a	14000 bar	2.03 x 10^5 psi
Annulus cross sectional area	0.021 m^2	0.226 ft^2
Friction factor in the drill string F_d	10000 bar m^{-6} s^2	116.3 psi ft^{-6} s^2
Friction factor in the annulus F_a	1250[a] m^{-5}	3.29 ft^{-5}
Lumped density per length parameter for the drill string - M_d	5700 x 10^{-5} kg/m^4	108.5 x 10^{-5} lb/ft^4
Lumped density per length parameter for the annulus – M_a	800 x 10^{-5} kg/m^4	15.22 x 10^{-5} lb/ft^4
Well depth (h_{bit})	2000 m	6562 ft
[a]For connection procedures, the values used for simulation were 0.01 (density) and 1000 (friction factor)		

Apr 2, 2014 Pressure Regulation and Kick Attenuation with Wired Drillpipe in MPD

MHE Estimator

Within 30 seconds of operation, the estimator has accurately determined the friction factor and has determined the annulus density within 1.5% of the actual value.

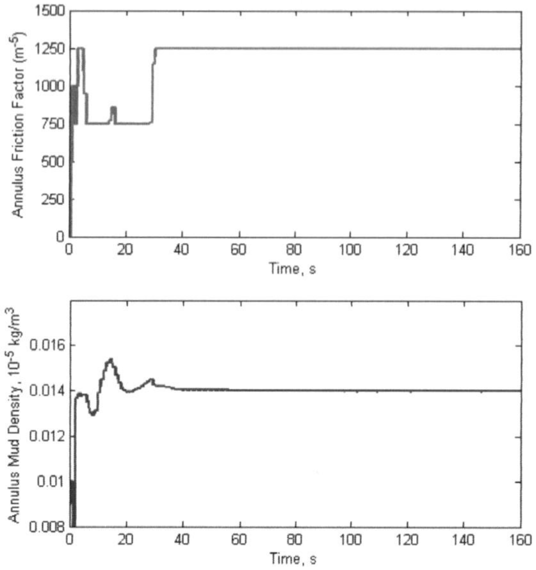

Case 1: Normal Operation

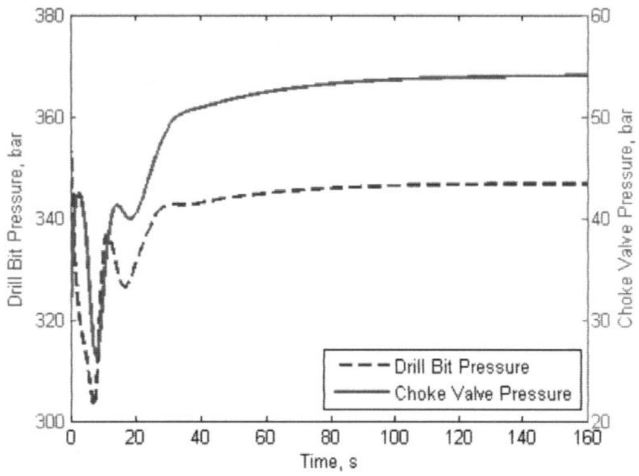

Within 30 seconds, the drill bit pressure has been controlled within 0.72% of its set point of 345 bar

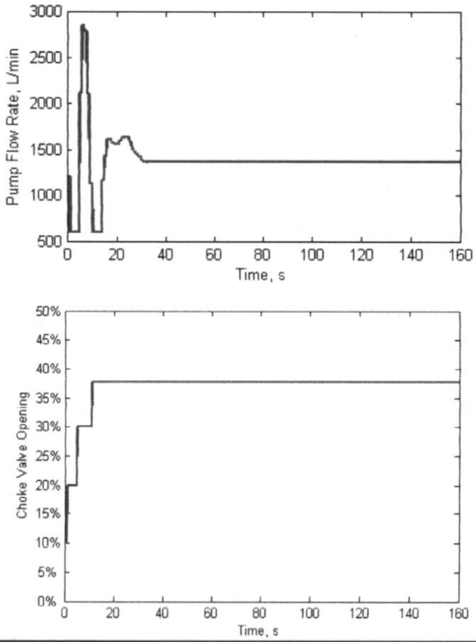

Case 2: Connection Procedure

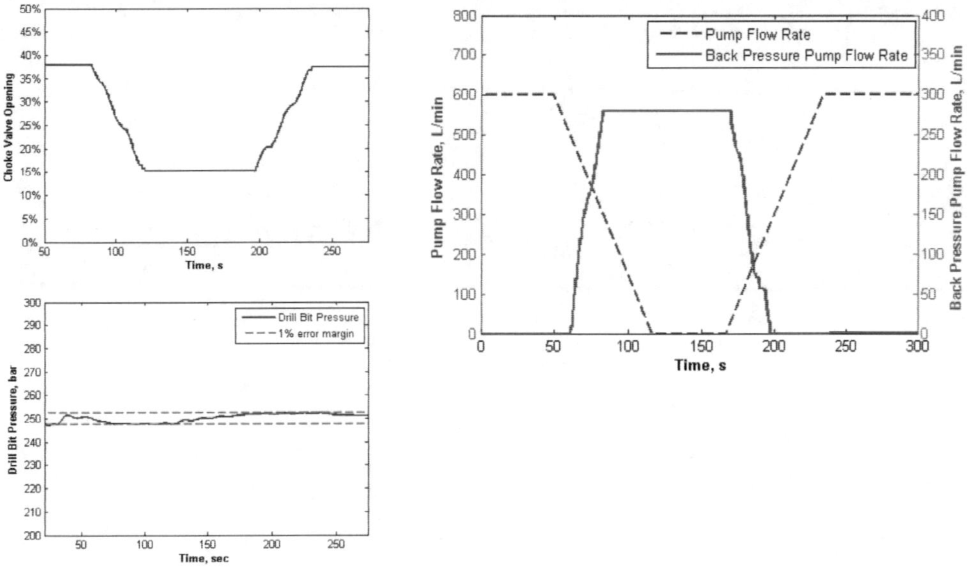

Apr 2, 2014 | Pressure Regulation and Kick Attenuation with Wired Drillpipe in MPD

Case 3: Unwanted Gas Influx

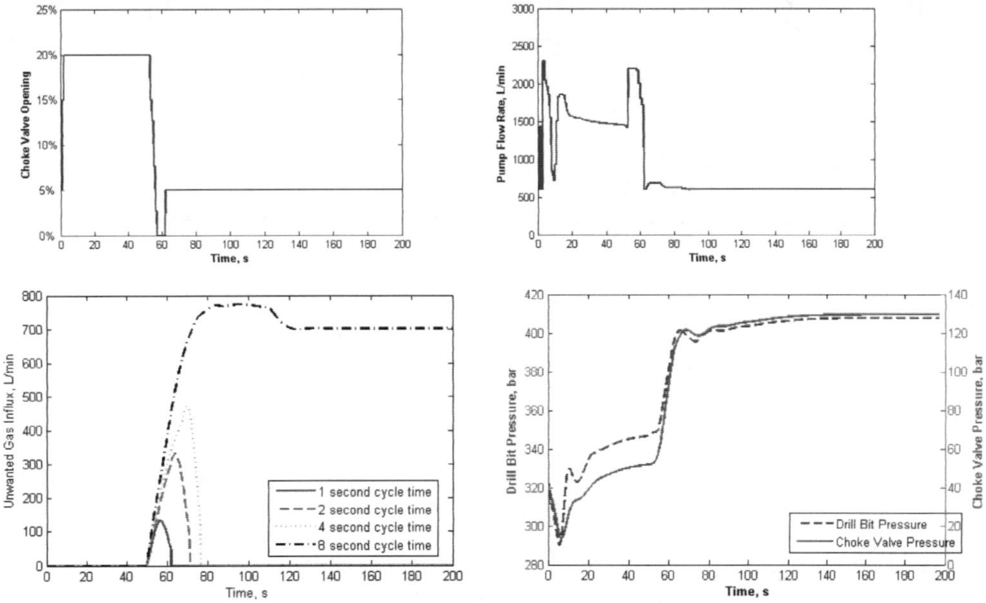

Apr 2, 2014 | Pressure Regulation and Kick Attenuation with Wired Drillpipe in MPD

Thank You for Your Attention

Apr 2, 2014 Pressure Regulation and Kick Attenuation with Wired Drillpipe in MPD

14AIChE®
Spring Meeting, New Orleans, LA

Risks from Unconventional Wells Gasification

Dennis B. Butts
Americas Risk Practice
Environmental Resources Management (ERM)
CityCentre Four
840 West Sam Houston Parkway North, Suite 600, Houston, TX 77024
dennis.butts@erm.com

Prepared for Presentation at
American Institute of Chemical Engineers
2014 Spring Meeting
Topical 2: Shale Gas and Tight Oil
New Orleans, LA
March 30 – April 2, 2014

UNPUBLISHED

Risks from Unconventional Wells Gasification

Dennis B. Butts
Americas Risk Practice
Environmental Resources Management (ERM)
CityCentre Four
840 West Sam Houston Parkway North, Suite 600, Houston, TX 77024
dennis.butts@erm.com

Keywords: LNG, Risk Management, Process Safety, Unconventional Wells, Shale

Abstract

With the continued differential between diesel and natural gas costs, onshore well operators have been working to convert both drilling rigs and hydraulic fracturing pumps to utilize natural gas as fuel. Associated with this conversion is a change in the risk profile that may not be fully recognized or understood in the field. Risk management philosophies and assumptions need to adjust to the difference in hazard characteristics between the fuels, as well as the variety of delivery methods.

Specific topics will include:

- Motivations for fuel conversion and the differences in scale for rig power, natural gas substitution on diesel hydraulic fracturing pumps, and dedicated natural gas powered hydraulic fracturing pumps. Risk mitigation and field training requirements may not fully accounted for in investment decisions.
- Scale of conversion and location specific availability drives natural gas sourcing, which can then drive personnel risk and well pad layout issues depending on whether the gas is sourced from local wellheads, pipeline, Compressed Natural Gas (CNG), or Liquefied Natural Gas (LNG). Separate Simultaneous Operations (SIMOPS) issues can arise based on the specifics of each sourcing method.
- Associated risks need to be appropriately managed in a manner that involves the personnel directly exposed. And there is a need to ensure that well sites which did not previously present a threat to surrounding activities do not become a threat due to either direct risk from the fuel conversion itself or from escalation risks not previously of concern.

1. Introduction

Over the past few years there has been a significant increase in efforts to substitute natural gas for diesel to provide fuel for drilling and fracturing onshore wells. This is an expansion of the conversion efforts that were already underway to convert transportation fleets away from diesel fuel. While the reasons for this may vary, and do include environmental benefits, the most significant reason is cost savings.

Natural gas has been averaging about $4.50/mmbtu [1] over the past 6 months, which is about $0.59 per diesel gallon equivalent (DGE) [1]. Even with added distribution costs, if pipeline gas is available, the economics are very attractive. Wellhead gas is even more attractive. And if LNG is the only alternative at a well site, at $4.50/mmbtu, the natural gas component of LNG is about $0.34 per gallon [1], with a markup of about $0.15 per gallon for liquefaction and storage [3] plus transport costs of about $0.15 per gallon (up to 100 miles) [4], giving a breakeven price of $0.64 per gallon or $1.10 per DGE. With typical supplier pricing to bulk customers as a discount to diesel typically around $1.50 per gallon, and more substantial discounts for large scale contracts, there are significant financial benefits for both LNG suppliers and drillers when diesel fuel is in the $4 per gallon range.

And as a bonus to the economic justification, drillers can tout emissions reductions of as much as 20 – 30% relative to diesel [5]. Rarely in these discussions is the subject of added costs or effort for process and technical safety brought up.

Substitution of natural gas for diesel has been underway for some time in heavy vehicle transportation fleets. With added competition in engine conversion markets and among storage tank suppliers, conversion costs have been reduced to the point where heavy vehicles can recover costs in less than 2 years [6]. Efforts to convert drill rigs originated in the Marcellus region, due to environmental concerns, but the economics have driven natural gas rigs into most of the major markets in the US, as well as into western Canada. Since hydraulic fracturing pumps use even more diesel than rigs, recent efforts have been in this direction, though technical challenges regarding power delivery have resulted in two types of conversion. One is the substitution of gas for diesel in existing engine drivers to ensure sufficient horsepower. The other is to use high powered turbine pumps, which are very well suited to gas power. Large scale conversion of fuel sources is not unprecedented, as the US heavy truck fleet went from less than 10% diesel powered to nearly 100% diesel powered from 1950 to 1980 [6].

So what does this mean in terms of risk management? Aside from the business and financial risk of converting equipment to a fuel in natural gas that is historically more volatile in pricing than diesel fuel at a time of historically high oil to gas price differentials, which is a subject for another paper, there are practical technical safety considerations that come into play as drilling sites convert from diesel to natural gas. These can result in increases to the risk profile of a well pad that is not generally accounted for, as well as changes in the risk profile within the drilling operation itself.

This paper will discuss some of the issues and options regarding conversion and highlight specific risk issues associated with each.

2. Sources of Natural Gas

2.1 Typical Volume Requirements

A heavy truck may use 20,000 gallons of diesel per year [4], with significant potential savings being the result of a conversion to natural gas fuel. Conventional fluid drill rigs use an average of 425,000 gallons of diesel per year [5], creating a fuel cost savings equivalent to converting a fleet of over 20 heavy trucks, but at an investment cost that is much lower. But even these volumes pale in comparison to the amounts of diesel used in hydraulic fracturing. Fracking crews can use 5,000 to 10,000 gallons of diesel per day [7], and it is common for the crews to be in operation for as many as 300 days per year, accounting for mobilization and demobilization time. That means 1.5 to 3 million gallons of diesel per year can easily be used by a single crew, often going well to well for the same operator. An unfortunate outcome of this is that conversions become a foregone conclusion, without due consideration to the resulting process safety risks.

2.2 Options for Sourcing and Delivery

Natural gas for a well site can be sourced as gas directly from the wellhead or gathering system, or it can be piped in from a commercial distribution line, when those are available. In more remote locations, distribution lines may not be available, and if the gas is too wet, wellhead or gathering gas may also be unsuitable for use. In those situations the options are to utilize either CNG or LNG. CNG would require considerably more transportation than LNG due to the lesser amount of gas that can be moved and comes with higher costs for storage and handling equipment. Therefore, CNG has limited utility for well sites, though its lower fuel cost makes it useful in some narrow applications. The use of LNG can be cost effective when other options are not available, as it is more available than in the past and although containing less energy than diesel per gallon, an onsite tank containing 15,000 gallons of LNG is still equivalent to over 8600 gallons of diesel [2] and is about 2.4 times the energy density of CNG [9].

2.3 Issues Associated with Each Option

Use of wellhead gas or gas from gathering lines requires either very dry gas or extensive treatment, as most engines converted to utilize natural gas are designed for gas approximating domestic gas. This is in contrast to offshore facilities where engines are often optimized to take advantage of fuel gas diverted from separation trains. Pipeline gas can be the best option, for those sites with ready access to a distribution line, and will provide sufficient flow and pressure for most natural gas drilling rigs. However, pipeline gas may have to have compression added for certain types of fracking pumps, and available flow may not be sufficient for some fracking operations. CNG can be useful in areas that do not have dry gas or ready pipeline access and may not have easy access to LNG. In such areas, CNG facilities can be set up to provide the product to users in the surrounding region much more readily than LNG facilities. To utilize CNG does require more transportation of the gas and more onsite storage than using LNG. With LNG being the last supply option, but becoming very common in the field, this is an option that carries significant process and technical safety risks. LNG is a material than can bring many benefits, but needs to be treated with considerable respect, often more than is witnessed at well sites.

3. Associated Risks

Associated process and technical safety risks arise in several stages of the conversion from diesel to natural gas. In most cases there is an increase in those risk levels relative to diesel fuel, though this is not always fully contemplated by decision makers or well communicated to the field level staff. Whereas diesel fuel is classified as a combustible liquid, NFPA Class II [8], is difficult to ignite if spilled and primarily a localized threat, natural gas is highly flammable vapor and depending on delivery method could have the potential to disperse considerable distances if released.

In use in equipment, natural gas can create the need for hazardous area class or hot work permit zones that were not required around diesel equipment. This issue has been missed on well sites and often is not included in discussions about conversion costs until after being identified in PHA studies or other risk assessments. The same consideration extends to gas monitoring and shutdown systems. System risks are manageable, but need to be given due consideration and treated with the proper respect. Some expansion of this thought is provided below.

3.1 Source of Gas

Wellhead gas primarily introduces risk through the separation and dehydration process. Even relatively dry gas needs to be treated prior to use in most of the engine drivers used for onshore production. To utilize it during drilling usually means that at least one well has already been put into operation on a multi-well pad. The risks associated with production while drilling and completing adjacent wells have to be assessed. Controls that may need to be in place during certain phases of drilling could interfere with the supply of gas. An alternative that may be available within a field of well sites under one operator is to use gas from the gathering system. While treatment would also be required for that gas, the potential SIMOPS issues of a producing well on a pad with drilling activity. In either case the gas being treated will be a relatively high pressure and is likely to result in some amount of condensate or natural gas liquids (NGL) production and handling. This can increase the fire and explosion risks above those normally found during drilling activity.

Pipeline gas is only an option when a distribution line is nearby. But if available, this can provide a supply of natural gas that is already treated and can be adjusted to the necessary pressure at the takeoff location from the distribution network. In analysis of fire and explosion risks, pipeline gas supply most often has the smallest consequence zones and the least amount of equipment susceptible to failures. This is even the case if high horsepower fracking pumps are in use that requires some degree of compression of the natural gas to reach necessary operating pressures.

CNG introduces risk due to its high pressure, typically near or above 3000 psig [9]. This can both drive the spread of a gas cloud in the event of a release as well as provide the means through depressurization to cool the gas to a low enough temperature that it will disperse for some distance along the ground, further lengthening dispersion distances. Storage under pressure also introduces a finite risk of boiling liquid expanding vapor explosion (BLEVE) that could spread debris.

LNG usage at well sites is capable of imposing a larger potential impact zone than a well blowout scenario. And this is excluding a catastrophic failure of the storage medium, which is low likelihood event due to the dual tank construction of trailers that handle LNG. When released, LNG will initially disperse in a manner very similar to liquefied petroleum gas (LPG), but as the edges of the dispersing cloud warm, flammable vapors will loft upward. Assumptions regarding layout and spacing that are based on LPG can address the flammable breadth of a possible LNG cloud, an example of which is shown in Figure 1 from PHAST [10], going out about 850 feet (260 meters) to the 50% lower flammable limit (LFL). Maximum distance to a 1 psi overpressure, good for window breakage, trailer damage, and upsetting neighbors, would extend another 300+ feet (100 meters) beyond the 50% LFL distance. In general, the height of the 50% LFL section of the cloud will remain under 15 feet (5 meters), but this also needs to be confirmed relative to the drill rig flare height. And the failure of a LNG loading hose would impose an even larger impact zone. Though many operators end up using gas detection, this is only effective if appropriate emergency response procedures are in place.

Figure 1: 50% LFL Dispersion Distance for 1-inch LNG Leak [10]

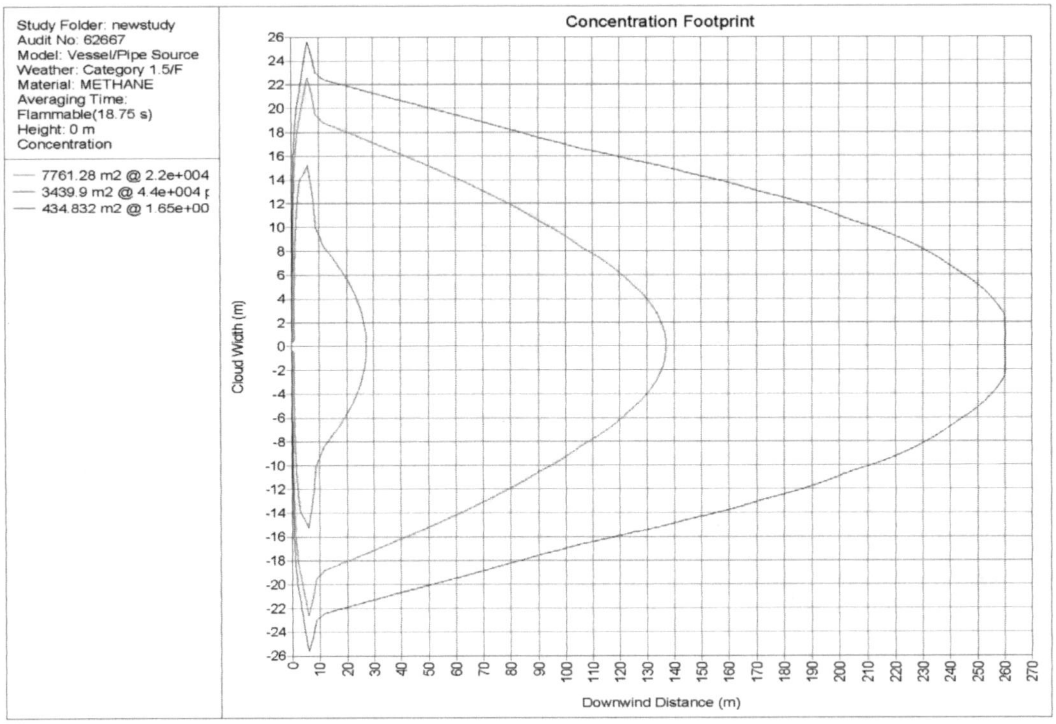

For LNG most releases behave as impinged releases, due to the low pressure environment. For CNG, the best case is for a free field release, which provide dispersion distances of less than half that of LNG. However, horizontal releases of CNG are capable of dispersion distances beyond those of LNG, in particular if they impact other equipment and become partially impinged, as shown for dispersion to 50% LFL for a 1-inch CNG release in **Figure 2**. An interesting phenomenon is how the LNG dispersion slumps outward from the beginning, due to release at temperatures low enough for the methane to be denser than air, but the CNG is initially released at warmer temperatures, then settles into a much lower temperature, resulting in a slumping spread.

Figure 2: 50% LFL Dispersion Distance for 1-inch CNG Horizontal Leak [10]

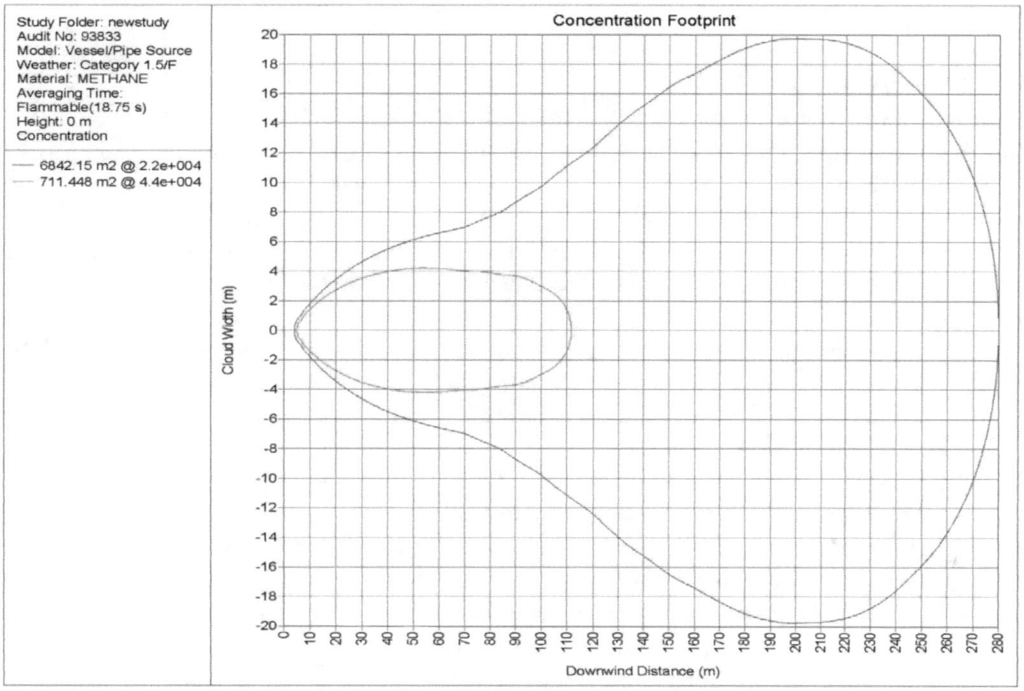

As a reference, an ignited well blowout, traditionally considered a worst case event at a well pad, will typically have an effect distance to 1500 Btu/hr-ft^2 (4.7 kW/m^2) of about 500 feet (150 meters).

3.2 Refueling at Site

When utilizing CNG or LNG as the source of gas, there are additional considerations for storage at the site and for refueling deliveries. Both require delivery by truck, just like diesel fuel, but the release of diesel from an overturned truck is not the same as a release of compressed or liquefied natural gas. This is why there are stringent design requirements on CNG and LNG transport vehicles. However, when assessing risks it is wise to keep in mind that those standards were developed for paved highway transportation, not gravel access roads to well sites. In addition to possible visibility issues, there are the impacts such roads could have on the expected behavior of safety features on the vehicles.

3.3 Compression and Delivery of Gas to Equipment

Further issues arise in situations where either pipeline gas or LNG is used to supply equipment such as hydraulic fracturing pumps that require higher pressure feed gas. In these circumstances, arrangements have to be made to provide compression at the site. A site that is typically already very crowded with equipment. In terms of risk from the equipment it is normally a localized threat, however the extra equipment adds a level of complexity that can complicate operations. The added complexity can add a human factors risk dimension that is rarely assessed in drilling situations.

4. Conclusion

Considerable financial and environmental benefits can be had by converting drilling operations from diesel fuel to natural gas, as long as the current differential in crude oil versus natural gas pricing exists. To achieve those benefits does entail some risk, with this paper having focused on the process and technical safety risks. In some instances, the benefits and risks can pull in opposing directions creating a challenge to find equilibrium between the two.

Depending on the source of natural gas for the drilling and completions operations, due consideration needs to be given to:

- Training needs for field staff that are used to dealing with combustible diesel fuel and not highly flammable natural gas;
- Flammable gas dispersion distances as well as resulting maximum potential zones for 1 psi overpressures, in particular for LNG or CNG applications;
- Review of drill rig flare location relative to natural gas supply point(s) to account for potential height of dispersion;
- Road conditions and their potential impact on transportation of fuels;
- Hazardous area classification zones and resulting equipment alterations, layout changes, or work permit requirements;
- Gas detection needs and the emergency response planning to accompany the detection; and
- Hazard assessment requirements and when they are activated to address changes made to drilling operations or well pad layouts once in service.

As we further explore this question, it is necessary to keep in mind not only the benefits of the ongoing process of converting away from diesel fuel to natural gas at well sites, but also the risk management challenges that can arise.

5. References

[1] US Energy Information Administration, 2014

[2] US Department of Energy, Fuel Energy Spreadsheet, 2014

[3] Quine, Tom, "Fundamentals in LNG", ASGMT White Papers, 2010

[4] Center for Energy Economics, University of Texas at Austin

[5] Oil & Gas Technology, "Gas Powers Fracking Plays", April 10, 2013

[6] Blomerus, Paul and Oulette, Patric, "LNG As a Fuel for Demanding High Horsepower Engine Applications: Technology and Approaches",

[7] Schutt, Bryan, "Diesel to Gas Switching Trend Takes Off in Drilling Sector", SNL Financial, September 3, 2013

[8] NFPA 30, National Fire Protection Association, 2012

[9] Envocare Ltd., "Liquefied Petroleum Gas (LPG), Liquefied Natural Gas (LNG), and Compressed Natural Gas (CNG)", 2007

[10] PHAST 6.7, DNV GL AS

353108 Supercritical CO_2 / Brine / Hydrocarbon Interactions in Shale Rock Fracture: A Geo-Microfluidics Approach

Abstract:

Shale oil and gas is a critical unconventional resource for the energy sector in the US and worldwide. The booming of its production in the recent decade, due to the hydraulic fracturing technique, has led to serious environmental concerns such as huge water demands and toxic additives into fracking water which may pollute groundwater. A proposed solution is to use supercritical CO_2 ($scCO_2$) as a more environment-friendly alternative fracking fluid to replace water for enhanced oil recovery (EOR) in oil shale. Many challenges lie in the way of this scheme such as to understand multiphase flow of $scCO_2$ at injection sites, its interactions with brines, hydrocarbons, and surrounding geologic matrix. We present here a microfluidic approach to investigate the micro-scale multiphase fluid flow and reactive transport within low-permeability shale rock. Unlike the conventional microfluidic techniques, the geo-microfluidic chip is directly fabricated from a thin-section of shale rock, hence provides the genuine surface properties. Simple fracture-like patterns are created on the shale rock surface by machining and femto-second laser etching to simulate the natural structures. Flow experiments in the geo-microfluidic chip are conducted within a high-pressure, high-temperature apparatus to simulate the in situ reservoir conditions. The interactions between $scCO_2$/brine/hydrocarbon will be quantified and the hydrocarbon extraction efficiency of using $scCO_2$ and using fracking water will be compared. The displacement patterns, the dissolution rate of $scCO_2$ in brine/hydrocarbon, and the surface wettability of $scCO_2$ and shale rock will also be investigated. These micro-scale experiments results will help to reveal the underlying mechanisms of multi-phase fluid flow under in situ temperatures and pressures.

352468 Model Shale Gas Properties from a Statistical Mechanical Perturbed Hard-Sphere Chain Cubic Equation of State

Abstract:

Shale gas is an increasingly important energy source in North America. Hence, an accurate description of the physical, transport and thermodynamic properties of shale gas mixtures are a key prerequisite for process design. A particularly advantageous feature is that multiple properties can be obtained from a single equation of state (EOS) for a model system. Traditional EOS's are cubic in molar volume and are derivatives of the basic van der Waals (1873) EOS. Redlich and Kwong (1949), Soave (1973) and Peng and Robinson (1976) are attempts to improve the accuracy of the basic vdW EOS; however, as is shown early in this talk, these equations of state overcorrect the attractive term to compensate for deficiencies in the repulsive term. On the other hand these EOS's have the advantages of a computationally simple form and *three well-defined molar volume roots*. Nonetheless, while these EOS's often give a reasonably accurate representation of the phase composition of simple binary and ternary alkane mixtures (model shale gas systems); poor results are obtained for other properties including liquid density, mixture second virial coefficient and diffusion coefficient.

On the other hand, a class of accurate molecularly-based but complex and computationally intense equations of state: the statistical associating fluid theory (SAFT) equations (Huang and Radosz, 1990; Chapman et. al., 1990), which account for individual molecular level effects (i.e., hard-sphere repulsion, London dispersion, covalent bond chain formation, multipolar forces and hydrogen bonding) have successfully emerged in the past few decades to account for multiple effects in pure fluids and mixtures including associating and/or polydisperse components. These EOS's capture the major molecular level effects; however, convergence difficulties and the existence of several molar volume roots complicate implementation of these EOS routines.

Consequently, considering the above difficulties, we have developed an approach which incorporates the rich molecular description of the SAFT equations into a cubic in molar volume form. Specifically, molecular simulation data for hard-sphere compressibility factor is nearly exactly fit to a simple empirical function in packing fraction, which ensures an overall cubic in molar volume equation of state and from which two universal parameters (applicable to all fluids and fluid mixtures) are obtained. Furthermore, London dispersion attractive forces are described by an extremely simple polynomial refit of data generated for Chen and Kreglewski's (1977) *thirty-six term* double sum in reciprocal absolute temperature and packing fraction. Finally, chain formation (covalent bonding) is described by a modified form of Wertheim's (1984) thermodynamic first-order perturbation theory (TPT1) for segment covalent bonding, in which a diimer reference radial distribution function is employed to obtain more accurate results. Conventional forms of the hard-dimer radial distribution function (Chiew, 1991; Yethiraj and Hall, 1992; Ghonasgi and Chapman, 1994; Sadus, 1999) were refit to a novel empirical functional form which results in an equation of state which is overall cubic in molar volume.

The resulting theoretically-based EOS was applied to predict several model shale gas properties including compressibility factor versus packing fraction isotherms, the vapor-liquid coexistence curves, the second-virial coefficient versus temperature and the self-diffusion coefficient versus pressure isotherms for the constituent pure fluids; and vapor-liquid equilibria and binary diffusion coefficients versus pressure for binary systems. The observed performance of the molecularly-based *cubic* EOS presented in this paper is generally better than the performance of conventional cubic equations of state and about equal to the performance of the more complex non-cubic SAFT equations of state for the same fluids but with much less computational effort. Ongoing work involves development of *cubic* EOS terms which account for multipolar and hydrogen bonding effect to include water and other species in the shale gas mixtures.

REFERENCES

Chapman, W. G., K. E. Gubbins, G. Jackson and M. Radosz, *New Reference Equation of State for Associating Liquids*, Ind. Eng. Chem. Res. **29**, 1709-1721 (1990).

Chen, S. S. and A. Kreglewski, *Applications of the Augment van der Waals Theory of Fluids: I. Pure Fluids*, Berichte der Bunsengesellschaft fur Physikalische Chemie, **81**, 1048-1052 (1977).

Chiew, Y. C., *Percus-Yevick Integral Equation Theory for Athermal Hard-Sphere Chains. II. Average Intermolecular Correlation Functions*. Molec. Phys. **73**, 359-373 (1991).

Ghonasgi, D. and W. G. Chapman, *A New Equation of State for Hard-Chain Molecules*, J. Chem. Phys. **100**, 6633-6639 (1994).

Huang, S. H. and M. Radosz, *Equation of State for Small, Large, Polydisperse and Associating Molecules*, Ind. Eng. Chem. Res., **29**, 2284-2294 (1990).

Peng, D.-Y. and D. B. Robinson, *A New Two-Constant Equation of State*, Ind. Eng. Chem., **15**, 59-64 (1976).

Redlich, O. and J. N. S. Kwong, *On the Thermodynamics of Solutions. V. An Equation of State. Fugacities of Gaseous Solutions*, Chem. Rev. **44**, 233-244 (1949).

Sadus, R. J., *A Simple Equation of State for Hard-Sphere Chains*, AIChE J., **45**, 2454-2457 (1999).

Soave, G. *Equilibrium Constants from a Modified Redlich-Kwong Equation of State*. Chem. Eng. Sci., **27**, 1197-1203 (1972).

van der Waals, J. D., *Over de Continuiteit van den Gas- en Vloeistoftoestand (On the Continuity of the Gas and Liquid State)*. Ph.D. thesis, Leiden, The Netherlands, (1873).

Wertheim, M. S., *Fluids with Highly Directional Attractive Forces I. Statistical Thermodynamics*, J. Stat. Phys., **35**,

19-34 (1984).

Yethiraj, A. and C. K. Hall, *Monte Carlo Simulations and Integral Equation Theory for Microscopic Correlations in Polymeric Fluids*, J. Chem. Phys., **96**, 797-809 (1992).

352032 Methanol to Olefins – a New and Profitable Route for the Production of Light Olefins

Abstract:

Forecasts predict that the world will need to add approximately 100 million metric tons per year of additional light olefin production capacity over the next ten years to meet the world's demand for plastics and other products. Currently, nearly all of the world's ethylene is produced via steam cracking of various hydrocarbons from ethane to heavy gas oils. Propylene, on the other hand, is mainly produced as a byproduct from steam crackers processing propane and heavier feeds or as a byproduct from fluid catalytic cracking units. With the increased availability of low cost ethane, the percentage of ethane fed steam crackers is growing. Since ethane crackers produce very little propylene, this is creating the need for alternative technologies that produce "on-purpose" propylene.

UOP has now commercialized the Advanced MTO process for producing propylene and ethylene from methanol. This process utilizes feedstocks such as methane, coal and other inexpensive carbon sources, to enable a new route for large scale production of light olefins. The process, which was jointly developed with INEOS and Total, produces polymer or chemical grade propylene and ethylene at carbon yields approaching 90%. The first commercial unit was successfully started up by Wison Clean Energy in Nanjing China.

In addition to providing an overview of the Advanced MTO technology, this presentation will focus on the economic and market forces that will drive further adoption of the technology.

Efficient Propylene Production by Catalytic Cracking of Light Naphtha with MFI-Zeolite/Metal-Oxide Composites

Dr. Shinya Hodoshima, Azusa Motomiya, Shuhei Wakamatsu and Dr. Fuyuki Yagi,*
Research & Development Center, Chiyoda Corporation, Yokohama, Japan
* E-mail: shodoshima@ykh.chiyoda.co.jp

Introduction

Propylene is quite important as a raw material building block in the petrochemical industry and the global demand for propylene has been increased steadily. A majority of the propylene (ca. 60%) has been produced as a byproduct by thermal steam-cracking of hydrocarbon feedstock such as naphtha, but this technology requires a huge amount of thermal energy (temperature: 850-900°C) as well as its low propylene/ethylene weight ratio (ca. 0.5). The recent shift to ethane-based feedstock for producing ethylene also might affect the propylene supply, because this technology produces little propylene. Therefore, any new method for producing propylene efficiently from light-naphtha, being widely available feedstock, should be developed to meet future propylene demand and utilize various light paraffinic hydrocarbons (e.g., light liquid hydrocarbons obtained from refinery processes of shale gas and other unconventional crude oil, GTL-naphtha) by converting them into valuable light olefins.

Catalytic cracking of naphtha-based hydrocarbons over solid-acid catalysts, e.g., ZSM-5, have been actively investigated as an alternative method for on-purpose propylene production to conventional thermal steam-cracking [1,2]. In particular, catalytic cracking using fixed-bed reactor has the following advantages compared with steam-cracking: (1) high propylene/ethylene weight ratio; (2) energy-saving. Furthermore, the cracking process in fixed-bed mode is superior to fluid catalytic cracking (FCC) processes in terms of operational simplicity in reaction unit. However, catalytic cracking process using fixed-bed reactor has not been commercialized, because practical cracking catalyst with long lifetime, being applicable to fixed-bed reactor, has been still undeveloped.

From the above-mentioned viewpoints, we have focused on the development of highly active and stable catalyst for light-naphtha cracking in the fixed-bed mode in order to produce propylene efficiently under mild conditions at around 550°C. The proprietary zeolite-based catalysts for cracking, consisting of modified Al-MFI zeolites and metal oxides (e.g., aluminum oxide) [3], were developed in the present research. This paper presents both characteristics of these catalysts and their excellent performance on the experimental basis.

Experimental

Preparation of zeolite-based catalysts

The desired MFI-type zeolites containing some heteroatom (Al-MFI(ZSM-5), A-Al-MFI, B-Al-MFI, A-B-Al-MFI, A, B: heteroatom) were hydrothermally synthesized in autoclave at 120-180°C for 24-72 h under adequate stirring conditions. These sodium-type zeolites were converted into proton-type ones by ion-exchange with NH_4NO_3 aqueous solution and subsequent calcinations at 550°C. The obtained zeolites were characterized by XRD, XRF and SEM measurements. Acid densities of these zeolites (Si/Al, Si/(A+Al), Si/(B+Al), Si/(A+B+Al)) were in the range of 19.4-20.3 (ref. **Table 1**).

Finally, the resultant powdered-zeolites were pressed, crushed and sorted into sizes of 350-600 μm for reaction tests.

The composite catalysts, consisting of the A-B-Al-MFI-type zeolites and aluminum oxide (Al_2O_3), were prepared at the mixed ratio of 65 wt%-zeolite/35 wt%-Al_2O_3 as follows. Firstly, the A-B-Al-MFI zeolites with different Si/(A+B+Al) ratios (12.0-177.5) were prepared by the above hydrothermal synthesis. Secondly, the powder mixture of these zeolites and aluminum oxide were formed into cylindrical shape with diameter of 1.0 mm by extrusion. After these samples were converted into proton-type ones with NH_4NO_3 aqueous solution, calcined at 550°C and sorted into length of 1.0-2.0 mm, the A-B-Al-MFI/Al_2O_3 composites were finally obtained as catalysts for light-naphtha cracking.

Catalytic reaction test

Catalytic cracking of *n*-hexane (*n*-C_6H_{14}) as model compound of light-naphtha over zeolites only and zeolite/metal-oxide composites were carried out for 5-24 h in a flow-type continuous reactor to measure activities. The *n*-hexane diluted with nitrogen (molar ratio of nitrogen to *n*-hexane: 15/1) were supplied to the reactor at 550°C with the Liquid Hourly Space Velocity (LHSV) of *n*-hexane at 1.0 h^{-1}. Both catalytic conversions of *n*-hexane and yields of products were calculated by gas-chromatographic analysis. Catalytic cracking of *n*-hexane without dilution over zeolite/metal-oxide composites were performed in order to examine catalytic performance under practical conditions. The *n*-hexane without dilution was supplied to the reactor at 565°C with the LHSV of *n*-hexane at 5.0 h^{-1}. In addition, a long-term reaction with the A-B-Al-MFI/Al_2O_3 composite (Si/(A+B+Al) of zeolite: 121.3) was also carried out to evaluate catalyst lifetime under the same conditions (vide infra).

Results and discussion

Modified MFI-zeolites as matrix component for catalytic light-naphtha cracking

The comparison of catalytic performance among four kinds of zeolite species (Al-MFI, A-MFI, B-MFI and A-B-Al-MFI(A, B: heteroatom)) with the almost same acid density (Si/T ratios measured by XRF: 19.4-20.3, T=Al, A+Al, B+Al, A+B+Al) was carried out under the same conditions. **Fig. 1** and **Table 1** show product selectivity and catalytic activities in cracking of *n*-hexane diluted with nitrogen at 550°C over these zeoiltes synthesized in the present study, respectively. As shown in **Fig. 1**, the product distribution was clearly changed by introducing heteroatom into the Al-MFI zeolite(ZSM-5) as a reference sample. Though the conventional Al-MFI exhibited comparatively high selectivity to light olefins, it also gave high selectivity to aromatics. In the case of the A-Al-MFI, aromatics yield became ca. 50 wt% (**Table 1**) by predominant formation of aromatics, resulting in the rapid deactivation due to coking. On the other hand, the B-Al-MFI and the A-B-Al-MFI remarkably enhanced selectivity to light olefins by suppressing aromatics formation, so that high propylene yields of 24.7-27.0 wt% were obtained in spite of low temperature of 550°C for cracking reactions and these activities were kept constant during reaction time (5 h). In general, catalytic cracking of paraffinic hydrocarbons over solid-acid catalysts proceeds via light olefins successively to generate aromatics and selectivity to light olefins strongly depends on acidity of catalysts. Since heteroatom (A) and heteroatom (B) in the zeolite have been known to accelerate dehydrogenation of alkanes and reduce acid strength, respectively, the combination of two kinds of heteroatom (A, B) or the addition of heteroatom (B) was considered to cause a drastic change of acidity. As a consequence, the A-B-Al-MFI-zeolite containing two kinds of heteroatom at optimized ratio (A/(A+B+Al)=0.3, B/(A+B+Al)=0.4, Al/(A+B+Al)=0.3) gave the best catalytic performance for light-naphtha cracking in the terms of selectivity and yield of propylene.

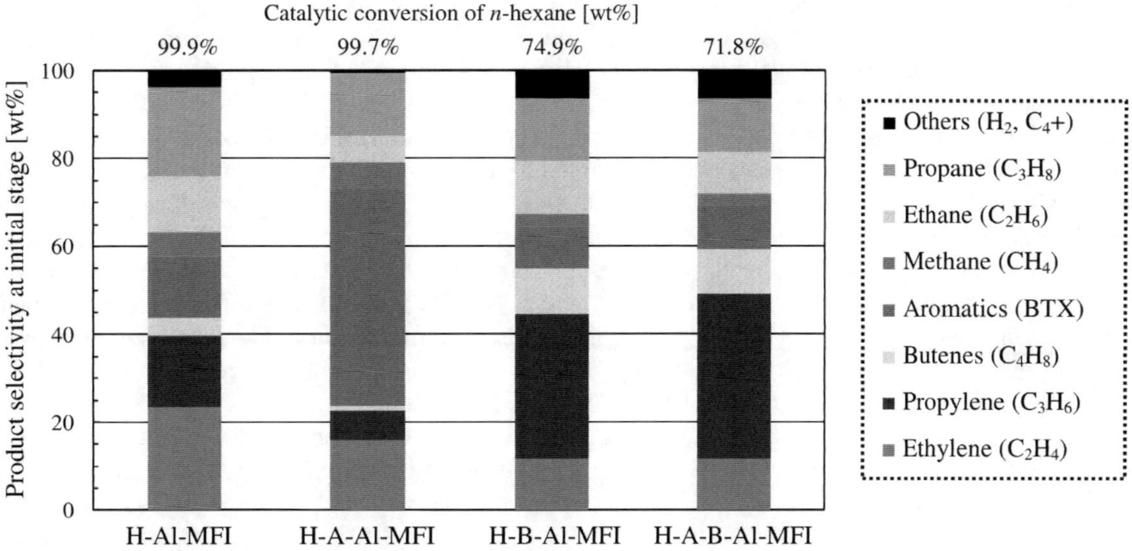

Zeolite species(catalyst); H-Al-MFI(Si/Al=20.0), H-A-Al-MFI(Si/(A+Al)=20.3, A/(A+Al)=0.3, Al/(A+Al)=0.7), H-B-Al-MFI(Si/(B+Al)=19.4, B/(B+Al)=0.5, Al/(B+Al)=0.5), H-A-B-Al-MFI(Si/(A+B+Al)=19.4, A/(A+B+Al)=0.3, B/(A+B+Al)=0.4, Al/(A+B+Al)=0.3),
A, B: Heteroatom contained in MFI-type zeolite
Reaction conditions; Temperature: 550°C, Pressure: 0.1 MPa, Charged volume of catalyst: 1.0 mL,
Feedstock: n-C_6H_{14} diluted with N_2(molar ratio of n-C_6/N_2: 1/15), LHSV of n-C_6: 1.0 h^{-1}, Reaction time: 5 h

Fig. 1 Initial product distribution in n-hexane cracking over various MFI-type zeolites

Table 1. Initial activities for n-hexane cracking over various MFI-type zeolites

Zeolite species	Acid density* [mol/mol]	Conversion [wt%]	Yield [wt%] C_3H_6	Yield [wt%] Aromatics
H-Al-MFI	20.0	99.9	16.3	14.2
H-A-Al-MFI	20.3	99.7	6.9	49.5
H-B-Al-MFI	19.4	74.9	24.7	7.3
H-A-B-Al-MFI	19.4	71.8	27.0	7.1

Cat.: MFI-type zeolites containing heteroatom (Al, A, B)
Acid density: Si/T molar ratio measured by XRF technique(T= Al, A+Al, B+Al, A+B+Al)
Reaction conditions; Temp.: 550°C, Press.: 0.1 MPa, Charged volume of catalyst: 1.0 mL, Feedstock: n-C_6H_{14} diluted with N_2(molar ratio of n-C_6/N_2: 1/15), LHSV of n-C_6: 1.0 h^{-1}, Reaction time: 5 h

Table 2. Initial activities for n-C_6H_{14} cracking over zeolite only and zeolite/Al_2O_3 catalysts

Catalyst	Conv. [wt%]	$C_3^=$ yield [wt%]	Space time yield [g-$C_3^=$/g-zeolite h]
H-A-B-Al-MFI only	71.8	27.0	0.43
H-A-B-Al-MFI/Al_2O_3	80.9	28.3	0.53

Cat.: H-A-B-Al-MFI(Si/(A+B+Al)=19.4, A/(A+B+Al)=0.3, B/(A+B+Al)=0.4, Al/(A+B+Al)=0.3) and H-A-B-Al-MFI/Al_2O_3 composite(Si/(A+B+Al)=19.4, A/(A+B+Al)=0.3, B/(A+B+Al)=0.4, Al/(A+B+Al)=0.3, Mixed ratio of zeolite/Al_2O_3: 65 wt%/35 wt%)
Reaction conditions; Temp.: 550°C, Press.: 0.1 MPa, Charged volume of catalyst: 1.0 mL, Feedstock: n-C_6H_{14} diluted with N_2(molar ratio of n-C_6/N_2: 1/15), LHSV of n-C_6: 1.0 h^{-1}, Reaction time: 5 h

Modified MFI-zeolite/metal-oxide composite catalysts for light-naphtha cracking

In general, zeolite-based catalysts for industrial use are often formed into certain shapes by use of binder (e.g., metal oxide) in order to improve mechanical strength or reduce pressure loss. In the present study, cylindrical composite catalyst with diameter of 1.0 mm, consisting of A-B-Al-MFI zeolite and aluminum oxide (Al_2O_3) binder, were prepared at the mixed ratio of 65 wt%-zeolite/35 wt%-Al_2O_3 by extrusion and its catalytic performance was examined. **Table 2** compares catalytic activities for cracking of *n*-hexane diluted with nitrogen at 550°C over the A-B-Al-MFI/Al_2O_3 composite (65 wt%/35 wt%) and the A-B-Al-MFI only. Both *n*-hexane conversion and apparent propylene yield were improved by utilizing the composite catalyst. Moreover, weight of propylene that can be generated per one hour with zeolite of one gram, i.e. space time yield of propylene, was also enhanced, suggesting that zeolite species were well dispersed by use of metal oxides as a binder. **Table 3** presents the change in catalytic activities for *n*-hexane cracking over the composite catalysts, consisting of A-B-Al-MFI zeolites of 65 wt% with different acid densities (Si/(A+B+Al)=12.0, 19.4 and 31.3) and aluminum oxide of 35 wt%. Higher propylene yields than 20 wt% were obtained in all cases. Both conversions of *n*-hexane and yields of light olefins increased at lower Si/(A+B+Al) ratios (larger amounts of heteroatom), indicating that cracking reaction proceeded over acid sites of zeolite. In particular, the composite catalyst with the Si/(A+B+Al) ratio of 12.0 exhibited high propylene yield of 31.2 wt%, being competitive to thermal steam-cracking or other technologies (e.g., FCC processes). Moreover, it is noteworthy that this reaction temperature (550°C) was much lower than those (850-900°C) in the steam-cracking. With respect to the reaction test attaining 31.2 wt%, it was also found that the activity was maintained for 24 h at least, as shown in **Fig. 2**. This long lifetime was due to the size effect of zeolite as well as its acidity. Zeolites listed in **Table 3** were in the form of nano-ordered particles with average sizes of below 1.0 μm. Average size of the A-B-Al-MFI with Si/(A+B+Al) ratio of 12.0 was ca. 570 nm by SEM measurement. On the other hand, coarse A-B-Al-MFI zeolite with both micro-ordered average size of ca. 6.3 μm and the almost same compositions as that of the nano-sized zeolite was prepared by controlling synthesis conditions in order to examine the size effect of zeolite on durability. Although this microsized-zeolite gave almost the same *n*-hexane conversions at initial stage as those with the nanosized-zeolite, its performance decreased rapidly in several hours (see **Fig. 2**). Fine zeolite particles was considered to increase the number of effectively working micropores for cracking reactions, so that the zeolites was highly resistant to deactivation due to coke formation or accumulation.

In terms of highly catalytic performance as well as enhancement of mechanical strength required for industrial use, it was thus confirmed that zeolite species combined with metal oxides were utilized efficiently as catalysts for light-naphtha cracking.

Excellent performance of MFI-zeolite/metal-oxide composite catalysts under practical conditions

For the purpose of applying the present composite catalysts to practical cracking process, catalytic performance of the present composites under severe conditions were examined by performing cracking of *n*-hexane without dilution at 565°C. Catalytic cracking of light-naphtha without dilution is superior to catalytic cracking with diluents such as nitrogen or steam, in terms of both no thermal energy for heating diluents and avoidance of irreversible catalyst deactivation derived from dealumination of zeolite. **Fig. 3** presents the relationship between acid density of A-B-Al-MFI/Al_2O_3 catalyst and catalytic activities for cracking of *n*-hexane without dilution. In contrast to the cracking of *n*-hexane diluted with nitrogen (see **Table 3**), the A-B-Al-MFI/Al_2O_3 catalysts with low Si/(A+B+Al) ratios (e.g., 19.4, 31.3) gave high aromatics yields in the cracking of *n*-hexane without dilution. As amounts of acid sites decreased (higher the Si/(A+B+Al) ratios), selectivity to light olefins became higher by suppressing aromatics formation

Table 3. Initial activities for *n*-hexane cracking over MFI-zeolites/Al_2O_3 catalysts with different acid densities

Catalyst	Acid density [mol/mol]	Conversion [wt%]	Yield of light olefins [wt%]	
			C_2H_4	C_3H_6
H-A-B-Al-MFI/Al_2O_3	31.3	46.1	3.7	20.0
H-A-B-Al-MFI/Al_2O_3	19.4	80.9	10.0	28.3
H-A-B-Al-MFI/Al_2O_3	12.0	85.6	9.8	31.2

Catalyst: H-A-B-Al-MFI/Al_2O_3 composites(Si/(A+B+Al)=12.0-31.3, A/(A+B+Al)=0.3, B/(A+B+Al)=0.4, Al/(A+B+Al)=0.3, Mixed ratio of zeolite/Al_2O_3: 65 wt%/35 wt%)
Reaction conditions; Temp.: 550°C, Press.: 0.1 MPa, Charged volume of catalyst: 1.0 mL, Feedstock: n-C_6H_{14} diluted with N_2(molar ratio of n-C_6H_{14}/N_2: 1/15), LHSV of n-C_6H_{14}: 1.0 h^{-1}, Reaction time: 5 h

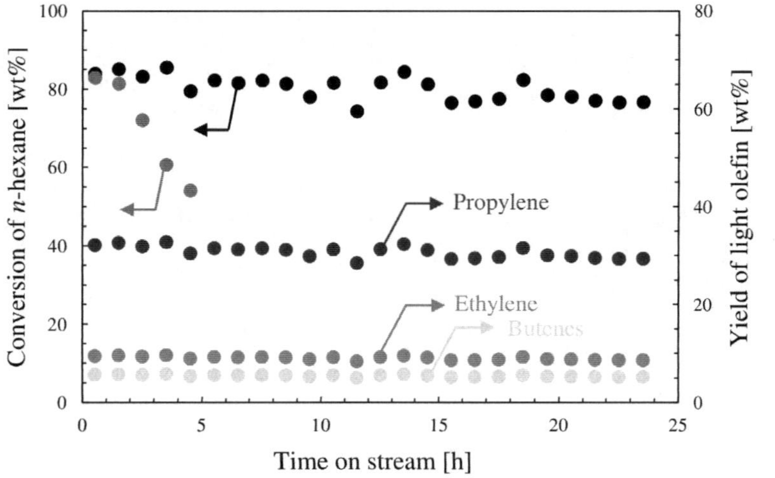

Fig. 2 Time courses of catalytic activities by A-B-Al-MFI/Al_2O_3 composites with high acid density

and the highest propylene yield (ca. 18 wt%) was obtained at the Si/(A+B+Al) ratio of ca. 120. At higher Si/(A+B+Al) ratios than 150 (smaller amounts of acid sites), yields of light olefins decreased because of lower *n*-hexane conversions. **Table 4** shows catalytic performance of the A-B-Al-MFI/Al_2O_3 catalysts in the *n*-hexane cracking with or without dilution. Here the propylene productivity was ca. 1.1 [g-propylene/g-zeolite h], being ca. twice of the value in the case of diluting with nitrogen.

In the cracking of *n*-hexane without dilution, aromatization proceeds much easier than cracking of *n*-hexane diluted with nitrogen due to bimolecular reaction of *n*-hexane and/or re-adsorption of light olefins on acid sites, indicating that aromatics would be generated quite selectively in case of using zeolites with high acid densities. In other words, zeolites with low acid densities would give high selectivity to light olefins by suppressing aromatics formation.

The lifetime of the A-B-Al-MFI/Al_2O_3 composite catalyst with the adequate Si/(A+B+Al) ratio, giving the highest propylene yield of ca. 18 wt%, was also estimated by performing *n*-hexane cracking without dilution at 565°C. **Fig. 4** shows time courses of yields of propylene and ethylene. Propylene yields of ca. 18 wt% obtained at initial stage (ethylene yield: ca. 8.0 wt%) were maintained for ca. 80 h at least in spite of severe conditions to cause coke formation easily. This long lifetime was achieved due to high resistance to coke formation. In addition, thermal energy required for producing propylene from *n*-hexane was calculated on this experimental basis. **Fig. 5** presents the comparison of energy required for propylene production in reaction unit with the conventional steam-cracking technology. As shown in **Fig. 5**, thermal energy required for producing propylene of unit weight was estimated to be reduced by ca. 70% compared with the steam-cracking, because no thermal energy for heating diluent such as steam (i.e., apparent heat) was needed as well as its low reaction temperature. The supply of hydrocarbon feedstock without dilution was thus confirmed to be advantageous in terms of energy consumption in reaction unit.

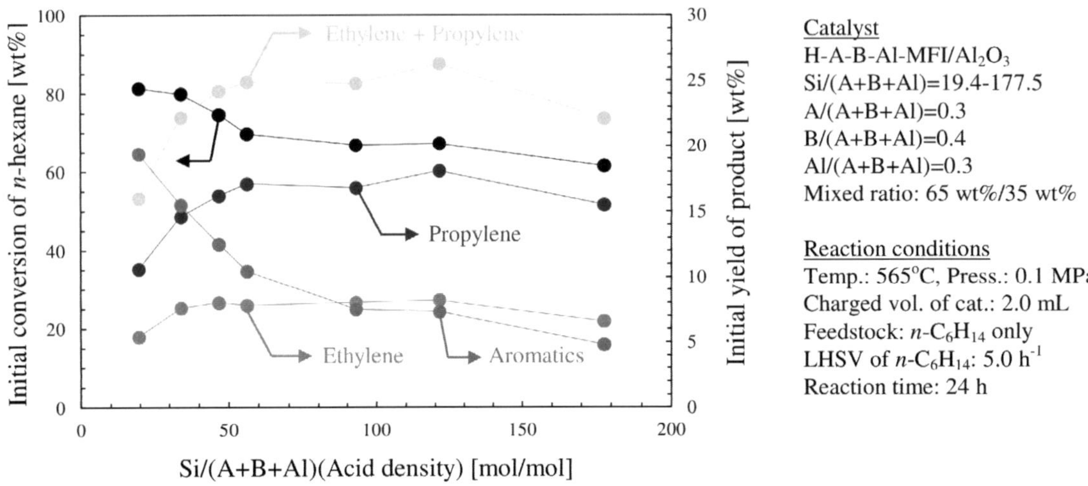

Fig. 3 Both conversion and product yield in n-C_6H_{14} cracking without dilution vs. acid density

Table 4. Catalytic activities for *n*-hexane cracking over H-A-B-Al-MFI/Al_2O_3 composites in the presence and the absence of diluent

Feedstock	Mixed ratio [mol/mol]	LHSV of n-C_6 [h^{-1}]	Temp. [°C]	Conv. [wt%]	Propylene yield [wt%]	Space time yield of $C_3^=$ [g-$C_3^=$ / g-zeolite h]
n-C_6H_{14}/N_2	1/15	1.0	550	80.9	28.3	0.5
n-C_6H_{14}/N_2	1/15	1.0	550	85.9	31.2	0.6
n-C_6H_{14} only	-	5.0	565	67.3	18.1	1.1

Cat.: H-A-B-Al-MFI/Al_2O_3 composite(Si/(A+B+Al) ratio: 19.4-121.3, A/(A+B+Al)=0.3, B/(A+B+Al)=0.4, Al/(A+B+Al)=0.3, Mixed ratio of zeolite/Al_2O_3: 65 wt%/35 wt%
Reaction conditions; Temperature: 550-565°C, Pressure: 0.1 MPa, Charged volume of catalyst: 1.0 mL(n-C_6/N_2), 2.0 mL(n-C_6 only), Reaction time: 5.0 h(n-C_6/N_2), 24.0 h(n-C_6 only)

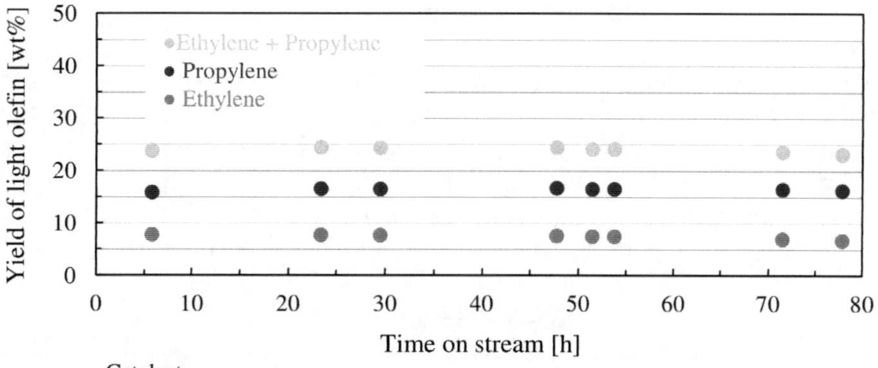

Catalyst
H-A-B-Al-MFI/Al_2O_3
Si/(A+B+Al)=121.3, A/(A+B+Al)=0.3, B/(A+B+Al)=0.4, Al/(A+B+Al)=0.3
Mixed ratio of zeolite/Al_2O_3: 65 wt%/35 wt%

Reaction conditions
Temperature: 565°C, Pressure: 0.1 MPa, Charged volume of catalyst: 2.0 mL,
Feedstock: n-C_6H_{14} only, LHSV of n-C_6H_{14}: 5.0 h^{-1}

Fig. 4 Time courses of n-C_6H_{14} cracking without dilution over A-B-Al-MFI/Al_2O_3 composite catalyst

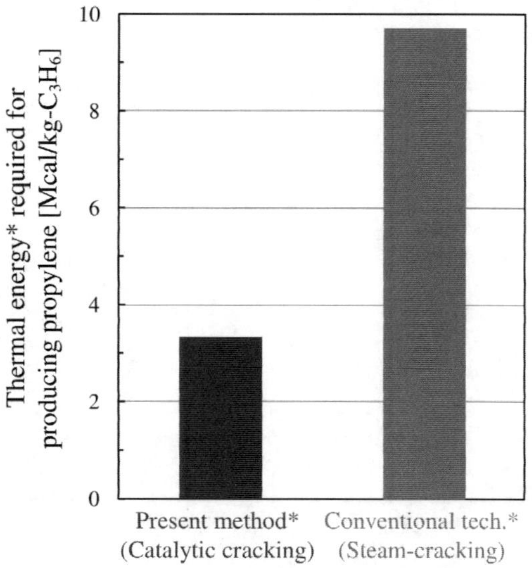

* Reaction heat + apparent heat
* Catalytic cracking of n-hexane without dilution at 565°C
* Thermal steam-cracking at 850°C (Non-catalytic)

Fig. 5 Comparison of energy required for producing propylene in reaction unit between present catalytic-cracking and conventional steam-cracking

Summary

The proprietary composite catalysts, consisting modified Al-MFI-zeolite and metal-oxide, developed in the present study made it possible to produce propylene quite efficiently from light-naphtha at mild reaction temperatures of 550-565°C and gave long lifetime that is applicable to cracking process using simplified fixed-bed reactor. It was suggested on the experimental basis that the energy-saving process for producing propylene efficiently from light-naphtha could be realized by using the present catalysts as an alternative on-purpose technology to conventional thermal steam-cracking.

Acknowledgements

A part of the present study was financially supported by the Japan Oil, Gas and Metals National Corporation (JOGMEC). We would like to express our deep gratitude here. The authors acknowledge Prof. Sachio Asaoka (The University of Kitakyushu) for guidance regarding zeolite synthesis.

References

[1] H. Konno et al., J. Jpn. Petrol. Inst., 55(4), 267-274(2012).
[2] A. Corma et al., Appl. Catal. A: General, 421-422, 121-134(2012).
[3] S. Hodoshima et al., Extended Abstract of 7th Int. Symp. on Acid-Base Catal., PB-011, 12-15 May 2013, Tokyo, Japan.

351465 A Reactive Distillation Process for Separation of Propylene-Propane Mixtures

Abstract:

Propylene has been historically produced as a by-product of ethylene in the refinery gas oil or naphtha steam-cracking process, or as a by-product of motor gasoline in the fluid catalyst cracking (FCC) process. Prior to 2005, propylene market supply/demand was in balance and no ***on purpose production*** of propylene was necessary. However, the propylene market has continued to expand and traditional refinery-based modes of co-production can no longer meet demand. Global consumption of propylene in 2012 was about 79 million metric tons per year (MMTA) with growth at 7% annually. It is estimated that US production of propylene must increase by 2 million tons per year by 2020 to meet the increasing demand. Expanded shale gas production in North America has recently sent a supply shock throughout the world energy picture, and is beginning to transform the global ethylene and propylene markets. Production of ethylene by steam cracking of naphtha or gas oil is becoming non-competitive, and U.S. producers are shifting their steam crackers to lighter feedstock such as shale gas-derived liquids. Energy-efficient separation and recovery of high-purity propylene is a crucial step in its ***on-purpose*** production.

The design study in this project focuses on the development of a novel, reversible reaction-based process to separate and recover propylene from propylene/propane mixtures in ***on-purpose*** propylene production. Separating light olefins from paraffinic feed stocks by conventional distillation is highly energy and capital intensive because of small boiling points differences. In order to improve energy efficiency and reduce capital, alternate separation processes have been developed and implemented. This design study focuses on an advanced reactive distillation concepts including heat integration and side reactors. In the process, propylene is hydrated to form isopropanol or esterified to form an isopropyl ester that is immediately separated from propane. The reverse reaction regenerates propylene of high purity.

139d

Fuel Flexibility in Low NO_x Industrial Gas Turbines

Hans Mattsson and Jenny Larfeldt

hans.mattson@siemens.com and jenny.larfeldt@siemens.com

Siemens Industrial Turbomachinery AB, Finspong, Sweden

Abstract

Siemens experiences a continuously expanding request from customers to utilize varying gaseous fuels for power production. From such requests follows an increased need for testing new fuels and fuel systems limitations. In recent years Siemens has performed a number of fuel flexibility tests on the medium size range industrial gas turbines. This paper is a survey of Siemens experience on diluting natural gas with;

- Up to 50 volume -% nitrogen (inert gas),

- Heavy hydrocarbons with testing up to 23 volume-% pentane and 100% ethane

- Up to 40 volume-% hydrogen.

A novel approach for fuel testing has been developed, where a single burner rig testing is combined with a full scale engine test to give a cost effective and flexible solution. The combination of the two approaches is accomplished by using a separate feed of testing fuel to one or more burners in a standard gas turbine installation where the other burners use standard fuel.

Also the function and principles of Siemens power producing industrial gas turbine, particularly the combustion system, are described. It is proved that the Siemens 3rd generation DLE (Dry Low Emissions) burner is a very fuel flexible and reliable, low emissions solution for both oil and gas upstream applications and in various chemical processes in the industry.

INTRODUCTION

The requirements on medium sized gas turbines are becoming more demanding as the requirement for profitability and low environmental impact increases. The basic requirement for plant profitability is high reliability and availability since every hour the gas turbine is out of operation results in loss of production for the plant owner.

Fuel-flexible gas turbines can also improve plant profitability. A fuel-flexible gas turbine operating on flare gas can increase plant profitability and lower environmental impact. If the gas turbine can operate directly on the flare gas, complex and sometimes expensive fuel treatment systems can be avoided. A complex fuel-treatment system will definitely have a negative impact on the plant reliability.

Low environmental impact is achieved by designing high efficiency plants to minimize CO_2 in combination with low-NO_x emission combustion systems. One market perception is that DLE combustion systems are less reliable than conventional combustion systems. Furthermore, the DLE combustion systems are also considered to be less fuel flexible. However, Siemens DLE system has proven over 6 million operating hours to have the same reliability as conventional combustion systems.

The complete Siemens gas turbine program ranges from 5 MW to 375 MW in size and can be used both in mechanical drive and power generation applications (Figure 1). The equipment is supplied to a variety of solutions such as pipelines, power generation or mechanical drive, combined heat and power, combined cycle power plants or marine propulsion in onshore and offshore applications. This paper focuses on the medium-sized gas turbines SGT-700 [1] and SGT-800 [2], their DLE performance and fuel flexibility.

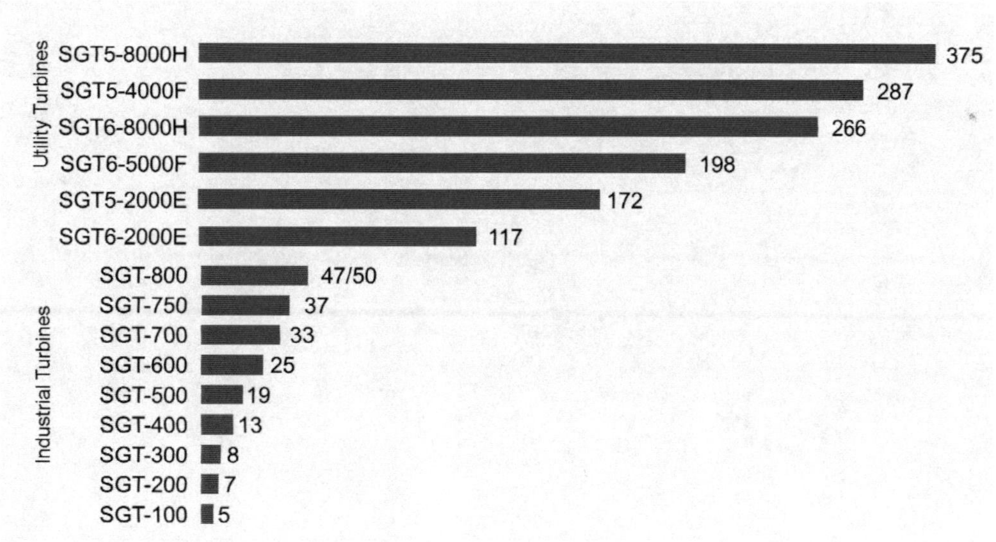

Figure 1. Siemens complete gas turbine range from 5 to 375 MW

FUEL FLEXIBILITY

Fuel flexibility is very important for industrial gas turbines and Siemens is constantly requested to make judgments on a variety of gaseous fuels and how it could be used in an industrial gas turbine. The fuel enquires Siemens received during last year are seen Figure 2. Each dark blue dot corresponds to one fuel specification from a customer, shown as the mass based lower heating value versus the Wobbe Index (WI). This index is a key gaseous-fuel parameter from the gas turbine perspective and includes the heating value and the density of the fuel. It is defined according to the equation 1 below. Standard conditions for a normal cubic meter are at pressure 101.325 kPa and temperature 273 K.

$$WI^0 = \frac{LHV}{\sqrt{\rho_{rel}}} \qquad (1)$$

$$\rho_{rel} = \frac{\rho_{gas}}{\rho_{air}} \qquad (2)$$

For premixed low NO_x systems such as the DLE system described below, varying fuel means that expected emissions and risks for combustion related problems as flashback and flameout must be evaluated on a case by case basis. The wide spread in fuel qualities give a clear need for combustion testing as the fuel quality influences:

- Flame stability
- Flame position
- Combustion dynamics
- Flame out
- Flash back
- Combustion temperature field
- Hardware temperature
- Emissions

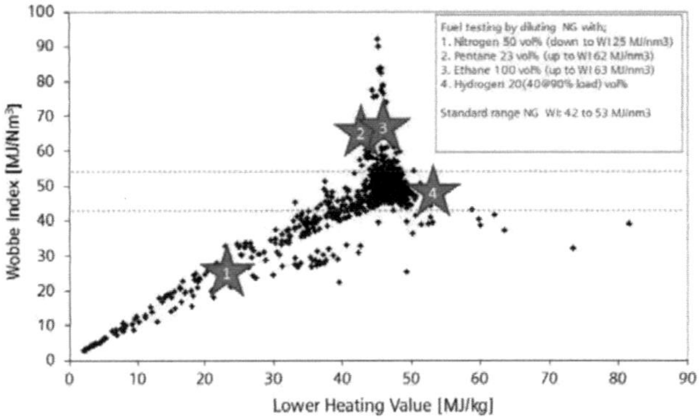

Figure 2. Gaseous fuel enquires given in the Lower Heating Value and Wobbe Index- plane. Fuel testing by diluting NG with nitrogen (1), pentane (2), ethane (3) and hydrogen (4). Standard range WI 42 to 53 MJ/nm3 (dotted lines).

SGT-700/800 COMBUSTION SYSTEM

The SGT-700 (33 MWel) and SGT-800 (47/50MWel) gas turbines are equipped with 18 and 30 3rd generation DLE burners respectively, in an annular combustor. The burners use pilot fed at the burner outlet for flame stabilization (see Figure 3). Typically for DLE combustion, fuel is burnt with the double amount of air than what is needed for stoichiometric combustion. The large amount of air keeps the flame temperature low which reduces emissions of nitrous oxides but on the other hand increases the risk of combustion instabilities and incomplete combustion leading to emissions of CO as illustrated in Figure 4.

In Siemens 3rd generation DLE burner robust operation is secured by the pilot flames during e.g. changes in fuel composition as well as during transients in engine operation such as rapid load changes. The pilot fuel ratio can be programmed in the control system to be increased in milliseconds as a response to sudden/unexpected transients and thereby avoiding tripping the engine - but the penalty for this stability is a slight increase in NO_x emissions.

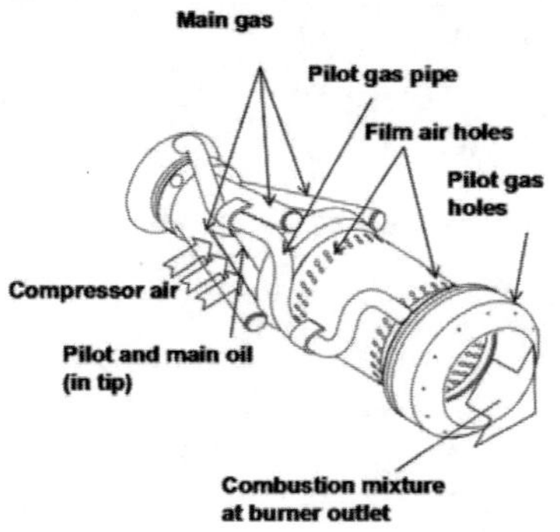

Figure 3. The 3rd generation burner used in the annular combustor of SGT-700 and SGT-800.

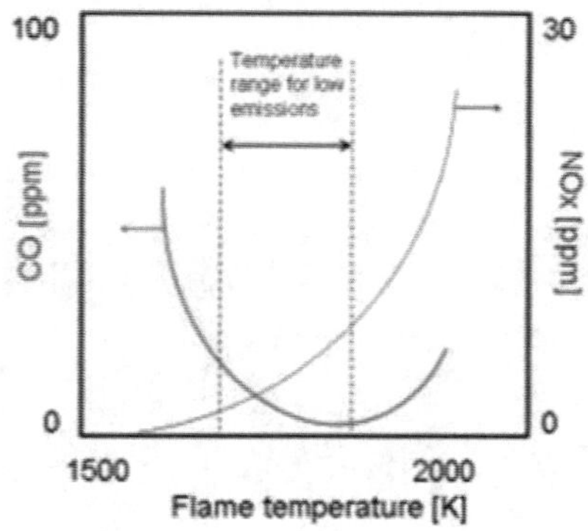

Figure 4. Typical behaviour of NO_x and CO versus flame temperature.

NATURAL GAS WITH HIGH INERT CONTENT

Natural gas with high contents of inert gas, such as N_2 or CO_2, may increase the tendency for lean blow out in a DLE combustion system.

The standard SGT-800 was tested on nitrogen rich fuel by simply injecting nitrogen into the engine gas fuel feed system during normal operation [3]. A nitrogen truck (see Figure 5), was rented. The connection and mixing of nitrogen into the natural gas was situated in the meter point cabinet of Öresundskraft AB , prior to the gas turbine fuel skid delivery point.

No hardware modifications were made to the engines. The control system was adjusted to allow for higher fuel flows since the engine is normally protected by limiters for too high fuel flows using standard natural gas fuels. These modifications were made on-line as the heating value changed. The control-loops themselves were unchanged and it was only the limiters that were adjusted.

The loads tested were 20, 30, 35 and 40 MW electric output from the engine. The nitrogen feed was also in this case gradually increased to a constant natural gas flow as illustrated in Figure 6 for the 40 MW load case. The green curve shows the load, and load swings can be seen as nitrogen content increase in steps. The red curve shows the total fuel flow to the engine, about 2,8 kg/s when running on natural gas increasing to about 4,3 kg/s.

In the SGT-800 test the pilot fuel ratio was set to about 14 % before the nitrogen feed was started. This is a much higher pilot flow applied as a precaution to secure stability, normally pilot is operated between 2 and 5 % of the total fuel flow at high load. During the test it was seen that the emissions of NO_x decrease as the nitrogen content in the fuel increases, however no specific emission tuning was performed during this relatively short test.

Fuel transients

As can be seen in Figure 7, the nitrogen feed was abruptly cut off after about 43 minutes of nitrogen feeding for the 20 MW case. Within 2 minutes pure natural gas was again fuelling the gas turbine.

The nitrogen rich fuel thus has a WI of about 25 MJ/Nm^3 shifted back to natural gas with a WI of about 48 MJ/Nm^3 implying a rate of change of about 45%/minute or 0,19 MJ/Nm^3/s.

Figure 5. The mobile nitrogen supply unit.

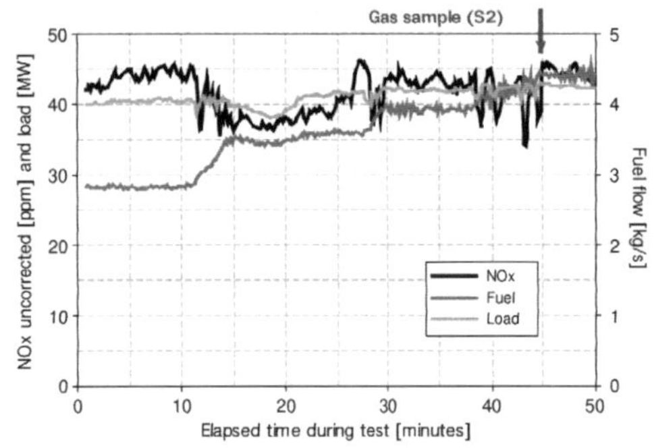

Figure 6. Emissions of NO_x, load and fuel flow to SGT-800 versus time at about 40 MW with increased N_2 content in NG.

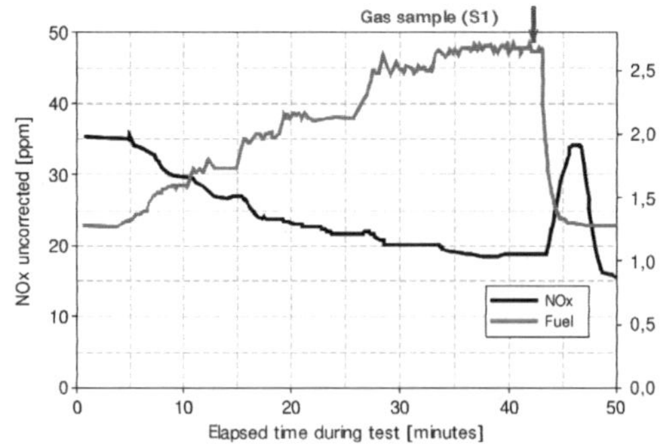

Figure 7. Emissions of NO_x and fuel flow to SGT-700 versus time at 20 MW with increased N_2 content in NG.

The testing has proved that the SGT-700 and SGT-800 are capable of operating on natural gas fuel with up to 50% by volume of nitrogen.

NATURAL GAS WITH HIGH HHC

Using natural gas with higher contents of heavy hydrocarbons, such as ethane, propane, butane and even heptane, as gas turbine fuel it is the ignition delay time that is of main concern. The larger the hydrocarbon chain is the lower the auto ignition temperature is (see Figure 8), meaning that ignition happens easier but also faster. Too fast ignition in a DLE combustion system means that the mixture of fuel and air is ignited upstream in the burner before the gas mixture enters the combustion chamber. Such undesired ignition will cause a flashback with potential burner hardware damages and gas turbine failure.

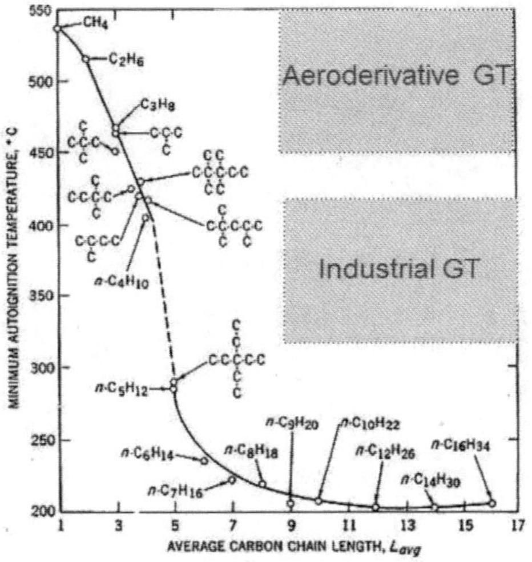

Figure 8. Auto ignition temperatures at atmospheric pressure for hydrocarbons versus carbon chain length [4].

In order to test our capability of using natural gas with high amounts of heavy hydrocarbons a test was performed in 2010 [5]. Pentane was sprayed into the natural gas feeding pipe at Siemens test bed in Finspong, Sweden on a SGT-700. Since the achievable gas temperature in the test bed is limited to about 60 °C, the pentane content was in this test limited to maximum 10 volume-% (33 weight-%) by its dew point.

In order to realize higher pentane concentrations in the fuel another approach using a separate feed of test fuel to one burner in a standard gas turbine, while the other burners use standard fuel for engine operation [6]. Testing was also this time performed in the delivery test bed and both the single fuel mixing line and the gas unit 1 (filters, etc.) are outdoor installations (see Figure 9). Vaporized LNG, i.e. natural gas, is bled from the main gas supply to the engine and then mixed with pentane into the experimental burner in the engine. As there are 18 burners in the SGT-700 engine it means that 1/18 of the gas turbine heat demand is supplied to the experimental burner. A heating was designed for the separate test fuel feeding line fuel gas temperature of 150 °C was used to avoid condensation of pentane (dew point).

Testing was performed in an SGT-700 engine such that all engine equipment and controllers are standard and all test conditions and transients are engine realistic. This approach has shown to be a very cost effective and flexible solution for fuel flexibility testing at actual engine conditions. A test with ethane in summer 2012 showed that the SGT-700/800 burner is capable of running on pure ethane.

Figure 9. Single burner feed unit and Finspong gas turbine delivery test beds.

The engine has extra instrumentation during a test with single burner feed as illustrated in Figure 10. A borescope camera is used for visual observation in the combustor at the test burner exit. An emissions measurement probe is located after the turbine, downstream of the test burner.

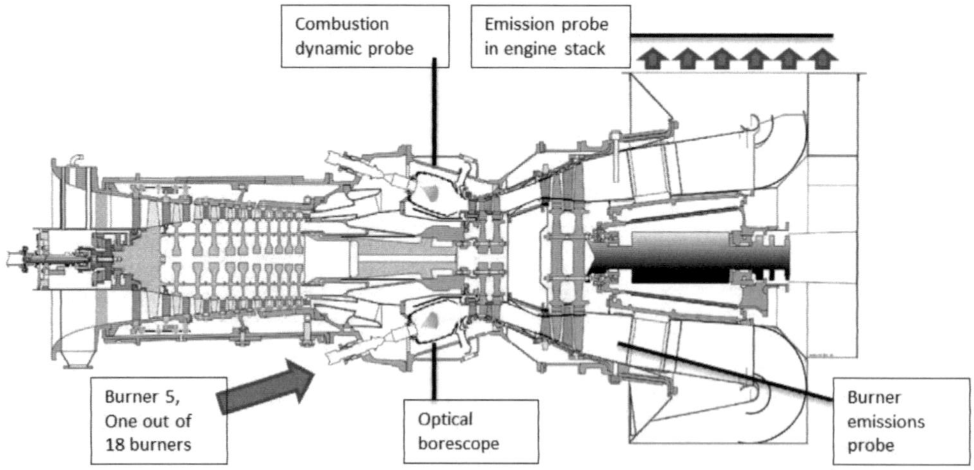

Figure 10. Survey of SGT-700 monitoring equipment during test with single burner feed.

Figure 11 shows the test with stepwise engine load increases and stable operation at 20, 25 and 30 MW. There are also stepwise changes in pentane feed to the single burner (burner 5). Pentane flow control was done manually and locally by adjusting the pump that fed the liquid pentane to the hot gas stream. About 400 kg of pentane was consumed in EBIT during the test. Total fuel consumption for the engine was approximately 17 tons of NG.

As can be seen in Figure 11 the maximum pentane content used was approximately 23% by volume. This corresponds to 53% of the heat input or 55% by weight of the fuel. The heating value of this fuel mixture is 63 MJ/Nm^3 and the Wobbe Index is 62 MJ/Nm^3 (Nm^3 at 0 °C and 1013 mbar).

Based on this test it is concluded that 20 volume-% pentane, or heavy hydro carbon equivalent, can be accepted in the 3rd generation DLE burner used in the SGT-700 and the SGT-800. Only a 20% increase in NO_x emissions is expected.

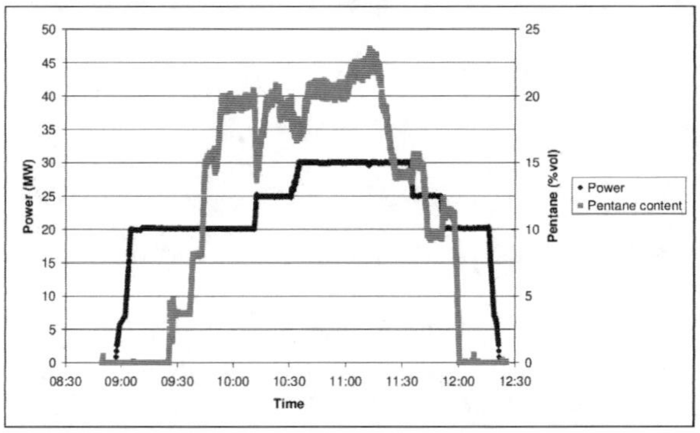

Figure 11. Running profile and pentane content in fuel during the test

NATURAL GAS WITH HIGH H2

Hydrogen is a very reactive fuel as the flame speed is about ten times higher compared to natural gas. In an atmospheric combustion test it is seen that the typical blue, premixed natural gas flame becomes denser and moves closer to the burner outlet (left hand side in Figure 12). This will increase the risk of flashback in the burner during gas turbine operation.

A test at engine like conditions was performed using the single burner feed in September 2012 [7].

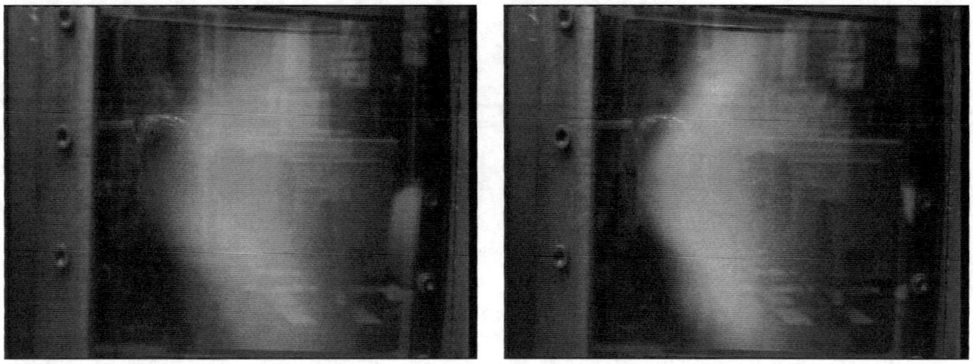

0 % by volume H_2 30 % by volume H_2

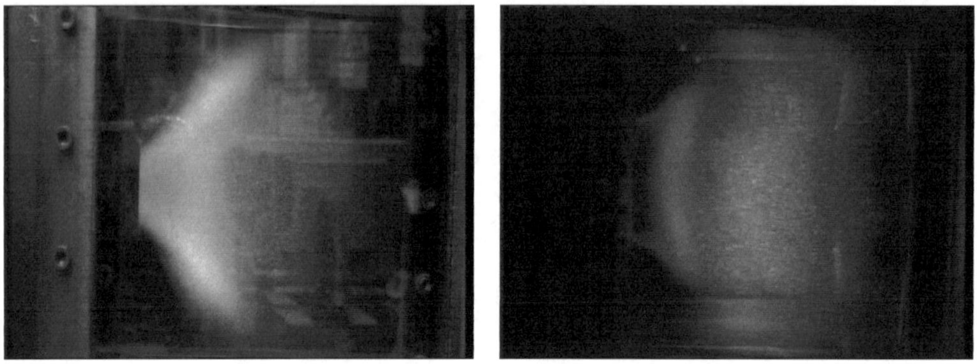

60 % by volume H_2 100 % by volume H_2

Figure 12. Flame from atmospheric combustion test rig as H_2 content in natural gas is increased.

Figure 13 shows a test where the compressor inlet temperature was controlled with electrical heaters. For constant electrical load a change in compressor inlet temperature corresponds to a change in relative load. Higher compressor inlet temperature means a higher relative load. Hydrogen was fed to a targeted concentration of about 30 volume % during the higher inlet temperature. Hydrogen feeding was lowered before the compressor inlet temperature was changed. During the last part of the test even higher hydrogen levels were tested, up to 40 % by volume. This test shows that a small load reduction allows higher hydrogen content to be used.

The increase in hydrogen content was stopped when indications of flame shifting was seen. Gas samples were also taken during the test. The time for sampling is marked with a vertical line in Figure 13. The analysis showed slightly lower hydrogen concentration, 34 % by volume, than expected, 38 % by volume, from the mass flow meter readings. Since hydrogen is very diffusive this error can be associated with the sampling procedure but also there were some instabilities in the hydrogen feeding system.

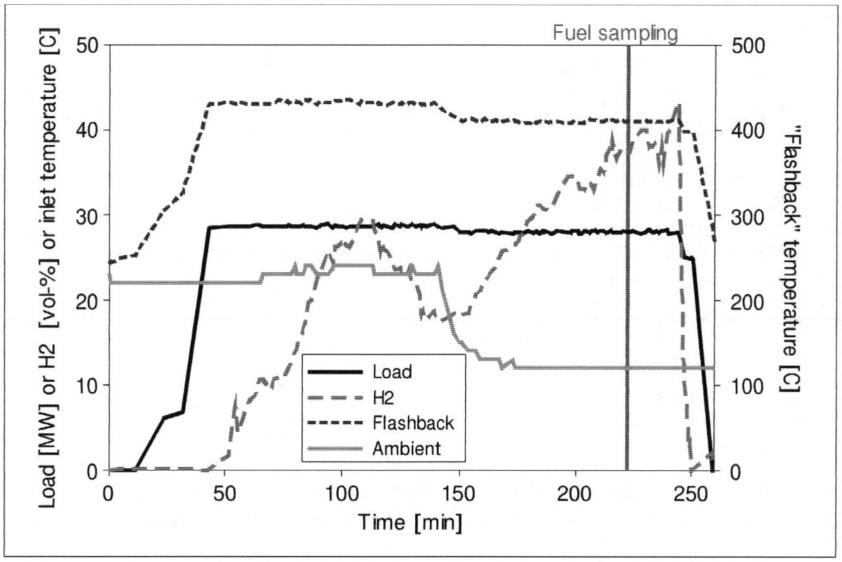

Figure 13. Stable operation at two different ambient temperatures. Fuel sample taken at around 220 minutes indicated by red arrow.

Based on this test the accepted level of hydrogen in the SGT-700 and SGT-800 was increased to 15 % by volume, well above the proposed level of 10 % H_2 by volume in the European natural gas net.

CONCLUSIONS

As a response to Siemens customer requests on utilizing various gaseous fuels for power production fuel flexibility testing has been performed. This paper surveys three types of fuels, all of them mixed into natural gas being the gas turbine standard fuel;

1. *Inert gases* such as N_2 and CO_2
   i. The challenge for a DLE combustion system is the increased risk of lean blow out, i.e. extinction, as the fuel heating value decreases.
   ii. Testing shows that the SGT-700 and SGT-800 are capable of operating on natural gas fuel with up to 50% by volume of nitrogen.
2. *Heavy hydrocarbons* such as ethane, propane, butane and pentane
   i. The challenge for a DLE combustion system is mainly the more rapid ignition compared to natural gas which might cause flash back.
   ii. Testing shows that the SGT-700 and SGT-800 can operate on pure ethane or up to 20 volume-% pentane, or heavy hydro carbon equivalent, in the fuel.
3. *Hydrogen*
   i. The challenge for a DLE combustion system is the high flame speed which can lead to a flash back in the burner.
   ii. Testing has resulted in updated fuel specifications of SGT-700 and SGT-800 allowing for 15 % by volume. Testing showed potential for higher hydrogen contents, up to 40 % H_2 by volume. Pending other parts of the fuel specification, this can be discussed on a case to case basis in projects.

REFERENCES

[1] A. Hellberg and G. Nordén, "Siemens SGT-700 gas turbine performance upgrade yields more power and higher efficiency", PowerGen Europe 2009, May 26-29, Cologne, Germany.

[2] S. Shukin et al, "Siemens SGT-800 industrial gas turbine enhanced to 47 MW. Design modifications and operation experience", ASME Turbo Expo 2008, June 9-13, Berlin, Germany.

[3] Larfeldt J., Larsson A. and Andersson M., "SGT-700 and SGT-800 fuel flexibility testing activities", IGTC 12, October 17-18, 2012, Brussels, Belgium.

[4] Zabetakis, Michael G., "Flammability Characteristics of Combustible Gases and Vapors ", Bureau of Mines Bulletin 627, 1965.

[5] Näsvall H. and Larfeldt J., "Power Generation Utilizing Process Gases to Avoid Flaring", SYS08-841, Värmeforsk,(http://rapporter.varmeforsk.se/Rapporter/DownloadReport.aspx?DocId=2458&Index=D:\INETPUB\varmeforsk12h63pv\Rapporter\pdf\index), 2010

[6] Andersson M., Larsson A., Lindholm A. and Larfeldt J., "Extended fuel flexibility testing of Siemens Industrial Gas Turbines: A novel approach", GT2012-69027, ASME Turbo Expo, Copenhagen, Denmark, 2012.

[7] Andersson M., Larfeldt J. and Larsson A., "Co-firing with hydrogen in industrial gas turbines", SGC report 2013:256

A model for the steam cracking of naphthas

Amérika Arévalo
Juan José Arias
Astrid Ramírez
Luis Carlos López
Alejandro Molina

July Vivas

2014 AIChE Spring Meeting
April 2, 2014

During the steam cracking of naphthas, a mixture of paraffins, olefins, naphthenes and aromatics enters a coil located in a gas-fired furnace

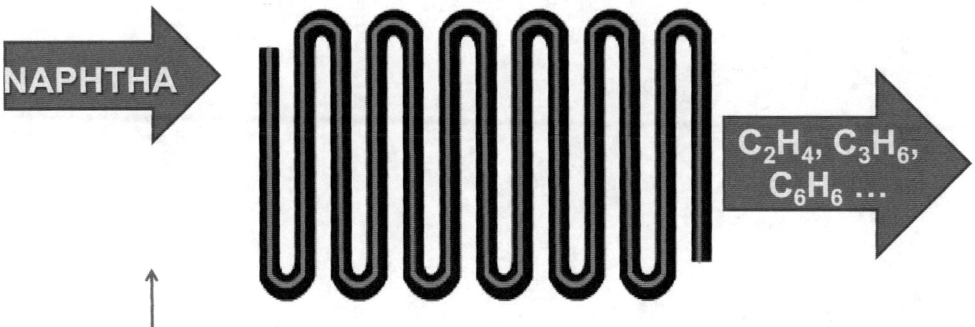

Steam is added to the mixture to control the reactive flow temperature and reduce the partial pressure of hydrocarbons

A model for the steam cracking of naphtha demands knowledge of: (1) Naphtha surrogate, (2) Reaction mechanism, (3) Reactor model, (4) Validation

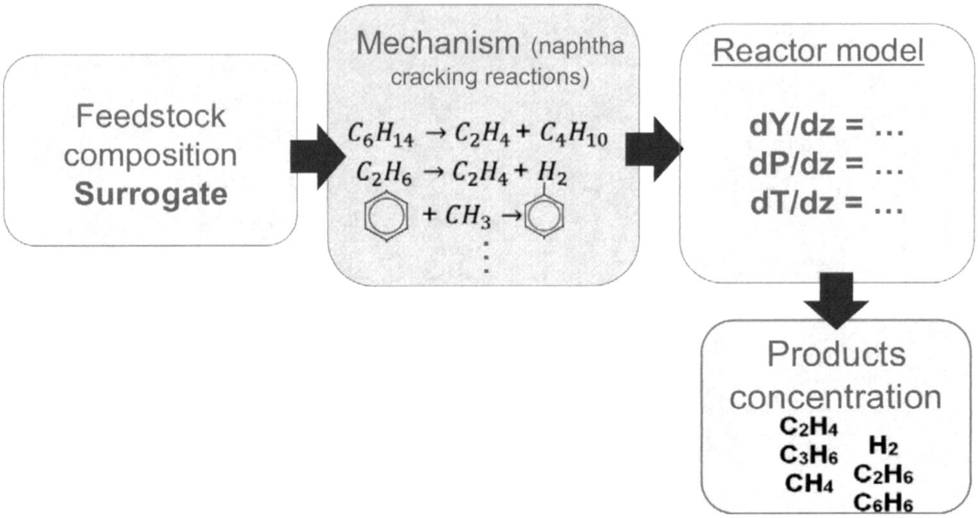

FRUN

UNIVERSIDAD NACIONAL DE COLOMBIA SEDE MEDELLIN

A model for the steam cracking of naphtha demands knowledge of: (1) Naphtha surrogate, (2) Reaction mechanism, (3) Reactor model, (4) Validation

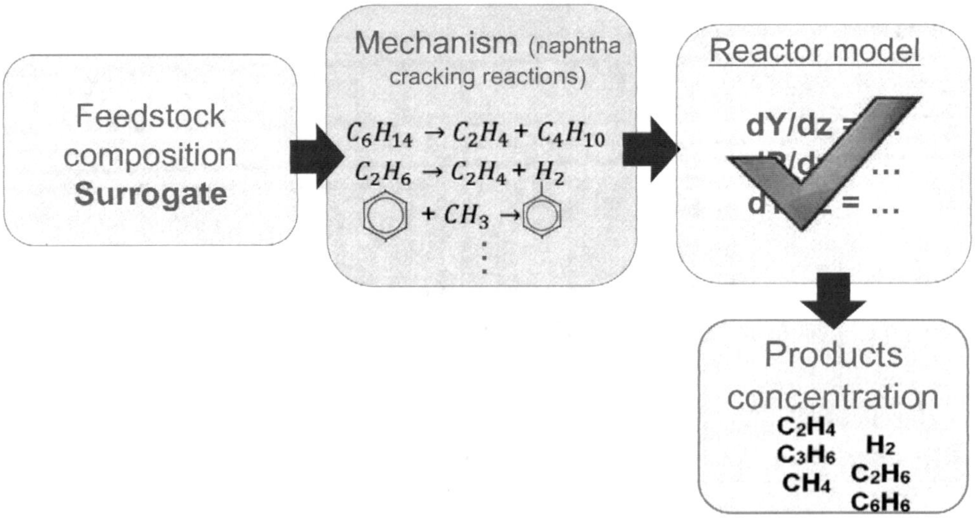

FRUN

UNIVERSIDAD NACIONAL DE COLOMBIA SEDE MEDELLIN

A model for the steam cracking of naphtha demands knowledge of: (1) Naphtha surrogate, (2) Reaction mechanism, (3) Reactor model, (4) Validation

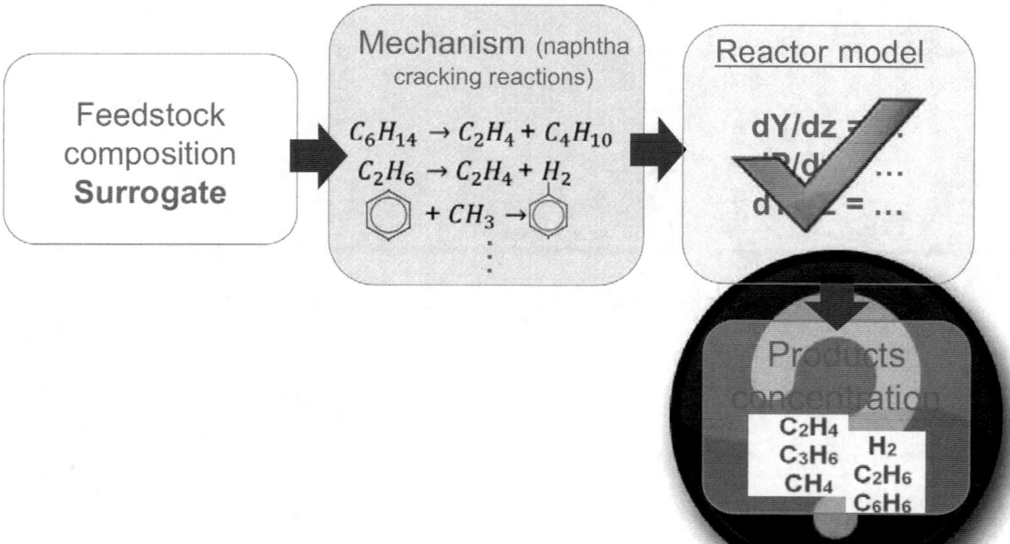

FRUN
Flujos Reactivos de la Universidad Nacional de Colombia

UNIVERSIDAD NACIONAL DE COLOMBIA SEDE MEDELLÍN

A model for the steam cracking of naphtha demands knowledge of: (1) Naphtha surrogate, (2) Reaction mechanism, (3) Reactor model, (4) Validation

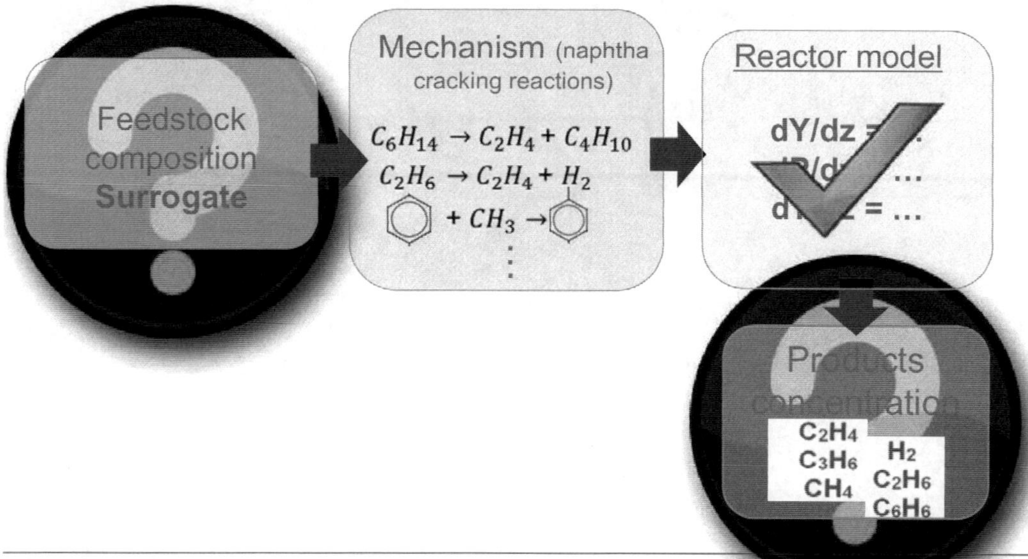

FRUN
Flujos Reactivos de la Universidad Nacional de Colombia

UNIVERSIDAD NACIONAL DE COLOMBIA SEDE MEDELLÍN

A model for the steam cracking of naphtha demands knowledge of: (1) Naphtha surrogate, (2) Reaction mechanism, (3) Reactor model, (4) Validation

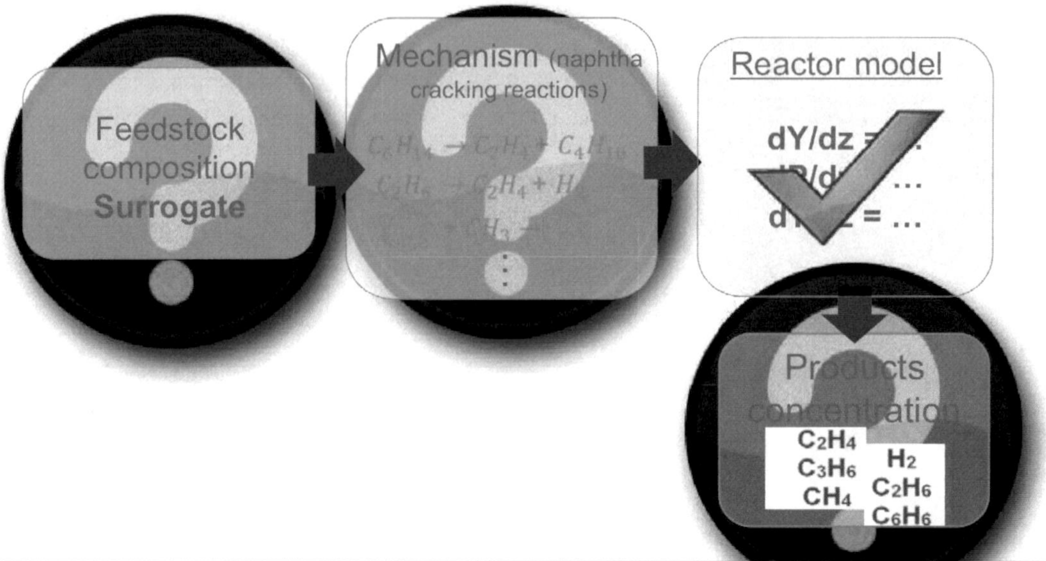

3FRUN

UNIVERSIDAD NACIONAL DE COLOMBIA SEDE MEDELLÍN

This presentation applies state of the art methods available in the open literature to simulate the steam cracking of naphthas

Feedstock composition Surrogate

Paraffin	N-propane, n-butane, n-pentane, n-hexane, n-heptane, n-octane, n-nonane, n-decane.	8
Iso-Paraffin	Iso-butane, 3-methyl pentane, 2-methyl pentane, 2,3-dimethyl butane, 2,2-dimethyl butane, 2-methyl hexano, 3-methyl hexano, trans-2,3-dimethyl pentane, 2,4-dimethyl pentane, 2-methyl heptane, 3-ethyl hexane, 3-methyl heptane, 2,4-dimethyl hexane, 2,5-dimethyl hexane, 2,3-dimethyl hexane, 3-methyl octane, 2-methyl octane, 2,5-dimethyl heptane, 2,2,3-trimethyl hexane.	19
Naphthene	Cyclo pentane, dimethyl cyclo pentane, cyclo hexane, methyl cyclo hexane, 1,2-dimethyl cyclo pentane, trans-1,3-dimethyl cyclo pentane, cis-1,3-dimethyl cyclo pentane, 1,1-dimethyl cyclo pentane, 2,2-dimethyl cyclo pentane, ethyl cyclo pentane, ethyl cyclo hexane, 1,4-dimethyl cyclo hexane.	12
Aromatic	Benzene, toluene, m-xylene, o-xylene, p-xylene, ethyl-benzene, 1,2,4-trimethyl benzene.	7

Shannon's entropy (van Geem et al. (2006))

Species	x_i
n-Pentane	0.01
n-Propane	0.03
n-Hexane	0.04
n-Octane	0.06
n-Heptane	0.21
Iso-butane	0.08
2,5 dimethyl-heptane	0.30
Ethyl cycle-pentane	0.07
Benzene	0.06
1,2,4 trimethyl benzene	0.14

3FRUN

UNIVERSIDAD NACIONAL DE COLOMBIA SEDE MEDELLÍN

This presentation applies state of the art methods available in the open literature to simulate the steam cracking of naphthas

Feedstock composition **Surrogate**

Paraffin	N-propane, n-butane, n-pentane, n-hexane, n-heptane, n-octane, n-nonane, n-decane.
Iso-Paraffin	Iso-butane, 3-methyl pentane, 2-methyl pentane, 2,3-dimethyl butane, 2,2-dimethyl butane, 2-methyl hexano, 3-methyl hexano, trans-2,3-dimethyl pentane, 2,4-dimethyl pentane, 2-methyl heptane, 3-ethyl hexane, 3-methyl heptane, 2,4-dimethyl hexane, 2,5-dimethyl hexane, 2,3-dimethyl hexane, 3-methyl octane, 2-methyl octane, 2,6-dimethyl heptane, 2,2,3-trimethyl hexane.
Naphthene	Cyclo pentane, dimethyl cyclo pentane, cyclo hexane, methyl cyclo hexane, 1,2-dimethyl cyclo pentane, trans-1,3-dimethyl cyclo pentane, cis-1,3-dimethyl cyclo pentane, 1,1-dimethyl cyclo pentane, 2,2-dimethyl cyclo pentane, ethyl cyclo pentane, ethyl cyclo hexane, 1,4-dimethyl cyclo hexane.
Aromatic	Benzene, toluene, m-xylene, o-xylene, p-xylene, ethyl-benzene, 1,2,4-trimethyl benzene.

Shannon's entropy (Van Geem et al. (2006))

Species	x_i
n-Pentane	0.01
n-Propane	0.03
n-Hexane	0.04
n-Octane	0.06
n-Heptane	0.21
Iso-butane	0.08
2,5 dimethyl-heptane	0.30
Ethyl cycle-pentane	0.07
Benzene	0.06
1,2,4 trimethyl benzene	0.14

Mechanism (naphtha cracking reactions)

$C_6H_{14} \rightarrow C_2H_4 + C_4H_{10}$

$C_2H_6 \rightarrow C_2H_4 + H_2$

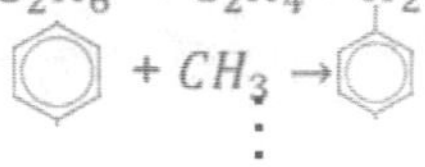

Automatic mechanism generator e.g. **(RMG)**

Combustion literature

FRUN Flujos Reactivos de la Universidad Nacional de Colombia

UNIVERSIDAD NACIONAL DE COLOMBIA SEDE MEDELLIN

A model for naphtha cracking demands a detailed knowledge of its composition

Naphthas are hydrocarbon mixtures with variable composition (around 200 components)

Paraffin	N-propane, n-butane, n-pentane, n-hexane, n-heptane, n-octane, n-nonane, n-decane.
Iso-Paraffin	Iso-butane, 3-methyl pentane, 2-methyl pentane, 2,3-dimethyl butane, 2,2-dimethyl butane, 2-methyl hexano, 3-methyl hexano, trans-2,3-dimethyl pentane, 2,4-dimethyl pentane, 2-methyl heptane, 3-ethyl hexane, 3-methyl heptane, 2,4-dimethyl hexane, 2,5-dimethyl hexane, 2,3-dimethyl hexane, 3-methyl octane, 2-methyl octane, 2,6-dimethyl heptane, 2,2,3-trimethyl hexane.
Naphthene	Cyclo pentane, dimethyl cyclo pentane, cyclo hexane, methyl cyclo hexane, 1,2-dimethyl cyclo pentane, trans-1,3-dimethyl cyclo pentane, cis-1,3-dimethyl cyclo pentane, 1,1-dimethyl cyclo pentane, 2,2-dimethyl cyclo pentane, ethyl cyclo pentane, ethyl cyclo hexane, 1,4-dimethyl cyclo hexane.
Aromatic	Benzene, toluene, m-xylene, o-xylene, p-xylene, ethyl-benzene, 1,2,4-trimethyl benzene.

Species	x_i
n-Pentane	0.01
n-Propane	0.03
n-Hexane	0.04
n-Octane	0.06
n-Heptane	0.21
Iso-butane	0.08
2,5 dimethyl-heptane	0.30
Ethyl cycle-pentane	0.07
Benzene	0.06
1,2,4 trimethyl benzene	0.14

The model benefit if naphthas are represented as **surrogates** (10-20 components)

FRUN Flujos Reactivos de la Universidad Nacional de Colombia

UNIVERSIDAD NACIONAL DE COLOMBIA SEDE MEDELLIN

The most used methodologies to generate a surrogate for naphthas imply (1) pseudocomponents or (2) stochastic methods

Pseudo-component representation →	Fast, limited application range and flexibility →	*(Dente et al.,1988; Joo et al. 2001)*
Stochastic →	High computational demand, wide application range, more flexibility →	*(Liguras and Allen, 1991; Hudebine and Verstraete., 2004)*

The Shannon's entropy maximization method is stochastic in nature

FRUN

Flujos Reactivos de la Universidad Nacional de Colombia

UNIVERSIDAD NACIONAL DE COLOMBIA SEDE MEDELLÍN

Following Van Geem et al. (2006) we selected Shannon's entropy maximization to generate the naphtha surrogate

$$MAX\ S(x_i) = -\sum_{i=1}^{N} x_i \ln(x) \quad \text{with} \quad \sum_{i=1}^{N} x_i = 1$$

The method finds the mole fraction composition (x_i)

Subject to constrains related to the desired naphtha properties (MW, H/C, ρ, PINA and TBP), known as commercial indices

$$\sum_{i=1}^{N} f_{i,j}\, x_i = f_j, \qquad j = 1, \ldots, J$$

With J the number of exact linear constrains

$$\sum_{i=1}^{N} x_i\, (f_k - f_{i,j})^2 \leq \sigma^2, \qquad k = 1, \ldots, K$$

With K the number of linear constrains with uncertainty

FRUN

Flujos Reactivos de la Universidad Nacional de Colombia

UNIVERSIDAD NACIONAL DE COLOMBIA SEDE MEDELLÍN

Exact linear constrains were applied to molecular weight, HC-ratio and specific density

$$\sum (M_{exp} - M_i) x_i = 0$$ → Molecular weight

$$\sum \left(\frac{H}{C}\right)_i M_i x_i = \left(\frac{H}{C}\right)_{exp} M_{exp}$$ → H/C

$$\sum \left(\frac{1}{d_i}\right) M_i x_i = \left(\frac{1}{d_{exp}}\right) M_{exp}$$ → Specific density

Simple mixture rules were applied to evaluate the mixture properties. The individual component values were taken from the ASPEN database using a CAPE-OPEN approach

3FRUN
Flujos Reactivos de la Universidad Nacional de Colombia

UNIVERSIDAD NACIONAL DE COLOMBIA
SEDE MEDELLÍN

Constrains that involved multiple responses were as well modeled as if they were linear

If component i ∈ PINA class K · *If component i ∉ PINA class K*

$$\sum \left(\frac{\%F_k^{exp}}{100} - 1\right) M_i x_i = \sum \left(\frac{\%F_k^{exp}}{100}\right) M_i x_i$$ → PINA analysis

If boiling point i < boiling poit K · *If boiling point i < boiling poit K*

$$\sum \left(1 - \frac{\%F_{exp}}{100}\right) \frac{M_i}{d_i} x_i = \sum \left(-\frac{\%F_{exp}}{100}\right) \frac{M_i}{d_i} x_i$$ → TBP

↑ Species that are part of the PINA or TBP evaluated

↑ Other species

As Shannon's entropy maximization has the tendency to favor the extreme MW values an uncertainty constrain was required

$$-\sum x_i (M_i - M_{exp})^2 = \sigma^2$$

3FRUN
Flujos Reactivos de la Universidad Nacional de Colombia

UNIVERSIDAD NACIONAL DE COLOMBIA
SEDE MEDELLÍN

The optimization considered 51 species that represent 95% of the naphtha composition

Paraffin	n-propane, n-butane, n-pentane, n-hexane, n-heptane, n-octane, n-nonane, n-decane.	8
Iso-Paraffin	iso-butane, 3-methyl pentane, 2-methyl pentane, 2,3-dimethyl butane, 2,2-dimethyl butane, 2-methyl hexane, 3-methyl hexane, trans-2,3-dimethyl pentane, 2,4-dimethyl pentane, 2-methyl heptane, 3-ethyl hexane, 3-methyl heptane, 2,4-dimethyl hexane, 2,5-dimethyl hexane, 2,3-dimethyl hexane, 3-methyl octane, 2-methyl octane, 2,5-dimethyl heptane, 2,2,3-trimethyl hexane.	19
Naphthene	cyclo pentane, dimethyl cyclo pentane, cyclo hexane, methyl cyclo hexane, 1,2-dimethyl cyclo pentane, trans-1,3-dimethyl cyclo pentane, cis-1,3-dimethyl cyclo pentane, 1,1-dimethyl cyclo pentane, 2,2-dimethyl cyclo pentane, ethyl cyclo pentane, ethyl cyclo hexane, 1,4-dimethyl cyclo hexane.	12
Aromatic	benzene, toluene, m-xylene, o-xylene, p-xylene, ethyl-benzene, 1,2,4-trimethyl benzene.	7

FRUN

UNIVERSIDAD NACIONAL DE COLOMBIA SEDE MEDELLIN

The commercial indices (physicochemical properties) selected for optimization are typical of a naphtha

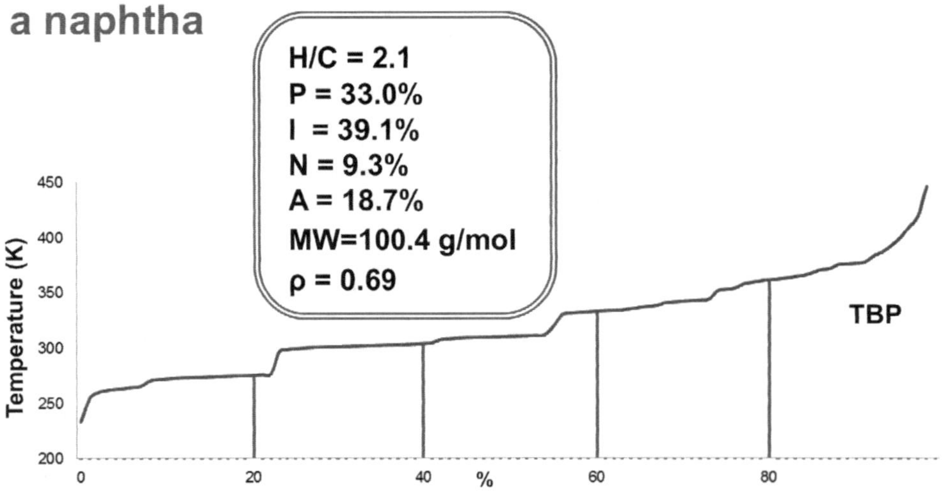

The fmincon Matlab tool (interior point algorithm) maximizes the objective function subject to the specified restrictions

FRUN

UNIVERSIDAD NACIONAL DE COLOMBIA SEDE MEDELLIN

The results obtained with the Shannon's entropy maximization method accurately (MAX ERROR > 1.1%) represent most naphtha properties

Species	x_i	Experimental	Simulated	Error %
n-Pentane	0.01			
n-Propane	0.03			
n-Hexane	0.04	P 33.0%	P 33.1%	0.3
n-Octane	0.06			
n-Heptane	0.21			
Iso-butane	0.08	I 39.1%	I 39.3%	0.5
2,5 dimethyl-heptane	0.30			
Ethyl cycle-pentane	0.07	N 9.2%	N 9.3%	1.1
Benzene	0.06	A 18.7%	A 18.6%	0.5
1,2,4 trimethyl benzene	0.14			
Molecular weight (kg/kmol)		100.4	100.4	0.0
Specific density		0.689	0.688	0.1
HC ratio		2.060	2.066	0.3

3FRUN
Flujos Reactivos de la Universidad Nacional de Colombia

UNIVERSIDAD NACIONAL DE COLOMBIA SEDE MEDELLÍN

The error for the TBP is slightly higher (13.8%) particularly for the 300 K – 340 K temperature range

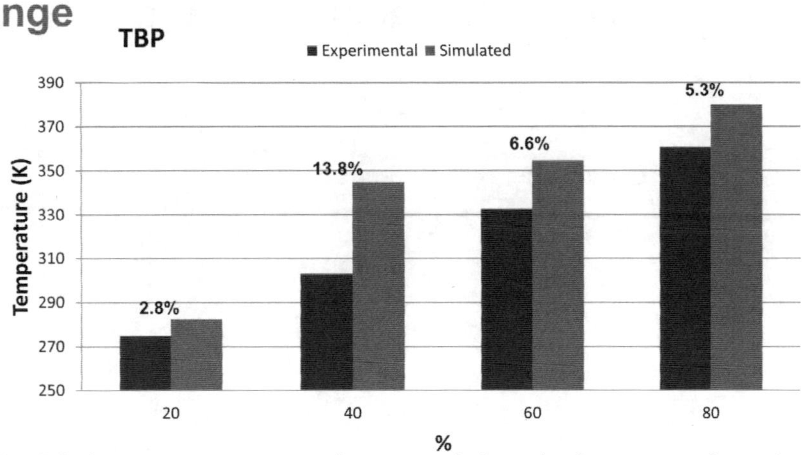

This higher error may be explained, in part, by the discrete representation of the TBP curve

3FRUN
Flujos Reactivos de la Universidad Nacional de Colombia

THM1: SHANNON'S ENTROPY CAN EFFICIENTLY GENERATE A NAPHTHA SURROGATE

UNIVERSIDAD NACIONAL DE COLOMBIA SEDE MEDELLÍN

The development of a reaction mechanism implies high costs (experiments) and a deep knowledge of reaction chemistry

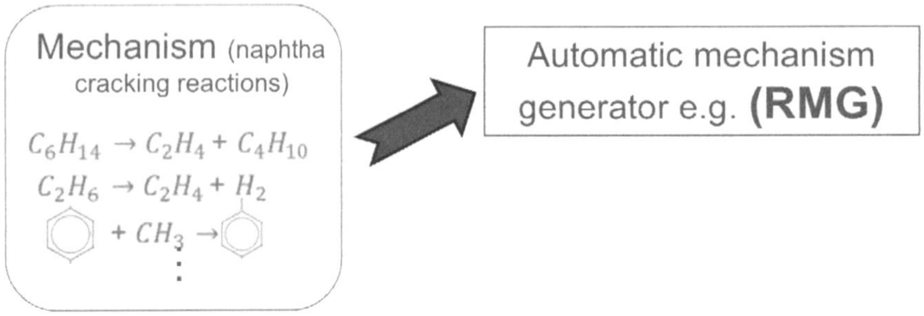

In the last decade, different groups have developed "automatic" mechanism generation softwares that may provide a simpler solution

3FRUN

UNIVERSIDAD NACIONAL DE COLOMBIA SEDE MEDELLIN

RMG is an automatic chemical reaction mechanism generator that designs a mechanism based on knowledge of chemical reactions

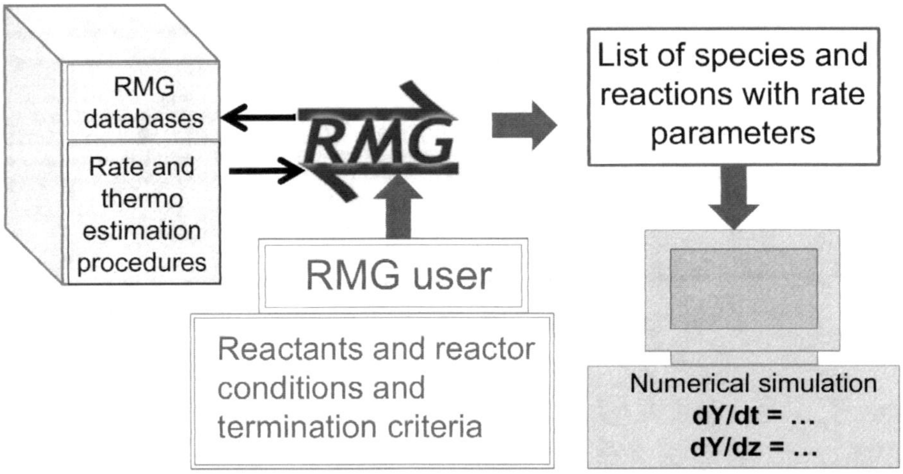

3FRUN

http://rmg.mit.edu/ Accessed (this morning). Green research group

UNIVERSIDAD NACIONAL DE COLOMBIA SEDE MEDELLIN

The input conditions to RMG were T, P and naphtha composition

Species	Concentration (mol/m^3)	
n-Pentane	0.17	From naphtha surrogate
n-Propane	0.77	
n-Hexane	1.19	
Benzene	1.59	
n-Octane	1.62	
Ethyl cycle-pentane	2.08	
Iso-butane	2.25	
1,2,4-trimethyl-benzene	4.07	
n-Heptane	6.06	
2,5-dimethyl-heptane	8.64	
Temperature (K)	1100	Typical conditions in industrial processes
Pressure (Pa)	260000	

The termination criteria were 0.1 for tolerance (recommended value for new systems) and n-heptane conversion (0.5, typical value)

3FRUN
Flujos Reactivos de la Universidad Nacional de Colombia

UNIVERSIDAD NACIONAL DE COLOMBIA SEDE MEDELLÍN

The mechanism generated has 48 species and 221 reactions, of the order of other well-known mechanisms in this area (Ranzi et al. and Gao et al.)

Mechanism	Number of species	Number of reactions
Ranzi et al. (2010)	85	1351
JetSurF (2012)	352	2083
Keyvanloo et al. (2010)	353	2083
Gao et al. (2009)	10	1
Kumar and Kunzru (1985)	14	22
Marin et al. (2006)	60	1178
Marin et al. reduced (2006)	24	58
RMG results	48	221

Reactions families: biradical recombination, disproportionation, H abstraction, internal disproportionation, internal H migration, multiple bond addition and recombination

3FRUN
Flujos Reactivos de la Universidad Nacional de Colombia

UNIVERSIDAD NACIONAL DE COLOMBIA SEDE MEDELLÍN

The predictions with the mechanism were compared to experimental results for naphtha steam cracking in a 1.5-m long pilot plant

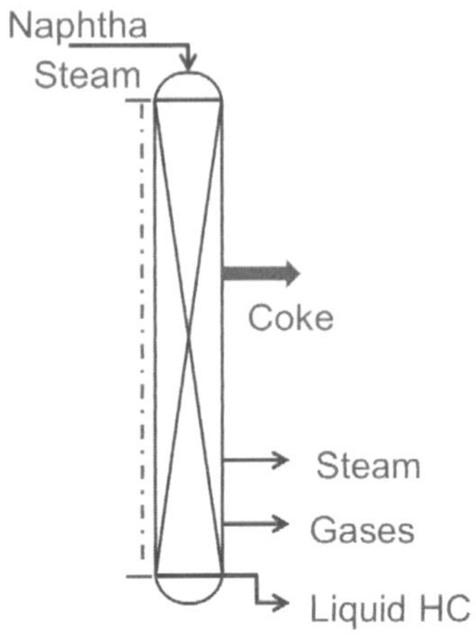

Inlet conditions	Value
HC mass flow (kg/h)	0.2
P (kPa)	206
Dilution factor (kg H_2O/kg HC)	0.8

The temperature profile was controlled at specific values

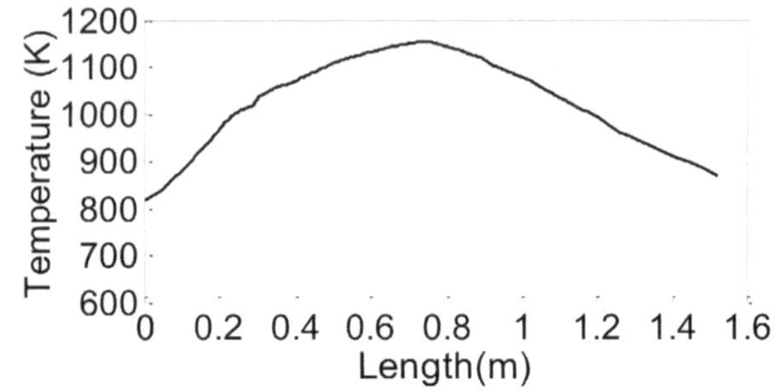

3FRUN

Flujos Reactivos de la Universidad Nacional de Colombia

UNIVERSIDAD NACIONAL DE COLOMBIA SEDE MEDELLÍN

The surrogate, mechanism and model approximately capture the experimental behavior

	Exper.	Model	Model/Exp.
C_2H_4	0.32	0.34	1.04
CH_4	0.25	0.32	1.30
C_3H_6	0.24	0.21	0.89
C_6H_6	0.10	0.10	0.96
C_2H_6	0.08	0.02	0.30
H_2	0.01	0.01	0.90

The most significant differences are for methane and ethane. Future work may need more iterations from the RMG software

3FRUN

Flujos Reactivos de la Universidad Nacional de Colombia

THM: RMG YIELDS, RELATIVELY EASILY, AN ACCEPLE MECHANISM

UNIVERSIDAD NACIONAL DE COLOMBIA SEDE MEDELLÍN

The development of a reaction mechanism implies very high costs (experiments) and a deep knowledge of reaction chemistry

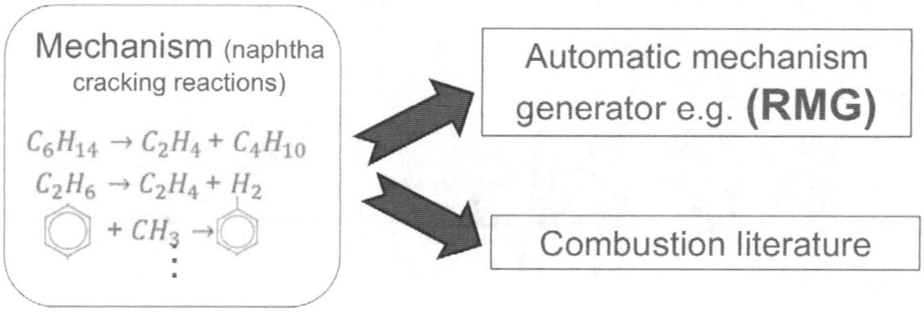

Motivated by environmental regulations and a very long history there have been continuous support to develop complex mechanism that predict HC combustion. The first step for combustion is pyrolysis (cracking)

FRUN

UNIVERSIDAD NACIONAL DE COLOMBIA SEDE MEDELLIN

We have previously taken advantage of combustion mechanism to model the steam cracking of ethane

	Ratio model/Industrial furnace				
Specie	Molecular model	S and F	ABF	Ranzi	USC
H_2	1.03	1.02	1.10	1.02	1.06
C_2H_4	1.02	1.00	1.10	1.10	1.09
C_2H_6	1.02	1.02	0.88	0.94	0.89
CH_4	0.63	0.48	0.29	0.72	0.65
C_2H_2	0.01	0.11	0.86	0.55	0.88
C_3H_8	7.69	0.043	-	2.09	0.80
Butadiene	0.03	0.053	-	0.40	1.07
Benzene	-	-	1.11	0.70	1.09
Butane	-	-	-	-	0.90

Combustion models such as ABF and USC (now JetSurF) yield better predictions that traditional cracking mechanism such as those by Sundaram and Froment and by Ranzi

FRUN

Ramírez. A.Y. M.Sc. thesis, Univ. Nac. de Colombia, Medellín, 2011

UNIVERSIDAD NACIONAL DE COLOMBIA SEDE MEDELLIN

Driven by concerns over pollution, there have been considerable gains in the understanding of soot formation

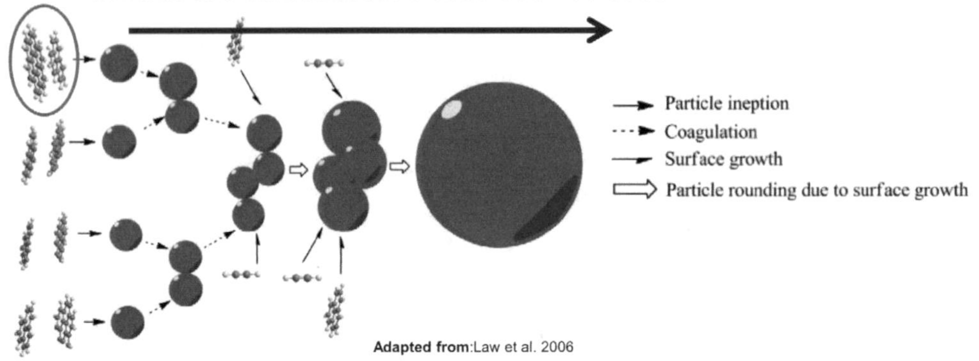

Because physical and chemical coalescence of PAHs is responsible for soot inception, any gas to solid mechanism should start with PAH formation

FRUN Ramírez. A.Y. M.Sc. thesis, Univ. Nac. de Colombia, Medellín, 2011 UNIVERSIDAD NACIONAL DE COLOMBIA SEDE MEDELLIN

In the combustion literature we identified two mechanisms: JetSurF (comb.) and Ranzi (comb./pyrol.) that could represent the steam cracking of naphthas

Mechanism	Number of species	Number of reactions	Hydrocarbon Range
JetSurF	352	2083	$C_1 - C_{12}$
Ranzi	85	1351	$C_1 - C_{10}$

FRUN UNIVERSIDAD NACIONAL DE COLOMBIA SEDE MEDELLIN

In the combustion literature we identified two mechanisms: JetSurF (comb.) and Ranzi (comb./pyrol.) that could represent the steam cracking of naphthas

Mechanism	Number of species	Number of reactions	Hydrocarbon Range
JetSurF	352	2083	$C_1 - C_{12}$
Ranzi	85	1351	$C_1 - C_{10}$

We arbitrarily selected JetSurF as it had a wider HC range* and represented a better test as it was designed for combustion

3FRUN

*Another mechanism by Ranzi's group that goes up to C18 with ~8000 reactions

The ability of JetSurF to predict the steam cracking of naphthas was tested in an industrial furnace as reported by Van Geem (2006)

Operating conditions

Parameter	Value
Sections	4/2/1
Length (m)	0.08/0.12/0.16
Inlet P (kPa)	300
Inlet T (K)	873

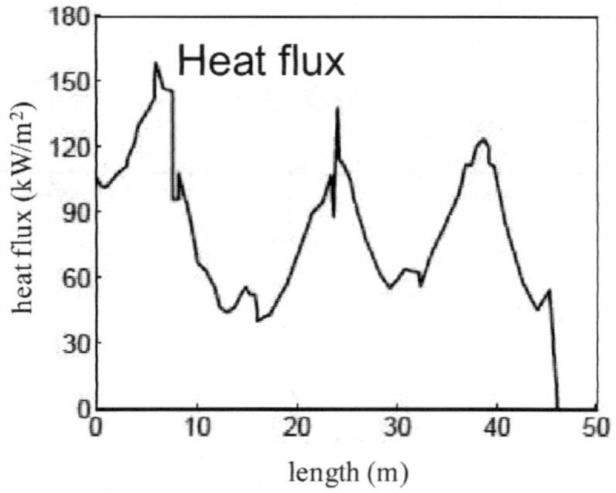

The naphtha composition was that of the surrogate

3FRUN

Van Geem Ph.D. thesis. Universiteit Gent, 2006

When the simulation with JetSurF is compared with the steam cracking of n-butane the results are at least encouraging

Species	van Geem* (%w)	This work (%w)
C_2H_4	0.22	0.48
CH_4	0.14	0.18
C_3H_6	0.20	0.17
C_2H_2	-	0.10
C_2H_6	0.03	0.03
H_2	0.01	0.03
C_6H_6	0.01	0.01
C_3H_8	-	0.00

* Simulation (n-butane steam cracking)

Future work is needed that provides a more robust comparison

3FRUN — Flujos Reactivos de la Universidad Nacional de Colombia

THM3: COMBUSTION MECHANISM GIVES PROMISING RESULTS

UNIVERSIDAD NACIONAL DE COLOMBIA SEDE MEDELLÍN

FINAL REMARKS

THM1: SHANNON'S ENTROPY CAN EFFICIENTLY GENERATE A NAPHTHA SURROGATE

$$MAX\ S(x_i) = -\sum_{i=1}^{N} x_i \ln(x) \quad \text{with} \quad \sum_{i=1}^{N} x_i = 1$$

RMG

THM2: RMG YIELDS, RELATIVELY EASILY, AN ACCEPLE REDUCED MECHANISM WHEN MODELING NAPTHA STEAM CRACKING

THM3: A COMBUSTION MECHANISM GIVES PROMISING RESULRTS WHEN MODELING THE STEAM CRACKING OF NAPHTHAS

Mechanism	Number of species	Number of reactions	Hydrocarbon Range
JetSurF	352	2083	$C_1 - C_{12}$

3FRUN — Flujos Reactivos de la Universidad Nacional de Colombia

UNIVERSIDAD NACIONAL DE COLOMBIA SEDE MEDELLÍN

ACKNOWLEDGEMTS

We acknowledge the following financial support:

(1) ECOPETROL through the project: *Desarrollo de modelos fenomenológicos de pirólisis de hidrocarburos líquidos y gaseosos para predicción de rendimientos y formación de coque*.

(2) ECOPETROL and Colciencias through project 423-2013

AA acknowledge as well Facultad de Minas and Colciencias for a traveling scholarship

FRUN

UNIVERSIDAD NACIONAL DE COLOMBIA SEDE MEDELLIN

END OF PRESENTATION

FACULTY OF MINES

U. NACIONAL DE COLOMBIA – MEDELLIN

FRUN

UNIVERSIDAD NACIONAL DE COLOMBIA SEDE MEDELLIN

WorleyParsons
resources & energy

To Shale or Not To Shale: Refinery Configuration Fit

AICHE 2014 Spring Meeting – April 2nd 2014

Odette Eng
Vice President

EcoNomics

Tight Oil Revolution

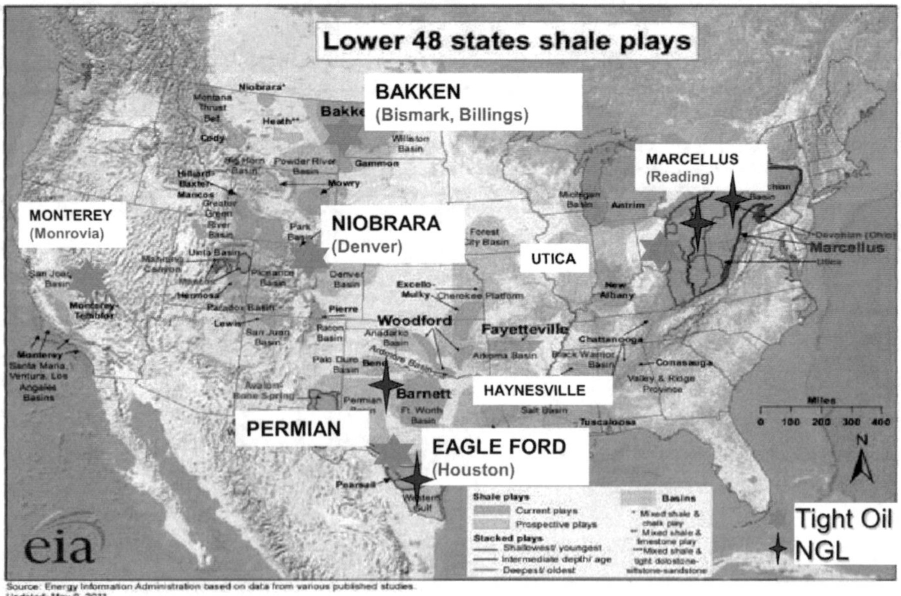

Source: Energy Information Administration based on data from various published studies.
Updated: May 9, 2011

Ever Increasing Projections

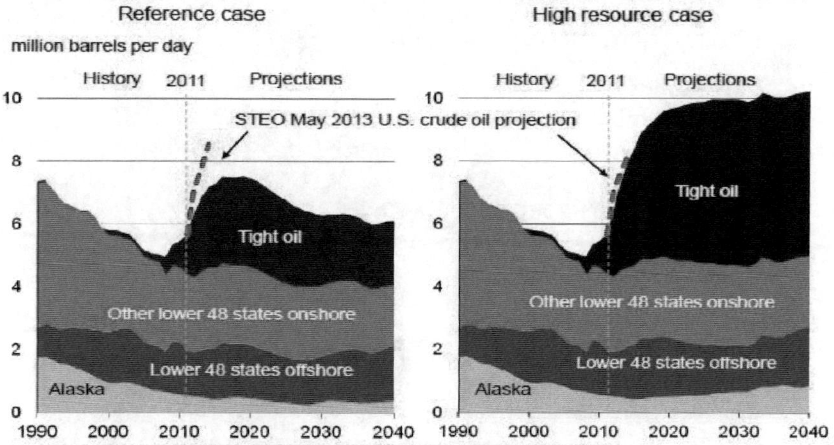

Source: EIA, Annual Energy Outlook 2013 and Short-Term Energy Outlook, May 2013

High Tight Oil Supply – Lowers Refiners' Crude Cost
Major Volume Impact on Refiners' Crude Diet

Tight Oil Brings Good News

- **Reversal:** Less Crude Import Dependency
  - Domestic crude production on the decline, tight oil will make up the difference.
  - Reduction in crude imports since 2008.
- **Reversal:** New opportunity for Refiners to lower feed cost.
- **Reversal:** Going light
  - Less processing required as compared to Heavy Oil.

Figure 3. Total U.S. petroleum and other liquids production, consumption, and net imports, 1970-2035
(million barrels per day)

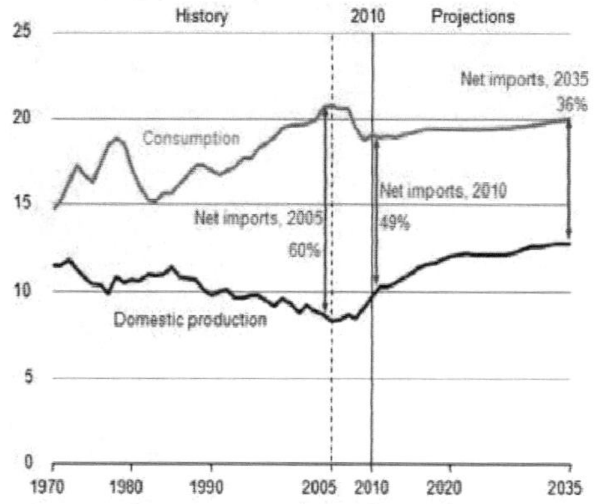

Source: EIA

High US Gas Production – Lowers Refiners' OpCost

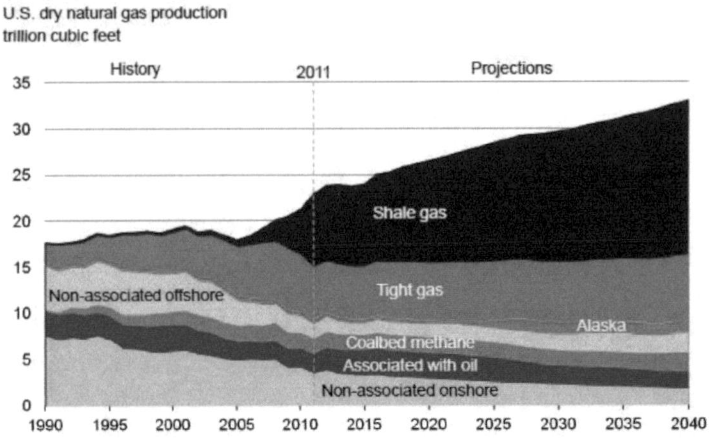

Source: EIA, Annual Energy Outlook 2013

To Shale or NOT To Shale?

Opportunity crude today includes Light Tight Oil

Light Tight Oil

Heavier Crude Oil

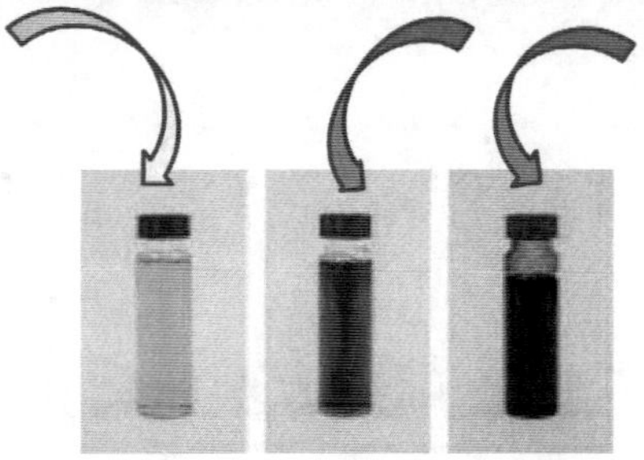

- High API, Low Sulur,
- Paraffinic
- High Natural Gasoline content

- Low API, High Sulur,
- High Aromatic content
- High Gas oil & Resid

Lighter Crudes Reversing Processing Trend

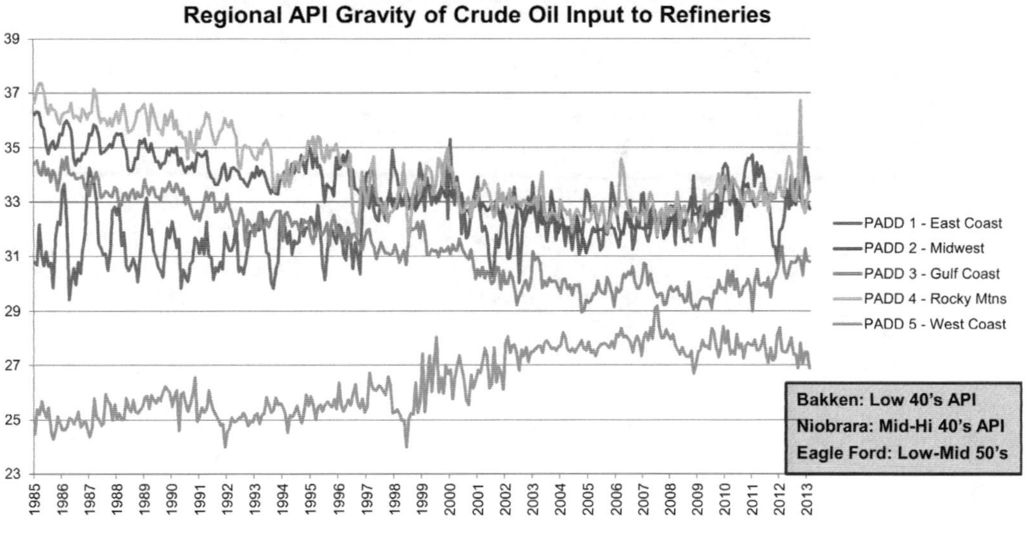

Source: EIA

A Macro Level View

of 20

Tight Oil Utilization Depends on Refinery Configuration…

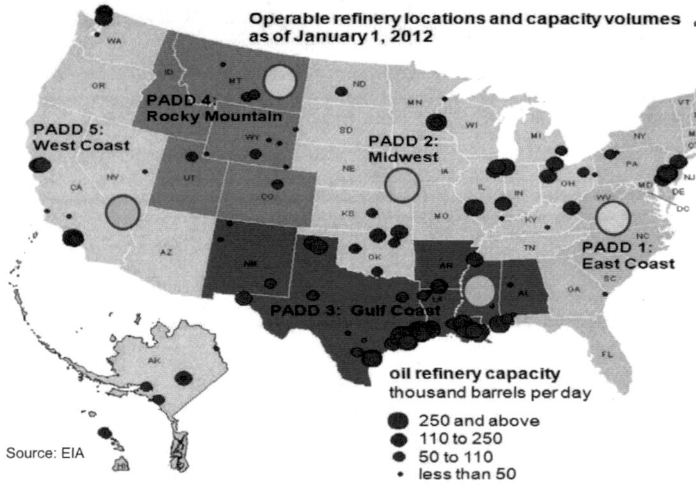

Source: EIA

(PADD: Petroleum Administration For Defense District)

PADD 1, 4= 2,000 MBPD (Relatively small)

REGION	Capacity, MBPD	Refineries	W/ Heavy	%
PADD 1	**1,400**	**10**	**2**	**20**
PADD 2	3,670	26	14	54
PADD 3	8,410	51	26	51
PADD 4	**600**	**14**	**5**	**36**
PADD 5	2,910	25	14	56

Currently Higher Density of Light Oil Refineries (>50%) → Preferential for Tight Oil
Currently Higher Density of Heavy Oil Refineries (>50%) → Balance out Crude blend with Tight Oil

…and All Potential Market

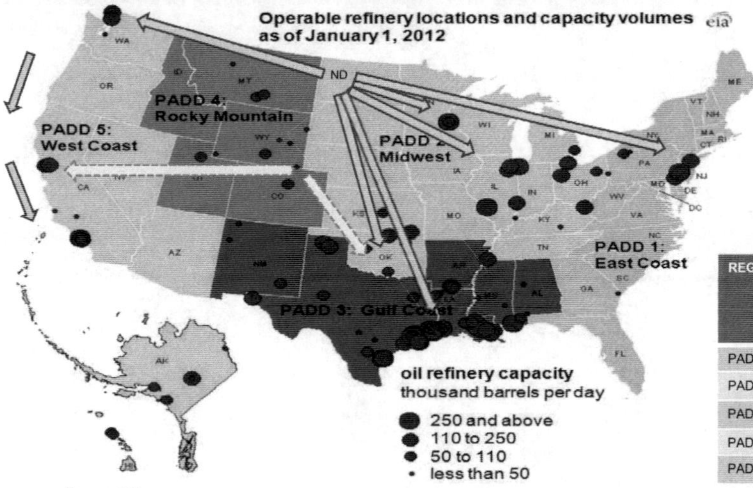

Bakken Tight Oil

Niobrara Tight Oil (potential destinations)

(PADD: Petroleum Administration For Defense District)

REGION	Capa city, MBPD	Refi neri es	W/ Heavy	%
PADD 1	**1,400**	**10**	**2**	**20**
PADD 2	3,670	26	14	54
PADD 3	8,410	51	26	51
PADD 4	**600**	**14**	**5**	**36**
PADD 5	2,910	25	14	56

Source: EIA

How Much Tight Oil Can Be Absorbed?

CRUDE RUNS IN 2011						
	PADD 1	**PADD 2**	**PADD 3**	**PADD 4**	**PADD 5**	**TOTAL**
Light Sweet Crude, KBPD	**687**	**313**	**881**	**24**	**221**	**2,126**
Light/Medium Crude, KBPD	**233**	**195**	**1,664**	**30**	**546**	**2,668**
Domestic Crude, KBPD	58	1,851	2,549	396	1,191	6,045
Heavy Crude, KBPD	119	1,013	2,380	180	387	4,079
	1,097	3,372	7,474	630	2,345	14,918

All Light Sweet Crude Components Can Be Displaced By Tight Oil

Mismatch

Source: Canadian Association of Petroleum Producers

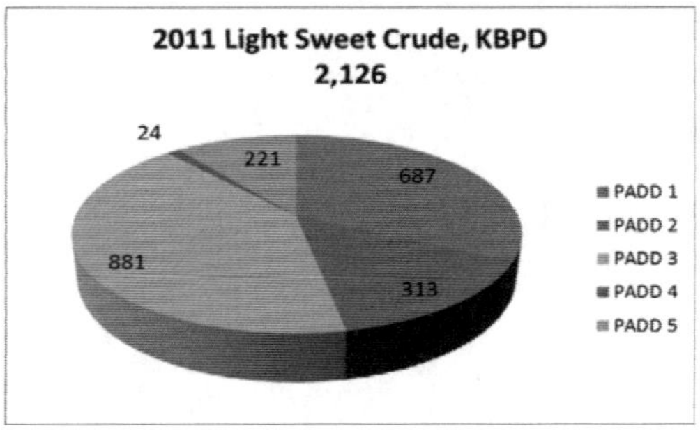

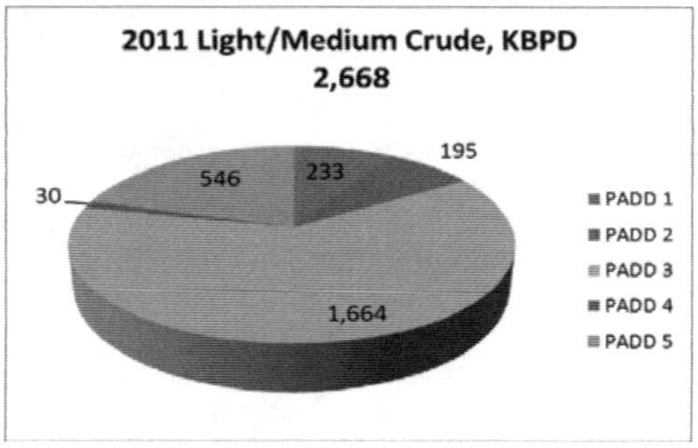

How Does Tight Oil Compare with Medium/Heavy Crudes?

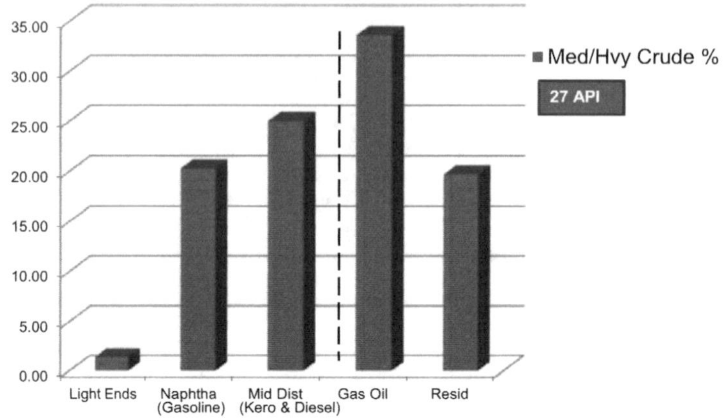

Tight Oil vs Med/Hvy Crude Refinery Yields

Excess Gasoline is expected from Light Tight Oil, based on crude yield distribution

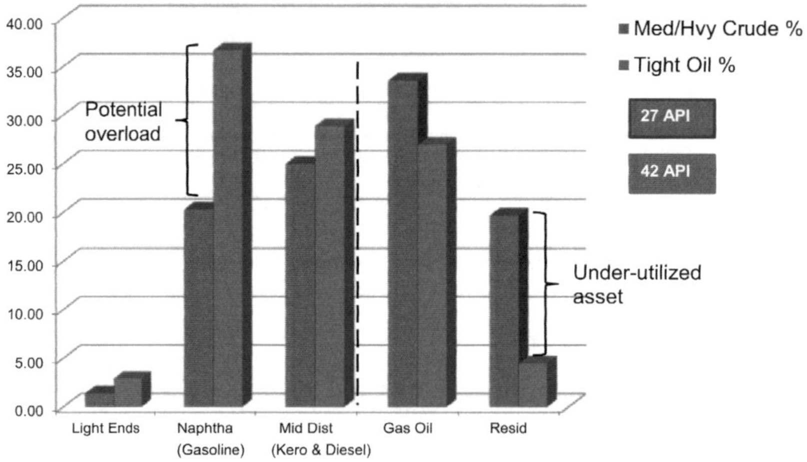

Potential Bottlenecks

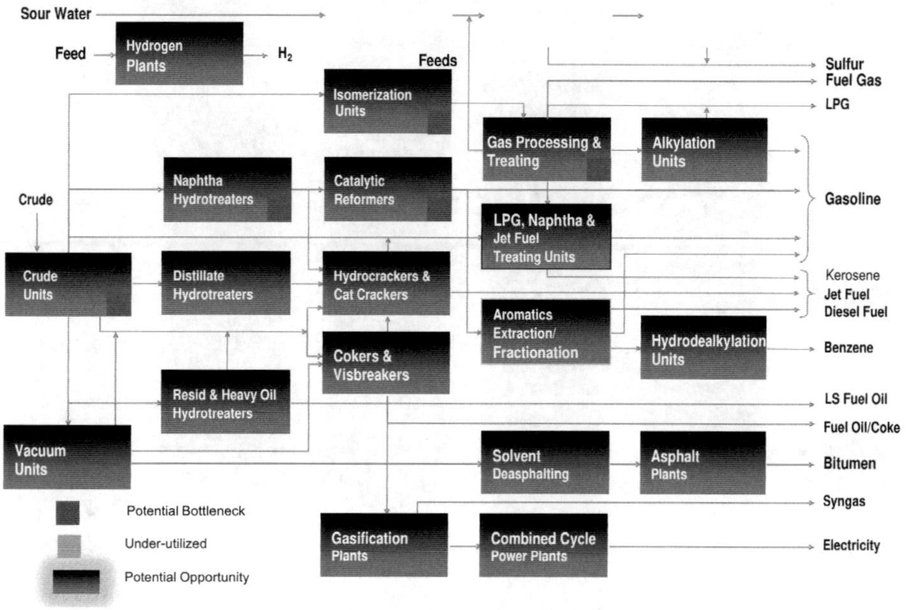

Potential Under-Utilization

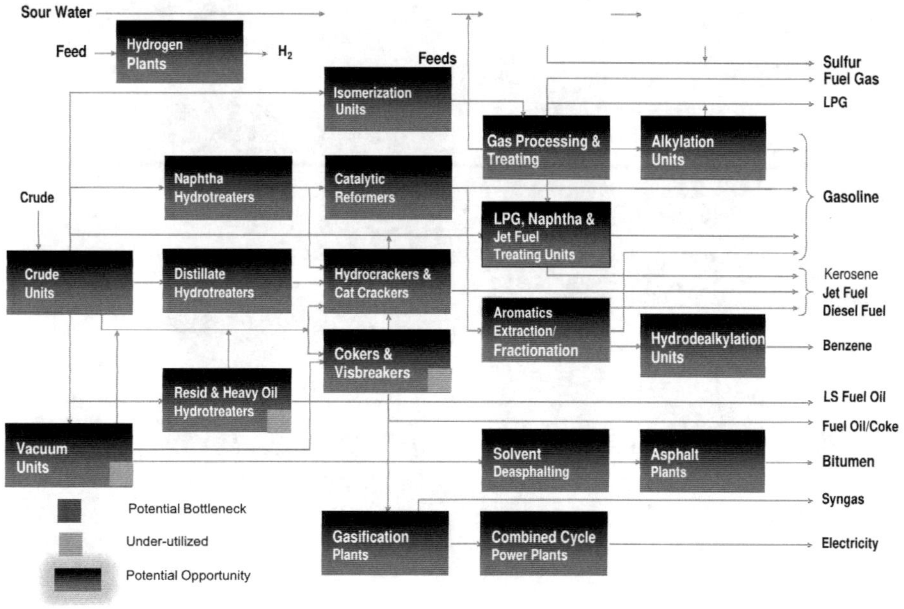

Potential Opportunities

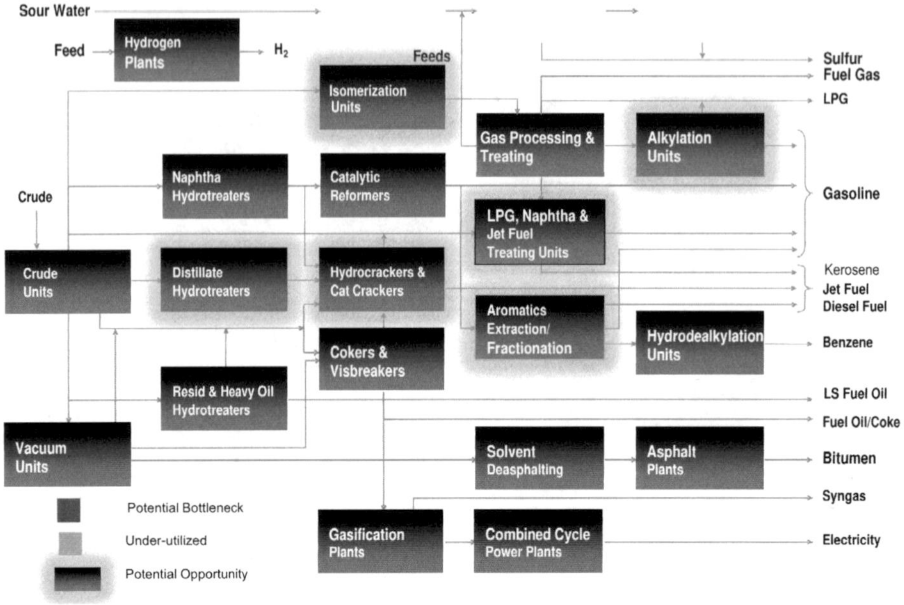

Overall Refinery Summary

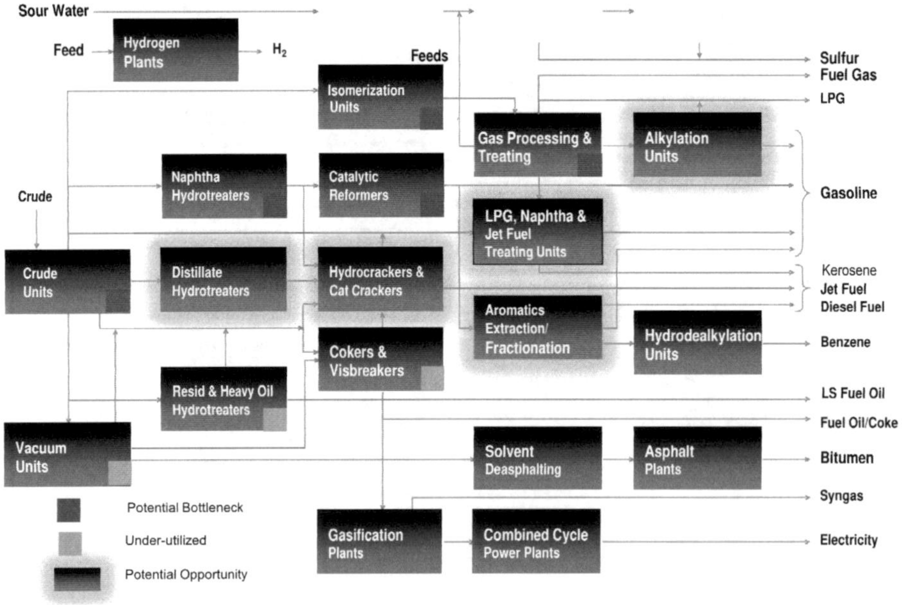

A Unit Level View

Taking Advantage of Tight Oil

- Simple Topping Refineries can take on more Tight Oil with minor changes

- More Complex refineries:
  - Can likely add 10-20% Tight Oil without problems
  - Tight Oil addition > 50% may require capex depending on amount and existing Refinery Configuration:
    - Retray Atmospheric Tower or use of Preflash Tower
    - Desalter review (Higher amount of Solids and Iron metals and Iron silicate)
    - Crude column Fired Heater and Reboiler – higher demand
    - Heat Exchanger network – Different Pinch Point – Heat availability is lower
    - Crude overhead system debottleneck
    - Naphtha HTU debottleneck
    - Catalytic Reformer debottleneck due to over capacity, higher H2 available

Taking Advantage of Tight Oil

- More Complex refineries (continued):
  - 100% Tight Oil addition could result in under-utilization of bottoms upgrading units:
    - Vacuum / FCC / Coker
  - Crude compatibility to be checked, risk of asphaltene agglomeration, fired heater and exchanger fouling
  - Product yield and quality need to be assessed:
    - Higher Gasoline Yield / Relatively lower Diesel Yield
    - Higher amount of straight run material, Octane number in Gasoline pool will likely drop

If Left Alone, The Following Trends Can Result

- Tight Oil carries higher Gasoline yield than Middle Distillate (Kerosene + Diesel).
- Processing more Tight Oil will further increase Gasoline & LPG (from Lt Ends) production, for export consideration.
- US became LPG exporter for the 1st time in 2012.
- Canada, L. America are the largest Gasoline importers

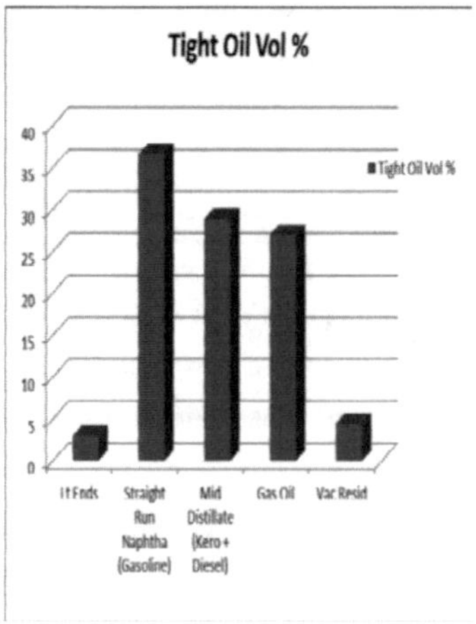

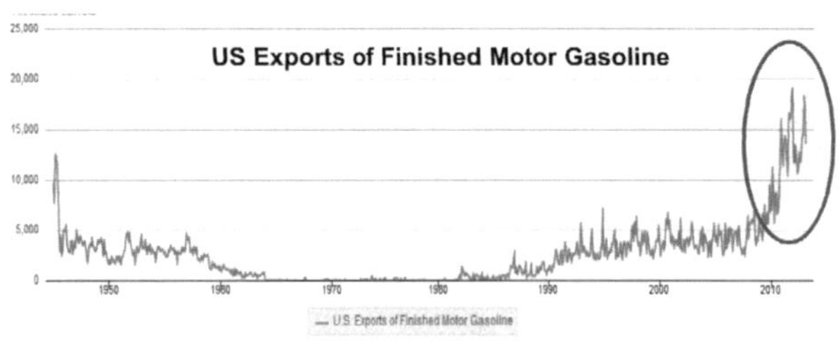

Source: U.S. Energy Information Administration

Other Product Options

of 20

What about Making Premium Gasoline?

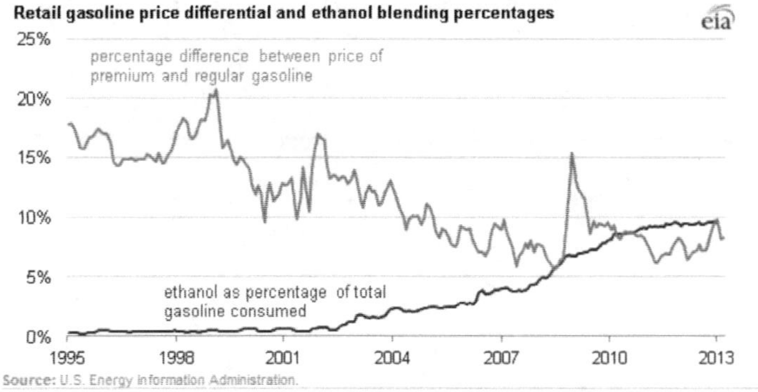

- E10 Ethanol blending helps narrow differential between Premium and Regular Gasoline.
- Premium Gasoline can be a niche opportunity in certain locations, but not universal.

What about Diesel?

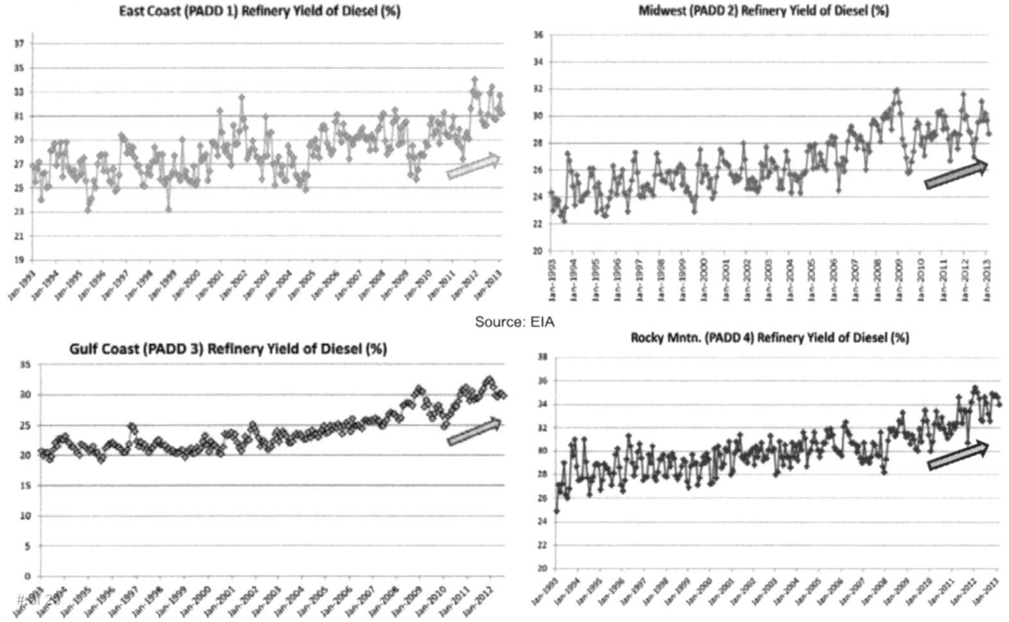

What about Diesel?

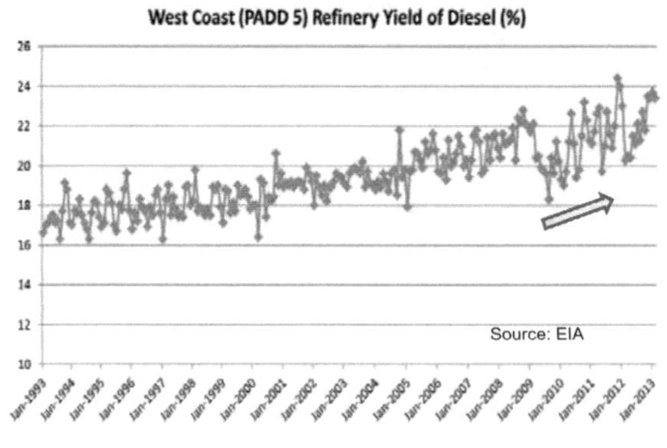

… While Gasoline production has been flat or declining.

What about Diesel?

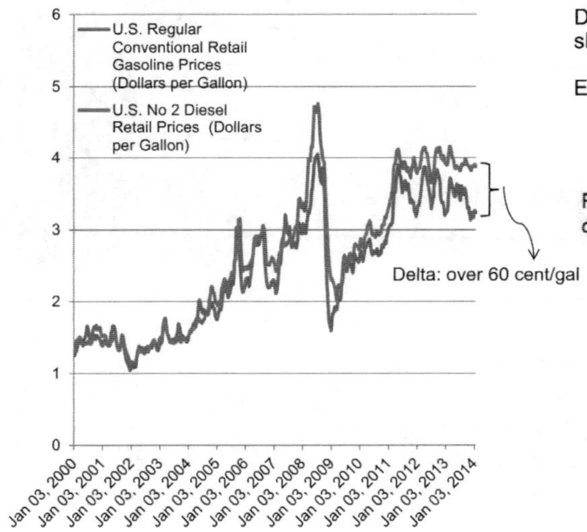

Differential between Gasoline and Diesel shows shift in higher demand towards Diesel.

EIA predicts the same.

Figure 102. U.S. motor gasoline and diesel fuel consumption, 2000-2040 (million barrels per day)

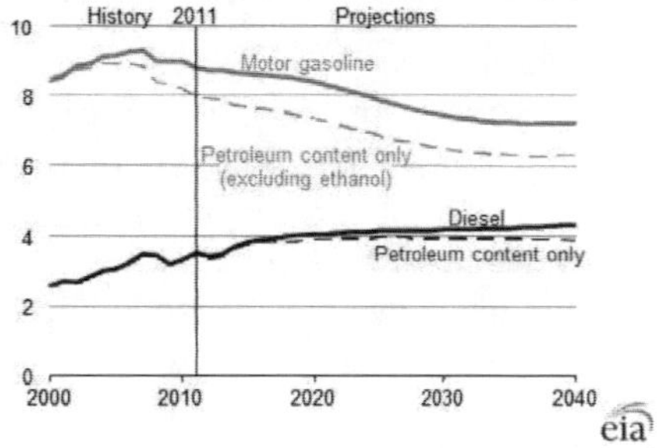

Other New Opportunities with Tight Oil

- Advantaged Natural Gas price for US means lower:
  - Hydrocracking Op cost
  - Hydrotreating Op cost
  - Fuel cost
  - Methanol production cost
- US can afford product diversification:
  - More Diesel (Hydrocracking) vs Gasoline (FCC)
    - Leveraging excess H2 availability
  - Light material needs to find home:
    - Excess Heavy Naphtha can be turned into BTX (Tier III requires lower Aromatics content)

World LNG Estimated November 2013 Landed Prices

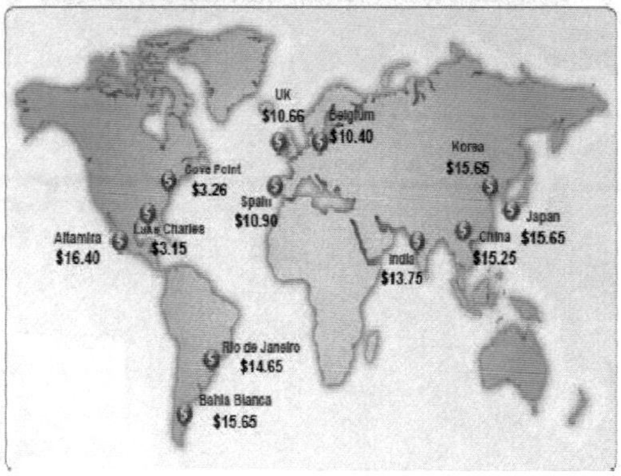

Source: Waterborne Energy Inc.

Summary

- Volume Impact from Tight Oil could be major.
- Lower Crude / Natural Gas costs will make US Refining industry more competitive, still subject to price differential volatility.
- New Opportunities for Refiners' diversification: new products, including Aromatics and Chemicals.
- Success factors for Refiners depend on:
  - Well balanced Crude Strategy.
  - Comprehensive asset utilization strategy, including transportation logistics and costs.
  - Production balance between Gasoline, Diesel and Aromatics for domestic and export markets.

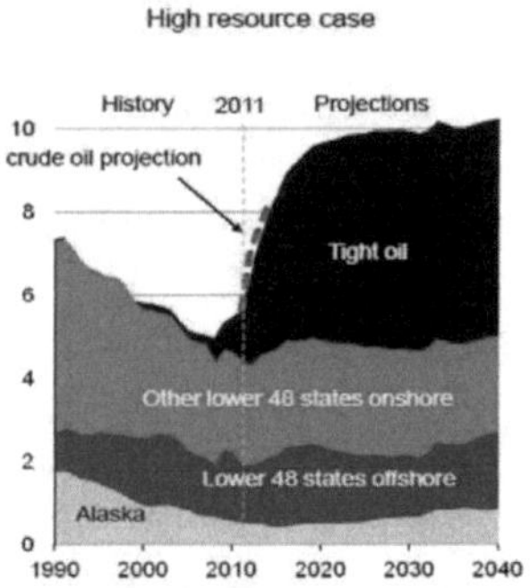

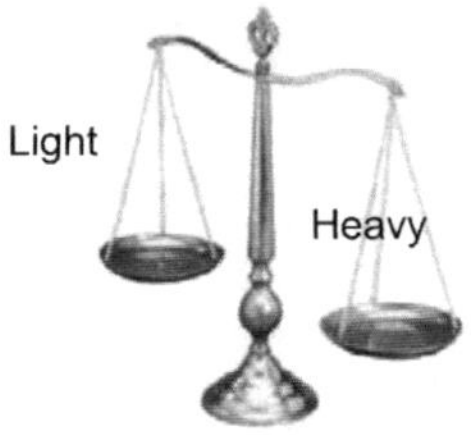

WorleyParsons

resources & energy

Getting Bitumen and Extra Heavy Oil to Market: Tight Oil Synergy, Blending, and Partial Upgrading Alternative Challenges Including Thermal Cracking Product Dienes, Gums, Self-Incompatible Asphaltenes, and Pricing Review

Stephen Pavel
Full Spectrum Consulting LLC

14AIChE
Spring Meeting, New Orleans, LA

2014 AIChE Spring Meeting & 10th Global Congress on Process Safety,
2014.04.02, 2:00pm, Belle Chase, Hilton NOLA Riverside. Paper (147b)

Full Spectrum Consulting LLC

Overview

ACKNOWLEDGMENTS

In this review, all of the paper and powerpoint are from the public domain, mostly newspapers.
See paper for power point slide references, use google to get current news.

Author thanks the organizers of the conference.

Author gratefully acknowledges those that have provided the foundation of knowledge in the public domain including the Fair Use Rules which have increased the dissemination of knowledge and are used in this review.

Author thanks those that skillfully reported topics that are frankly difficult to believe occur in developed counties, especially Canada and US, that required those topics be quoted rather than paraphrased.

Author acknowledges the US Patent and Trademark Office, with the new America Invents Act which improves the ability of individuals to pursue patents and to dispute patents that are in process and those that have been granted.

The author gratefully acknowledges generous support of friends, family and clients. Contact at pavel@full-spectrum-consulting.com

14AIChE

Full Spectrum Consulting LLC

Overview

- Getting Heavy Oil and Bitumen to Market
- Heavy Oil Locations, oil sands projects.
- Diluent sources, International Tight Oil
- Diluent source: hydrotreated, stable, fully upgraded synthetic crude oil, SCO
- Partial Upgrading field located technology options, in-situ and well head
- In-situ partial upgrading projects in various phases (20+)
- Well head partial upgrading technology proposals (40+)
- Partial upgraded synthetic crude oil, PUSCO, not hydrotreated for lower cost
- Stability, compatibility data notably very absent from published descriptions
- Emissions and waste quantity and composition also absent, but now known
- Instability & Incompatibility test results determine reduced marketability
- Dienes produced with short contact high temperature reactions create gums
- Precipitated semi-solid asphaltenes produced in high temperature
- Solids from feed and from thermal cracking reaction remain in product
- Self-Incompatible product defined by dienes, asphaltenes, solids, accurately
- Self-InCompatible Partially Upgraded Thermally Cracked Unstable Product: SICPUTCUP
- Compatible, stable merchant crude from bitumen by blending light or SCO
- Tight, light oil for diluent supply increasing, US exporting to Canada ...

See paper for powerpoint slide references

14AIChE Spring Meeting, New Orleans, LA

Full Spectrum Consulting LLC

Heavy Oil Basins

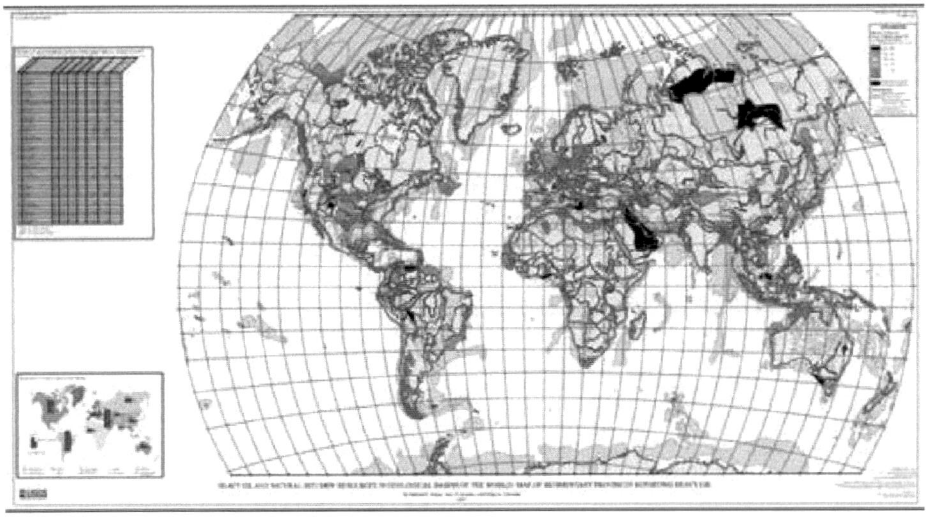

See paper for powerpoint slide references

14AIChE Spring Meeting, New Orleans, LA

Full Spectrum Consulting LLC

Bitumen Basins

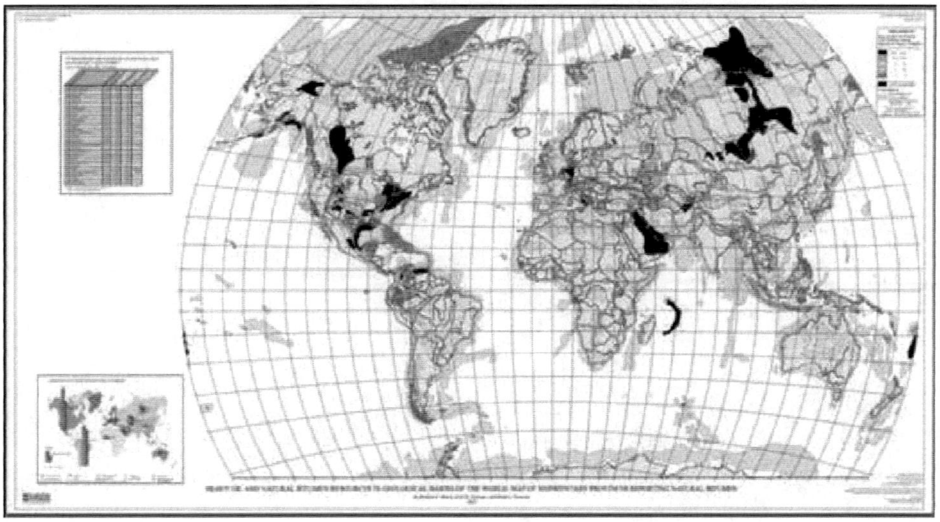

14AIChE
Spring Meeting, New Orleans, LA

Full Spectrum
Consulting LLC

Kleme basin types

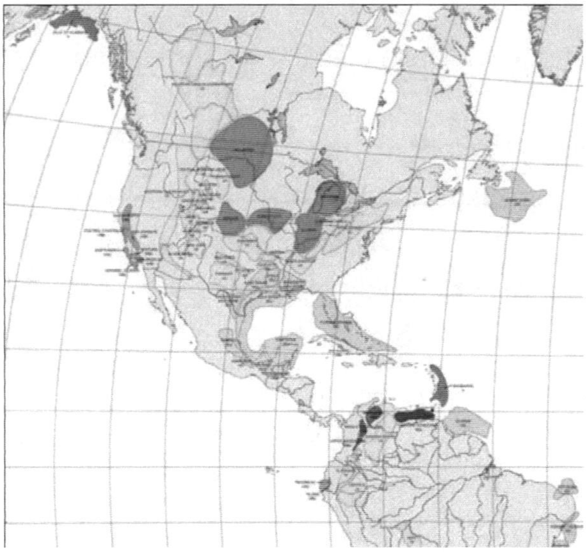

14AIChE
Spring Meeting, New Orleans, LA

Full Spectrum
Consulting LLC

Canadian Oil Sands Mining & In-Situ Projects

- " 1) Producing Projects: 1,853,650 bpd
- Mining - 921,000 bpd
- In-situ - 932,650 bpd
- 2) Under Construction: 714,905 bpd
- Mining - 295,000 bpd: 2projects
- In-situ – 419,905 bpd: 20 projects
- 3) Projects with Regulator Approval 2,199,650 bpd
- Mining-1,390,000 bpd: 12 projects
- In-situ - 809,650 bpd: 35 projects
- 4) Projects Under Regulator Review: 1,981,270 bpd
- Mining - 600,000 bpd: 7 projects
- In-situ-1,381,270 bpd: 55 projects
- 5) Projects Announced / Disclosed: 1,489,500 bpd
- Mining – 100,000 bpd: 1 projects
- In-situ–1,389,500 bpd: 45 projects"
- (Oilsands Developers Group, 2013.03)

14AIChE

Full Spectrum Consulting LLC

Oil Sands In-Situ Projects

- Projects continue and are provided for perspective:
- 1) 47 Producing Projects: 937,500 bpd
- 2) 17 Projects Under Construction: 532,000 bpd
- 3) 38 Projects with Regulator Approval: 1,236,000 bpd
- 4) 70 Projects Under Regulator Review: 1,925,270 bpd
- 5) 49 Projects Announced/Disclosed: 1,337,100 bpd
-
- (Oilsands Developers Group, 2013.04)

14AIChE Spring Meeting, New Orleans, LA

Full Spectrum Consulting LLC

Tight oil and gas plays, US

Source: Energy Information Administration based on data from various published studies. Updated: May 9, 2011

14 AIChE Spring Meeting, New Orleans, LA

Full Spectrum Consulting LLC

US domestic oil & gas to 2040

Figure 1. U.S. domestic crude oil production by source, 1990-2040
millions barrels per day

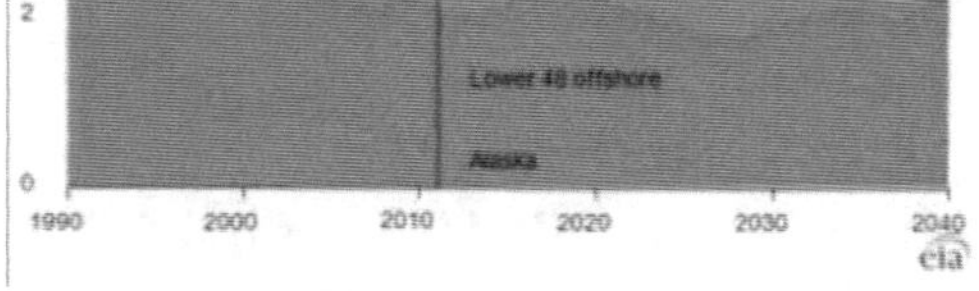

Figure 3. U.S. dry natural gas production by source, 1990-2040
trillion cubic feet

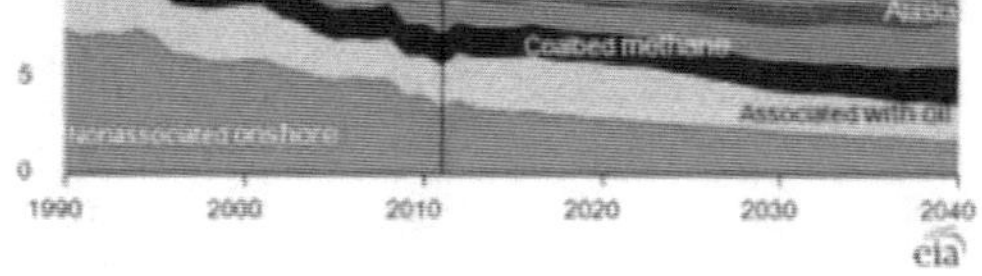

14 AIChE Spring Meeting, New Orleans, LA

Full Spectrum Consulting LLC

Basins assessed shale oil & gas, 5/2013

Figure 1. Map of basins with assessed shale oil and shale gas formations, as of May 2013

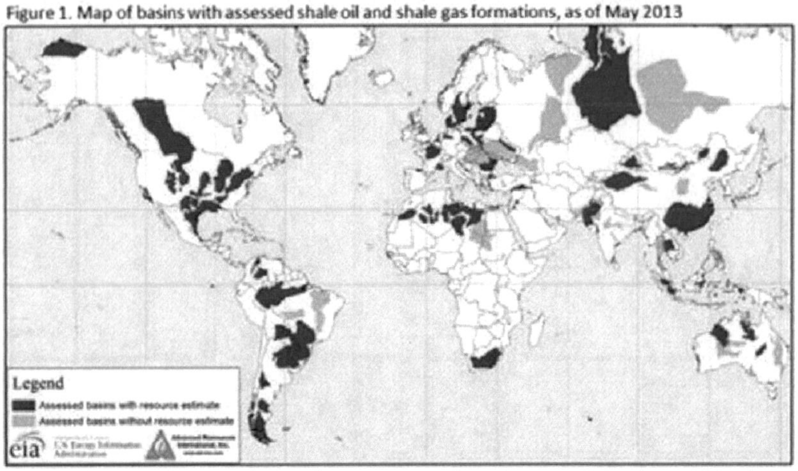

Source: United States basins from U.S. Energy Information Administration and United States Geological Survey; other basins from ARI based on data from various published studies.

14 AIChE Spring Meeting, New Orleans, LA

Full Spectrum Consulting LLC

Methods for Transportable Heavy Oil

- Previous presentation listed alternatives "to choose the best of them, regarding flow assurance and risk reduction associated with heavy oil production" :
- Dilution
- Heating
- Inverse emulsion
- Chemical Flow Improvers
- Upgrading
- Electrical traces
- Lubricated flow
- "**Pemex chose to apply the "Dilution at the surface" technique** to improve oil density and viscosity. … From the technologies evaluated, Dilution of heavy oil with a lighter one was selected; this one produces better profitability and lesser risks for the Project" Ayatsil-Tekel heavy oil field development.

Peregrino M.D.E., M.R., Mulato E., E.P., Tovilla P., I.C., (2013) Flow assurance Technologies for Ayatsil-Tekel heavy oil field development. Heavy ...nerica 2013. HOLA13-181.

14 AIChE Spring Meeting, New Orleans, LA

Full Spectrum Consulting LLC

Methods for Transportable Heavy Oil

- PEMEX, Shell, ExxonMobil, Cenovus, Canadian Natural, etc. have chosen dilution . There are no commercial partial upgraders.

Heavy oil, Bitumen Partial Upgrading In-situ, down hole Technologies

- 1. Aquathermolytic
- 2. Brine and supercritical conditions
- 3. Combination processes
- 4. Combined steam and vapor extraction process (SAVEX)
- 5. Downhole catalytic upgrading of heavy crude oil
- 6. Electrical heating elements
- 7. In-situ ethane / propane deasphalting (VAPEX)
- 8. Hot Fluid Injection Process for In-situ Upgrading
- 8. Naphtha / solvent deasphalting,
- 9. Hydrovisbreaking
- 10. Hydrocraking
- 11. In-situ heated annulus refining process
- 12. ICP Shell In-situ conversion process
- 13. ISU In-situ Upgrading with Nanocatalysts
- 14. Microwave down the wellbore
- 15. Natural gas addition to SAGD steam
- 16. Naturfrac
- 17. Paraffinic Froth Treatment, mining
- 18. Radiofrequency assisted gravity drainage
- 19. Others not listed above
- 19. SLP-SAGD Solvent, Low Pressure, low steam use
- 20. THAI Toe to Heel Air Injection
- 21. Upgrading heavy oil using aqueous base
- 22. EM-SAGD with SLP-SAGD and OTHERS

plus … others not included above

"There's a new wave of technology coming. Some people think we can do all this underground, said Cabrera." (Pratt, 2013)

Should know that
Down hole Technologies are in process to get oil out or improve oil quality, even adding hydrogen and upgrading underground. Appears Down hole Technologies are the New Wave of Technology coming.

Field Partial Upgrading Technologies-1

- A partial list of partial or integrated upgrading technologies:
- 1. Aquaconversion Process -- PDVSA
- 2. CCU Catalytic Crude Upgrading Process -- UOP
- 3. Canmet – UOP
- 4. Chattanooga Process -- Chattanooga Corp (i)
- 5. CPJ Process- - Wesco Energy Corp.
- 6. Eadiemac Process -- Eadie Oil Inc.
- 7. Ellycrack -- Ellycrack AS
- 8. EST Eni Slurry Technology -- Eni S.p.A. (i)
- 9. Genoil Process – Genoil (i)
- 10. GHU Heavy Oil Desulfurization Upgrader – Genoil
- 11. HCAT Process by Headwaters Technology Innovation (i)
- 12. HI-Q Process -- MEG Energy
- 13. H-Oil RC Process -- Axens North America
- 14. HOU – Ivanhoe Energy
- 15. HRH Process (TIPS Process) -- Mobis Energy (i)

SC High Conversion Soaker Cracking -- Toyo Eng. Corp

TL Process -- Ivanhoe Energy Inc.

14 AIChE Spring Meeting, New Orleans, LA

Full Spectrum Consulting LLC

Field Partial Upgrading Technologies-2

- A partial list of partial or integrated upgrading technologies:
- 19. IYQ Process -- ETX Systems Inc.
- 20. JetShear Process -- Fractal Systems Inc.
- 21. OrCrude Process – CNOOC (i)
- 22. PCUP Partial Crude Upgrading–ExxonMobil Upstream Res.
- 23. PetroBeam Process -- PetroBeam, Inc.
- 24. Petrosonic Heavy Oil Process -- Petrosonic Energy Inc.
- 25. Premium Upgrading & Gasification – OPTI (i)
- 26. ROSE -- KBR
- 27. Selex ASP – Well Resources
- 28. Sonocracking Process – SulphCo Ultrasound
- 29. Tarblaster -- Tarblaster AS
- 30. TRU Process -- Rival Technologies Inc.
- 31, Ultrasound and HOG Heavy Oil Upgrader-- NexGen
- 32. Uniflex -- UOP, from Canmet, Unicracking, Etc. (i)
- 33. Value Creation Technology -- Value Creation Group (i)
- 34. Viscositor -- FluidOil Ltd.
- 35. Wildcatter HCU -- Refinery Science
- 36. WRITE Process -- Western Research Institute
- 37. Catalytic Partial Upgrading – N.I.C.E.
- 38. Upgrading -- N.I.C.E.

Bitumen to Diesel – Ho, Sheenenergy, Bayshore Petr.

Ceramatec -- Ceramatec

14 AIChE Spring Meeting, New Orleans, LA

Full Spectrum Consulting LLC

Field Partial Upgrading Technologies-3

- A partial list of partial or integrated upgrading technologies:

- 39. Bitumen to Diesel – Ho, Sheenenergy, Bayshore Petr.
- 40. Ceramatec -- Ceramatec
- 41. FTcrude, FTupgrading, Ftetc. _Expander Energy
- 42. Others not listed above

14 AIChE Spring Meeting, New Orleans, LA

Full Spectrum Consulting LLC

Partial Upgrader Schemes do not provide enough data for accurate product evaluation, emissions, so valid economics are impossible.

- **Schemes not supported under scrutiny**
- Need full feed, product, waste, and emission details provided for accurate assessment
- Need full details and physical samples provided for independent analysis of all products
- Partial Upgrader schemes characterized by the **Next Big Thing, Game Changing, New Paradigm**, **Breakthrough Technology, and Hype Equivalents.**
- (Smith, 2013)

14 AIChE Spring Meeting, New Orleans, LA

Full Spectrum Consulting LLC

Partial Upgraders: many claim to be the Next Big Thing, Breakthrough Technology, Game Changing, Paradigm Shifting e.g. PSON

Tobin Smith, 2013, Petrosonic Partial Upgrading: Promo Material, 20 pages; no hard data

Partial Upgraders: many claim to be the Next Big Thing, Breakthrough Technology, Game Changing, Paradigm Shifting e.g. PSON

The pump on Petrosonic Energy Inc

(OTCMKTS:PSON) was quite a big affair – emails, video coverages, Mr. Tobin Smith's optimistic price targets plastered all over colorful brochures and lots of excited investors. It was a huge thing but, as you can see from the chart on the right, it was also quite tragic.

PSON not the only partial upgrader stock with hype price spikes

Partial Upgraders: many claim to be the Next Big Thing, Breakthrough Technology, Game Changing, Paradigm Shifting, etc.

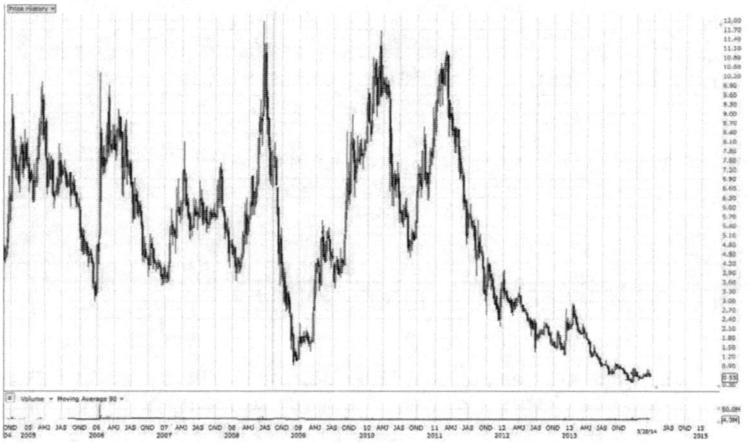

- "Large spikes in ... stock price can be traced to a potential major... gas play, exploration progress ..., a possible ... venture, advancements in unique [sic] ... technology, or an up- tick in reserves. All were heavily publicized and hyped, but few actually happened4. Others fell far short of expectations after the fact, ... The fallout from deals gone awry, dry wells, technology yet to be commercialized, or a downward adjustment of reserve size, took its toll ...."

14AIChE Spring Meeting, New Orleans, LA

http://amazonwatch.org/assets/files/2011-ivanhoe-energy-ecuador.pdf

Full Spectrum Consulting LLC

Upgrading Categories

- Upgrading technologies generally either takes carbon out, or puts hydrogen in, or both. Simple division to three categories: carbon out, hydrogen in, or both.
- More detailed General categories include
- catalytic/thermal cracking (CCU, Ellycrack, Tarblaster, Viscositor),
- thermal cracking (HTL, Fluid Coking, CPJ, Eadiemac, PCUP, HCU)
- thermal cracking coking (ETX, Genoil, HiQ, Write),
- cavitation/impingement (Viscositor)
- electron beams, electromagnetic waves,
- hydrogenation ,
- so-called integrated processes (OPTI, TRU, VCG) shown with an (i);
- SDA (Petrosonic, ROSE, SELEX),
- sonic waves (Sonocracking, Sonic Visbreaking),
- and all forms of hydrogen addition
- and any combination of categories.

14AIChE Spring Meeting, New Orleans, LA

Full Spectrum Consulting LLC

Partial Upgraders Promoted, but no partial upgraders built

- Partial upgrading technologies can be classified to those that have been built and commercialized or those that have not.
- Full Upgrading has been producing since 1967. No Partial Upgraders have been built, evidently for valid reasons.
- One partial upgrader claims no one has bought their technology because of resistance to change. (WHOC, 2014)
- However literature shows there have been many, including majors, Californian and Canadian heavy oil producers, national oil companies, and others that have looked at, and tested partial upgrading technology, one spent over 2 million, but none have gone forward after due pilot plant or demonstration plant testing. (NewTech, 2005).

Partial Upgrader Challenges

- History shows many technologies did not progress after pilot, demonstration, or commercial small scale trials.
- Major problems discovered in commercial testing resulting in abandoned schemes, e.g. full size FCCU converted to test run crudes, atmospheric bottoms and vacuum bottoms schemes terminated after no end of plugging. (Wiehe, 2009)
- Continuous plugging of reactors, transfer lines, overhead systems, distillation columns, guard beds, etc.

Two Upgrading Product Categories by determined by testing: Stable or Unstable

- **Stable, Compatible Product**
- Whole crudes are generally self compatible with some exceptions, e.g. YME.
- SCO – Synthetic Crude Oil from Full Upgrading including hydrotreating, e.g. Suncor, Syncrude, etc., then blended to specification. Low sulfur, low resid, high gravity.

- **Unstable, Self-Incompatible Product**
- Self-incompatible crude difficult to market, some crude production shut in.
- Naphthas, Jet Fuels, Diesels, and Fuel oils with dienes, asphaltene incompatible components, production and reaction solids required test methods to quantify.

- SICPUTCUPs--Self-InCompatible Partially Upgraded Thermal Cracked Unstable Products. Typically high sulfur, low gravity.

- All thermal cracked material must be assumed unstable and incompatible until irrefutably proven stable, that is without dienes, gum, and asphaltene precipitates, TIOM, and solids by extensive quantitative test results.

14AIChE Spring Meeting, New Orleans, LA

Full Spectrum Consulting LLC

Stability & Compatibility Tests – 1/3

- Optical Inspection for Self-Incompatibility, Wiehe
- Asphaltene content, Heptane (D3279)
- Asphaltene content, Hexane (D3279)
- Asphaltene content, Pentane (D3279)
- Toluene Insolubles (D4072)
- THF insolubles (D4072_modified)
- Flocculation Ratio, Peptizing Power (D7060)
- Stability and Compatibility (D7112)
- Intrinsic Stability (D7157)
- Automated Heithaus Titrimetry (D6703)
- Asphaltenes by nC7 phase separation (D7061)
- Asphaltenes (D6560)
- SHFT Total sediment (D4870a-96/ISO10307)
- SHFT Total sediment by thermal ageing (D4870b)
- SHFT Total sediment by chemical ageing (D4870c)
- SHFT Hot Filtration after 7 days, as is (D4870)
- SHFT Hot Filtration after 7 days, Heat (D4870)
- SHFT Hot Filtration after 7 days, Chemical (D4870)
- SHFT Hot Filtration after 28 days, as is (D4870)
- SHFT Hot Filtration after 28 days, Heat (D4870)
- SHFT Hot Filtration after 28 days, Chemical (D4870)
- Thermal Stability Tests (D873)

14AIChE Spring Meeting, New Orleans, LA

Full Spectrum Consulting LLC

Stability & Compatibility Tests – 2/3

- Spot test (D4740-95)
- Spot test and residue stability (IFP3024-82)
- Spot test (D102)
- P-Value, Peptization value Shell method
- Wiehe Compatibility Test Methods
- CII, Colloidal Instability Index, Weihe method
- Coking Index, Weihe, WRI calculation
- Fouling Test, Alcor Hot Liquid Process Simulator
- Fouling Tendency (D3711-95)
- Asphaltene dispersant tests by vendor
- Furnace tube fouling tests by vendor
- Viscosity with and without exposure to air
- Elemental Analysis: COHNS
- SARA, Sat, Arom, Resins, Asph (D2007/D2549)
- SARA Calculations: Asphaltene/Resin Ratio
- SARA Calculation: Saturate/Aromatic Ratio
- CSI Coking Stability Index re: furnace fouling
- Asphaltene Precipitation Onset, various methods
- Compatibility spot tests filter paper methods
- Elemental Analysis by ICP

14 AIChE Spring Meeting, New Orleans, LA

Full Spectrum
Consulting LLC

Stability & Compatibility Tests – 3/3

- Conjugated Di-olefins by Exxon Method
- Dienes UOP 326-82
- Dienes by 1hNMR published method
- Olefins by 1hNMR published method
- Bromine Number
- Water by Karl Fisher (D7377/D1744)
- BS&W by Centrifuge (D1796)
- Sediment by extraction (ISO 3735, IP53)
- Sediment content (D473)
- TAN, Total Acid Number (D664)
- Salt in crude (D6470M)
- Existent Gun in Fuel by Jet Evap (D381)
- Oxidation Stability of Naphtha (D525)
- Oxidation Stability of Kerosene (D873)
- Oxidation Stability of Distillate Fuel (D2274)
- JFTOT Thrm Oxyd. Stab. of Av. Turbine Fuel (D2274)
- Distillate Fuel Storage Stability (D4625)
- Endcut distillation published method
- Particle size by Fixed Beam Reflectance Measure
- Particle Vision Measurement, PVM Mettler-Toledo

- OTHER BULK TESTS NOT INDICATIVE OF STABILITY INCLUDE
- Simulated Distillation high temp (D2887)
- Viscosity at 0, 10, 40, 100, 150 C and pipeline specs
- Viscosity (D445, etc.)
- API (D287, etc.)
- ETC.

THESE ARE INADEQUATE TO DEMONSTRATE COMPATIBILITY

14 AIChE Spring Meeting, New Orleans, LA

Full Spectrum
Consulting LLC

Determination of incompatible product

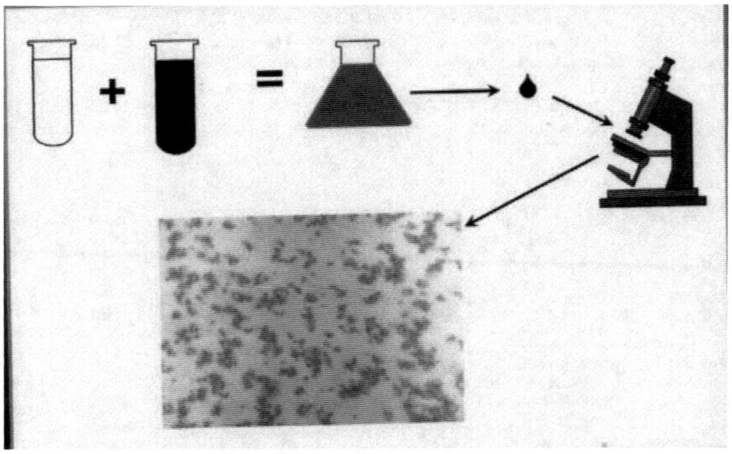

(Wiehe, 2008)

14AIChE Spring Meeting, New Orleans, LA

Full Spectrum Consulting LLC

Determination of self-incompatible product

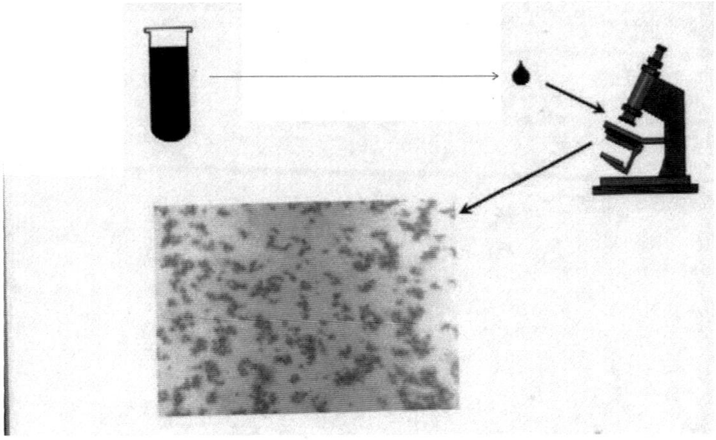

(Wiehe, 2008)

14AIChE Spring Meeting, New Orleans, LA

Full Spectrum Consulting LLC

Thermal cracked fuel oils instability documented from start of delayed coking, very unstable distillates,1929 and catalytic with thermal cracking at higher temperature temperature

Burk, R. (**1933**) U.S. Patent Application, Patent 2,081,218, May 25, 1932, METHOD FOR. STABILIZING MOTOR FUEL AGAINST GUM FORMATION. Washington, DC: U.S. Patent and Trademark Office. assigned to The Standard Oil Company, Cleveland, Ohio.

"In the customary methods of producing motor fuel by cracking petroleum, there is a formation of unstable gum-forming byproducts which subsequently require particular treatment if their results are to be avoided." (Burk, 1933).

- Laboratory Evaluation of Fuel Oil Stability
- 4. R. RESCORLA, J. H. CROMWELL, AND DANIEL RIILSO3I
- Cities Service Research and Development Co.,JYeu.York,.1'.Y.
- Analytical Chemistry, v24, no12, Dec. 1952
- WITH the introduction of catalytic cracking of the heavier crude oil fractions, the furnace oil produced from these cracked stocks was found to be generally very unstable. The furnace oil produced from the catalytic cracking stocks has a tendency to become more rapidly oxidized in storage to form an in- soluble precipitate and a dark colored oil which may have a high soluble residue. Although the fuel oils prepared from straight- run materials are more resistant to oxidation, when this material is mixed with the catalytic fuel oil the precipitation of the insoluble material is greatly accelerated. The consumer's first experience with the instability comes with the clogging of burner filters, and the dark oil removed from the system when clogging occurs.
- There have been cases reported where fuel oil stored for a period as short as 2 months has caused filter clogging. The cause of this instability has not been definitely established, but various investigators have attributed it to sulfur compounds, particularly thiophenols and thiocresols, unstable nitrogen compounds, such as pyrroles, quinoline, and isoquinoline, colloidal carbenes and carboids, or unstable diolefinic hydrocarbons. Actually the instability may be due to any one or a combination of these compounds resulting from oxidation and polymerization reactions in
- the fuel oil.

Full Spectrum
Consulting LLC

Thermal cracking naphtha gums, 1953

- **Gum Formation in Shale-Oil Naphtha**
- **G. U. DINNEEN AND JV. D. BICICEL' U. S. BureauofMines,Laramie, Wyo.,**
- **Industrial and engineering chemistry, vol 43, no 7, 1953.**

Existent gum (*2*) was determined on each of the raw naphthas as taken from storage which was in sealed containers under refrigeration. Results are shown in Table III. Gum content was also determined on the samples after accelerated oxidation for 60 minutes under an oxygen pressure of 100 pounds per square inch gage and at a temperature of 211° to 212° F. The procedure was the same as in the A.S.T.M. method for oxidation stability of aviation gasoline (*3*), except that filtration of the oxidized sample was omitted and the sample was transferred as completely as possible to the beaker used in the existent gum determination. The gums obtained from the oxidized naphtha were analyzed for nitrogen by a Kjeldahl method (unpublished), for oxygen by the Unterzaucher method (*1*), and for sulfur by a combustion-tube method.

Table III. Gums on Naphthas Produced by Thermal Processes

Sample No.	A.S.T.M. Gum, Mg./100 Ml.	A.S.T.M. Gum after 60 Min. Oxidation, Mg./100 Ml.	Elemental Analysis of Gum from Oxidized Naphtha, %			Elemental Analysis of Naphtha, %	
			Nitrogen	Oxygen	Sulfur	Nitrogen	Sulfur
419	494	5284	8.15	9.09	2.47	1.21	1.09
128	90	1922	8.57	9.19	1.36	0.90	0.57
130	274	3356	8.95	9.25	2.45	1.05	0.96
131	162	3770	8.40	10.28	3.28	0.99	1.04
132	322	4032	7.75	10.46	1.87	0.90	0.78
133	464	5956	7.12	9.59	5.14	1.15	1.26

The striking increase in gum content between the oxidized and original samples shows the importance of oxidation reactions in gum formation in shale-oil naphthas. However, the nitrogen

Full Spectrum
Consulting LLC

High butadiene with high temperature short contact time thermal cracking ENSYN RTP EXPERIMENTS shown

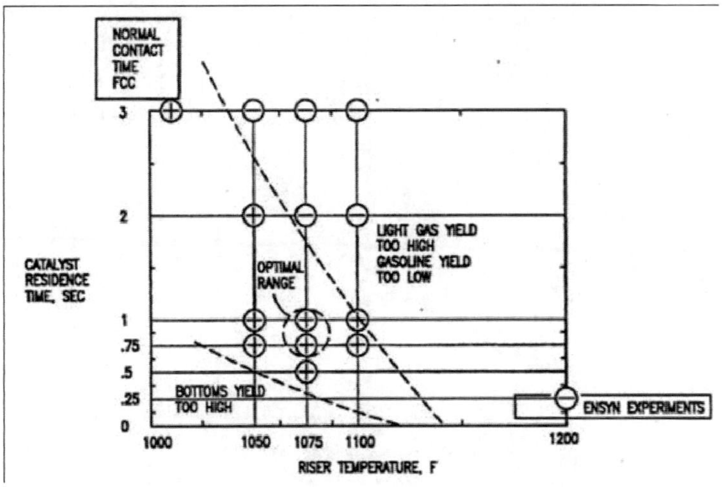

patent "FCC to Minimize Butadiene Yields" (Muldowney, 1994)

14AIChE Spring Meeting, New Orleans, LA

Full Spectrum Consulting LLC

High temperature, short contact time high 1,3-Butadiene from Ensyn data

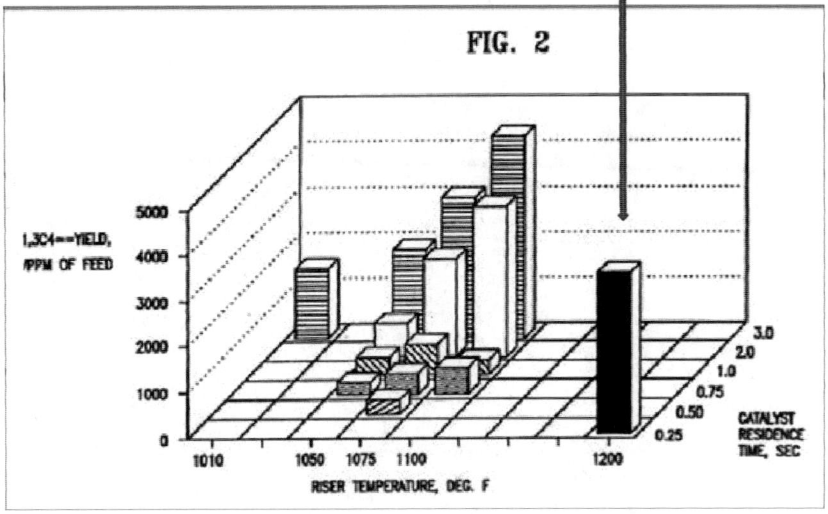

patent "FCC to Minimize Butadiene Yields" (Muldowney, 1994)

14AIChE Spring Meeting, New Orleans, LA

Full Spectrum Consulting LLC

Well established relationship Butadiene increase with Temperature

- FC

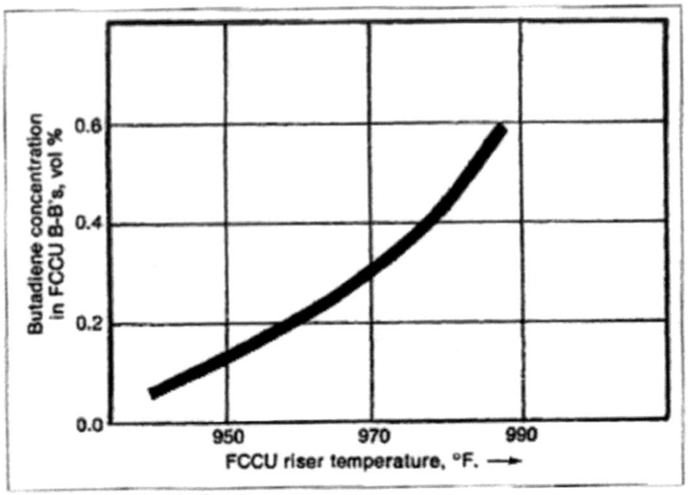

Lieberman, N.P. (2009). Troubleshooting Process Operations, Pennwell, Tulsa.

Butadiene vs Reactor Temperature

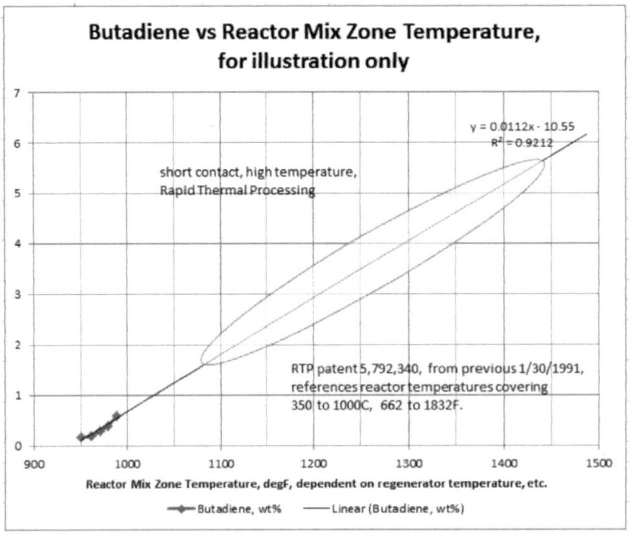

Highest reactor temperature makes least stable naphtha, e.g. FCC

Table 1 Physico-chemical characteristics of cracked naphthas and straight-run naphthas

Characteristic	FCC naphtha	Visbreaking naphtha	Straight-run naphtha A	Straight-run naphtha B
Density ($kg\,l^{-1}$ at 15°C)	0.7067	0.7240	0.6863	0.7444
Gravity (API at 60°F)	68.68	63.88	75.5	58.52
Reid vapour pressure ($kg\,cm^{-2}$)	0.46	0.29	0.69	0.35
H_2S (ppm)	Nil	Nil	Nil	Nil
Thiol sulfur (ppm)	0.8	3.3	3.1	1.6
Sulfur, total (ppm)	328.0	0.53 wt%	3.5	9.0
Nitrogen (ppm)	7.3	10.7	<1	<1
Basic nitrogen (ppm)	0.8	1.9	–	–
Hydrocarbon types (PONA):				
Aromatics (vol.%)	4.9	5.3	5.8	17.0
Olefins (vol.%)	61.9	42.9	–	–
Naphthenes (vol.%)	33.2[a]	51.8[a]	22.3	43.7
Paraffins (vol.%)			71.9	39.3
Diene value	4.9	2.6	–	–
Av. molecular weight	94.5	100.8	80.3	93.9
Potential gum, ASTM D 873 (4 h)	–	–	0.5	2.1
C/H ratio by weight	5.55	5.93	5.50	6.09
Distillation, ASTM D 86:				
IBP (°C)	47.0	48.0	42.0	57.0
10 vol.% at (°C)	62.0	70.0	51.0	74.0
50 vol.% at (°C)	82.0	113.5	63.0	98.5
90 vol.% at (°C)	117.5	155.0	102.5	125.5

[a] Naphthenes + paraffins

14 AIChE Spring Meeting, New Orleans, LA

Full Spectrum Consulting LLC

Highest reactor temperature least stable:

FCC least stable, coker most stable

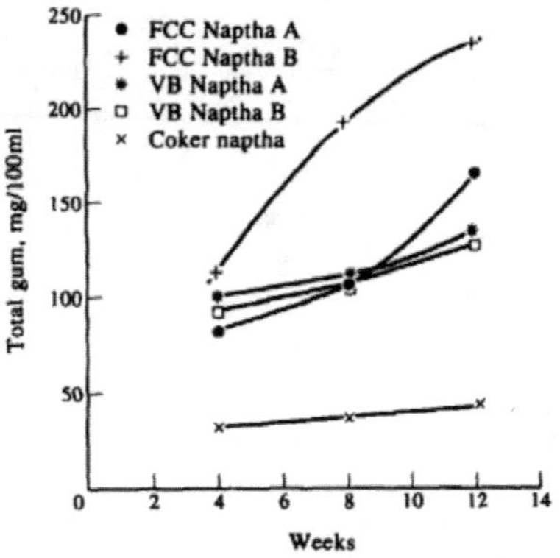

Figure 6 Three months stability test at 43°C, cracked naphthas as such. (●) FCC naphtha A; (+) FCC naphtha B; (•) visbreaking naphtha A; (□) visbreaking naphtha B; (×) coker naphtha

14 AIChE Spring Meeting, New Orleans, LA

Full Spectrum Consulting LLC

Steam Cracked Naphtha Dienes to C9s

Table 4. Conjugated Diene Analysis of Steam-Cracked Naphtha (SCN) before and after Hydrotreating

	molecular weight, MW	Analysis of SCN (wt %) before hydrotreatment	Analysis of SCN (wt %) after hydrotreatment
		Linear	
C_4	54	0.01	0.00
C_5	68	1.09	0.02
C_6	82	0.78	0.04
C_7	96	0.08	0.01
C_8	110	0.00	0.00
C_9	124	0.00	0.00
		Cyclic	
C_5	66	0.07	0.03
C_6	80	0.59	0.03
C_7	94	0.09	0.01
C_8	108	0.01	0.01
C_9	122	0.00	0.00
		Total	
		2.72	0.15

(Edwards, 2005)

14 AIChE Spring Meeting, New Orleans, LA

Full Spectrum Consulting LLC

Coker Product Dienes to C10s

Table 5. Conjugated Diene Analysis of Three Heavy Coker Naphthas (HCNs)

	molecular weight, MW	Coker Naphtha Analysis (wt %) HCN A	HCN B	HCN C
		Linear		
C_4	54	0.00	0.02	0.01
C_5	68	0.01	0.01	0.01
C_6	82	0.09	0.01	0.01
C_7	96	0.22	0.42	0.39
C_8	110	0.17	0.37	0.35
C_9	124	0.12	0.20	0.19
C_{10}	138	0.08	0.11	0.09
		Cyclic		
C_5	66	0.01	0.01	0.00
C_6	80	0.03	0.03	0.03
C_7	94	0.08	0.20	0.22
C_8	108	0.07	0.18	0.20
C_9	122	0.05	0.09	0.09
C_{10}	136	0.02	0.04	0.04
		Total		
		0.96	1.69	1.63

(Edwards, 2005)

14 AIChE Spring Meeting, New Orleans, LA

Full Spectrum Consulting LLC

Insufficient data for stability determination

Heavy Oil, not Bitumen TABLE 13

Properties and SimDist of liquid products prepared from Heavy Oil using the multi-stage Process (for feedstock properties see Tables 1 and 5).

Fraction	Temp (° C.)	R241* Day 0	R241* Day 30	R242** Day 30	R244*** Day 30
Density @15.6° C.*	—	0.9592	0.9597	0.9465	0.9591
API	—	15.9	15.8	17.8	15.9
Viscosity @40° C.**	—	79.7	83.2	25.0	49.1
Light Naphtha	<71	0.0	0.2	0.3	0.3
Light/med Naphtha	71-100	0.0	0.1	0.2	0.3
Med Naphtha	100-166	0.1	0.4	2.5	1.8
Naphtha/Kerosene	166-193	0.6	0.6	1.8	1.5
Kerosene	193-232	2.8	2.5	5.0	3.5
Diesel	232-327	21.8	21.0	23.1	18.9
Light VGO	327-360	10.8	10.2	9.9	8.8
Heavy VGO	360-538	51.1	45.0	44.9	43.2
Vacuum Resid.	>538	12.7	20.0	12.3	21.7

*reactor temp. 543° C.
**reactor temp. 557° C.
***reactor temp. 561° C.

Freel, Graham, Patent 7,270,743 "Products Produced From Rapid Thermal Processing of Heavy Hydrocarbon Feedstocks" provisional application filed on 09/18/2000

Vacuum resid apparently increases from 12.7 to 20.0, and from 12.3 to 21.7. Without accurate testing using the correct test methods, there is no support for a claim of stability.

14AIChE Spring Meeting, New Orleans, LA

Full Spectrum Consulting LLC

Insufficient data for stability determination, unsupported stability claim

TABLE 9

Stability of liquid products after single stage processing (reactor temperature 525° C.)

Fraction	Temp (° C.)	Feedstock	R232 Day 0	R232 7 days	R232 14 days	R232 30 days
Density @15.6° C.*	—	1.0095	0.979	0.980	0.981	0.981
API	—	8.5	12.9	12.7	12.6	12.6
Viscosity @40° C.**	—	30380	201.1	213.9	214.0	218.5
Light Naphtha	<71	0.0	0.1	0.1	0.1	0.1
Light/med Naphtha	71-100	0.0	0.1	0.1	0.1	0.1
Med Naphtha	100-166	0.0	1.5	1.5	1.5	1.4
Naphtha/Kerosene	166-193	0.1	1.0	1.0	1.0	1.1
Kerosene	193-232	1.0	2.6	2.6	2.6	2.7
Diesel	232-327	8.7	14.1	14.1	14.3	14.3
Light VGO	327-360	5.2	7.3	7.3	7.4	7.4
Heavy VGO	360-538	33.5	41.3	41.3	41.7	42.1
Vacuum Resid.	>538	51.5	32.0	32.0	31.2	30.8

*g/cm³
**cSt

Freel, Graham, Patent 7,270,743 "Products Produced From Rapid Thermal Processing of Heavy Hydrocarbon Feedstocks" provisional application filed on 09/18/2000

14AIChE Spring Meeting, New Orleans, LA

Full Spectrum Consulting LLC

Solid particle size in Athabasca Bitumen by FBRM

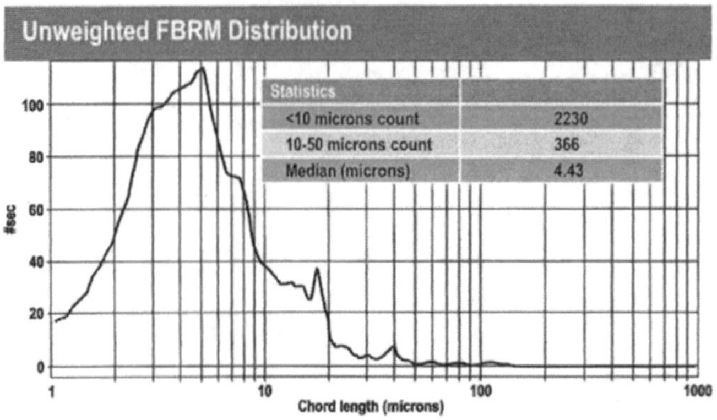

FIG. 21

(12) **Patent Application Publication** (10) **Pub. No.: US 2012/0167767 A1**
Pavel et al. (43) **Pub. Date: Jul. 5, 2012**

Solid particle size in Athabasca Bitumen by FBRM

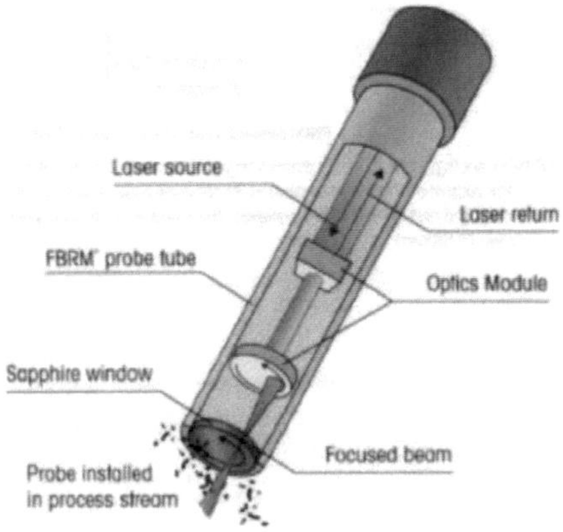

(12) **Patent Application Publication** (10) **Pub. No.: US 2012/0167767 A1**
Pavel et al. (43) **Pub. Date: Jul. 5, 2012**

Solids in Bitumen by PVM

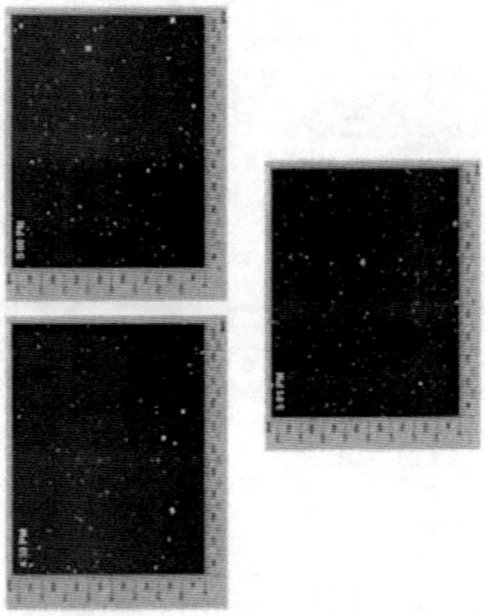

FIG. 20

(12) **Patent Application Publication** (10) **Pub. No.: US 2012/0167767 A1**
Pavel et al. (43) **Pub. Date: Jul. 5, 2012**

Solid particle size in Athabasca Bitumen

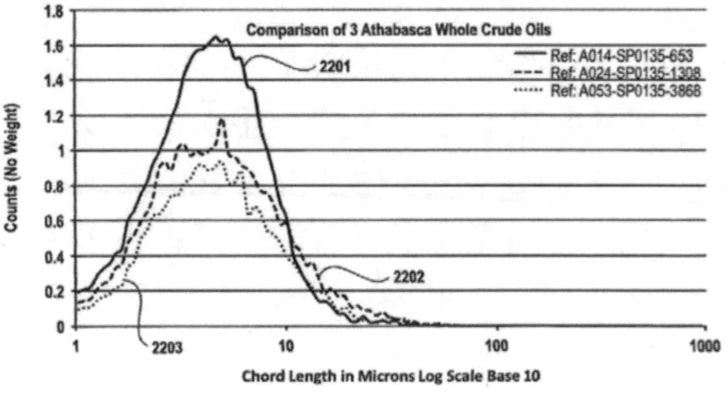

FIG. 22

(12) **Patent Application Publication** (10) **Pub. No.: US 2012/0167767 A1**
Pavel et al. (43) **Pub. Date: Jul. 5, 2012**

Solids removed from partial upgrading reactor product, patent figures

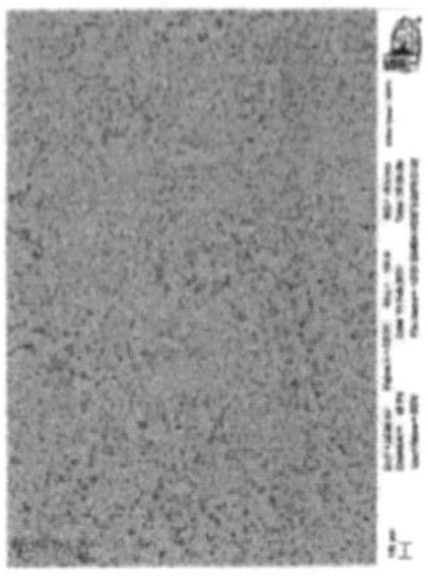

FIG. 10

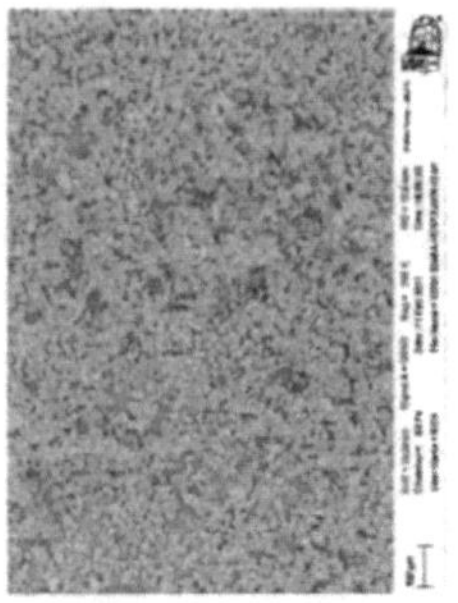

FIG. 11

FIG. 12

(12) **Patent Application Publication** (10) **Pub. No.: US 2012/0167767 A1**
Pavel et al. (43) **Pub. Date: Jul. 5, 2012**

14 AIChE Spring Meeting, New Orleans, LA

Full Spectrum Consulting LLC

FBRM note

- FBRM can also be used to for inline analysis detecting the formation and presence of asphaltenes and for laboratory analysis of compatible materials to show the incipient precipitation point of the asphaltenes and confirm the presence of Toluene Insoluble Organic Material, TIOM, solids.

Full Spectrum Consulting LLC

Assorted stability methods

Comparison of Methods					
Method	**Target Oil**	**1st Solvent**	**Titration Solvent**	**Detector**	**Report**
ASTM D7060	Residue >1wt% asphaltene	I methylnaphthalene Ratio = 1:1, 2:1, 3:1, 6:1	Cetane	Optical	FRmax, Po P-value
Wiehe Method (Manual)	Residue no asphaltene (Solvent oil/non-solvent oils	Toluene Ratio = 1:5 etc	Heptane	Spot test-Microscopy	I_N, S_{BN} Pa, Po, $FR_{5/1}$ P-Value =
DBR SDS Wiehe Method (Automated)	Residue no asphaltene (Solvent oil/non-solvent oils	Toluene Ratio = 0.33, 0.5, 0.66	Heptane	Optical	I_N, S_{BN} Pa, Po, $FR_{5/1}$ P-Value =
ASTM D6703	Asphalts, pyrolized heavy oil & resids	Toluene Ratio = 0.2, 0.3, 0.4	Isooctane	Optical	Pa, Po, P
Chevron Method	Crude Oil, resids and processed materials with asphaltene contents> 500 ppm	Methylene chloride (1000 ppm of sample; other concentrations might be used depending on asphaltene content). Toluene could be used for waxy samples	Heptane Gradient from Heptane to 90/10 methylene chloride/ methanol and finally methanol	ELSD (Evaporative Light Scattering Detector)	ΔPS : 0.5 to 6. Values > 2 indicate increasing unstability
ASTM D7112	Residue >0.05wt% asphaltene	Xylene Ratio = 1:2, 1:1, 2:1	Heptane	Optical	I_N, S_{BN} Pa, Po, $FR_{5/1}$ P-Value = 1.0
ASTM D7157 (Porla)	Residues, heavy fuel oils, crude oils >0.5 wt% asphaltene	Toluene Ratio = 3:1, 1:1, 1:3	Heptane	Optical	S, Sa, So FR_{max}, Po, P-Value I_N, S_{BN}

http://www.coqa-inc.org/20120619-20_CCQTA_Project_Updates.pdf

Assorted stability method results

Heavy Oil Compatibility Project - Test Method Results Summary							
	Sample Type	Low asphaltene synbit	High asphaltene dilbit	High asphaltene synbit	Conventional Heavy low asphaltene paraffinic	Conventional Heavy high asphaltene aromatic	Conventional Heavy waxy paraffinic
	Sample ID	Shell ARB	Cenovus Christina Lake	Suncor Mackay River	Brazilian	Bow River	ConocoPhillips Sample
ASTM D7060	FR_{max}	22	31	26		31	NA
	Po	71	80	59		71	NA
	P-ratio	3.27	2.58	2.27		2.29	NA
Wiehe Method (Manual)	S_{BN}	80.26	64.44	91.11		69.47	SSO = 42.22
	I_N	24.93	33	38.28		31.29	NA
	V_H	11.1	2.1	6.9		6.1	NA
	P-Value	3.22	2.02	2.38		2.22	NA
DBR SDS Wiehe Method Automated	Po	83.6	78.8	89.4		81.8	NA
	FR_{max}	22.4	29.2	29.6		30.8	NA
	I_N	22.44	29.22	29.65		30.76	NA
	S_{BN}	79.21	75.44	89.69		77.73	SSO = 79.62
	P-Value	3.53	3.68	3.02		2.53	NA
Chevron method	ΔPS	0.75	1.43			1.44	
ASTM D6703	Pa	74.24	67.38	71.42		68.71	
	Po	84.42	85.98	74.83		66.92	
	P	3.277	2.625	2.619		2.138	
ASTM D7157	Sa	0.81	0.71	0.74		0.7	NA
	So	0.79	0.87	0.8		0.96	NA
	S	4.26	3.05	3.09		3.25	NA
	I_N	18.8	28.1	26.8		29.7	
	S_{BN}	76.9	82.2	79.3		91.3	
	FR_{max}	18.8	28.1	26		29.7	NA
	Po	81	86	80		96	NA
	P-Value	4.33	3.07	3.08		3.28	
ASTM D7112	FR_{max}	27 / 22	33 / 33	26		38 / 27	
	Po	80 / 64	87 / 89	64		90 / 78	
	I_N	27	33	26.3		27	
	S_{BN}	78	86	63		79	
	P-Value	2.96 / 2.91	2.64 / 2.7	2.45		2.57 / 2.89	

http://www.coqa-inc.org/20120619-20_CCQTA_Project_Updates.pdf

Self Incompatible Streams
all about asphaltenes

- Self-incompatible material show asphaltenes precipitated out of solution floating in crude, or Self-InCompatible Partially Upgraded Thermally Cracked Unstable Product (SICPUTCUP) streams.
- Precipitated, agglomerated asphaltenes easily seen with microscope, sometimes seen with naked eye, and more quantitatively reported using heptane and toluene.
- (Weihe, 2008)

14AIChE Spring Meeting, New Orleans, LA

Full Spectrum
Consulting LLC

Self-incompatibility and Stability easily determined by Weihe's optical method using $79 microscope

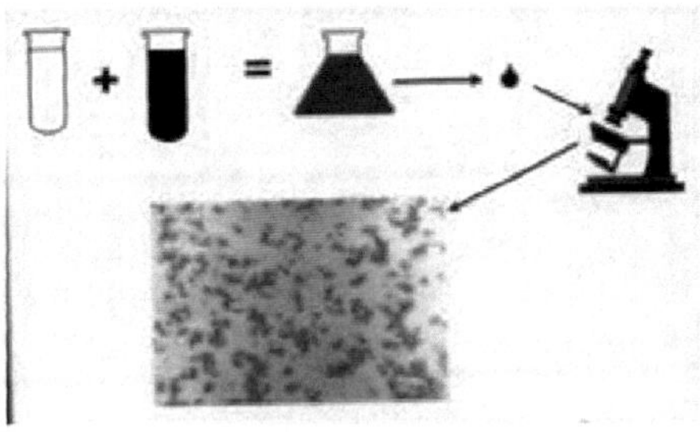

- This slide shows the precipitated and agglomerated asphaltenes floating in solution

(Weihe, 2009)

14AIChE Spring Meeting, New Orleans, LA

Full Spectrum
Consulting LLC

Canadian Heavies and Diluted Bitumen

TABLE 3-4 Comparison of Density, API Gravity, and Viscosity of Diluted Bitumen and Other Canadian Crude Oils

	Canadian Heavy Crude Oils					
	Bow River	Fosterton	Lloydminster Blend	Lloydminster Kerrobert	Smiley–Coleville	Western Canadian Blend
Density (kg/m^3)	914	927	927	930	932	929
API gravity (°)	23	21	21	20	20	21
Viscosity at 20°C (68°F) (cSt)	100	96	145	146	144	145
Viscosity at 40°C (104°F) (cSt)	37	36	52	52	51	52
	Diluted Bitumen					
	Access Western	Cold Lake	Peace River Heavy	Christina Lake	Wabasca Heavy	Surmount Heavy (Synbit)
Density (kg/m^3)	926	928	931	923	935	936
API gravity (°)	21	21	20	22	20	19
Viscosity at 20°C (68°F) (cSt)	150	153	113	178	134	131
Viscosity at 40°C (104°F) (cSt)	53	54	44	62	49	47

SOURCE: Data obtained from CrudeMonitor.com by Crude Quality, Inc. (http://www.crudemonitor.ca/tools/comp/crudecomparisons.php#results) and from Enbridge website (http://www.enbridge.com/DeliveringEnergy/Shippers/~/media/www/Site%20Documents/Delivering%20Energy/2012CrudeCharaceristics.ashx). Both websites accessed March 1, 2013.

14AIChE Spring Meeting, New Orleans, LA

Full Spectrum Consulting LLC

Diluents: Southern Lights, Suncor SCO

The specific diluents used in blending are selected on the basis of many factors, including their availability in bitumen production regions. Table 3-2 shows the chemical and physical properties of the common diluent Southern Lights, a condensate produced in the United States and piped to Alberta. Because of its low viscosity, this condensate and others can be mixed with bitumen at a ratio of about 30:70 by volume.[23] Table 3-2 also shows the chemical and physical properties of a Suncor synthetic crude oil. Because it has a higher density than condensate, this and other synthetic crude oils are usually blended in even (50:50) ratios with bitumen. Illustrative blending ratios and resulting density and viscosity values for synbit and dilbit are given in Table 3-3.

TABLE 3-2 Selected Properties of Two Common Diluents

Property	Southern Lights Condensate Diluent	Suncor Synthetic Crude Oil Diluent
Density (kg/m^3)	675	861
API gravity (°)	78	33
Sulfur (weight percent)	0.03	0.17
Viscosity at 20°C (68°F) (cSt)	<0.5	6.3
Sediment (parts per million by weight)	16	0

SOURCE: Data obtained from CrudeMonitor.com by Crude Quality, Inc. (http://www.crudemonitor.ca/condensate.php?acr=SLD; http://www.crudemonitor.ca/crude.php?acr=SYN) and from Enbridge website (http://www.enbridge.com/DeliveringEnergy/Shippers/~/media/www/Site%20Documents/Delivering%20Energy/2012CrudeCharaceristics.ashx). Both accessed March 1, 2013.

14AIChE Spring Meeting, New Orleans, LA

Full Spectrum Consulting LLC

Tight, Shale Oil and Gas Diluent Location

14 AIChE Spring Meeting, New Orleans, LA

Full Spectrum Consulting LLC

Tight oil & gas diluent opportunities for heavy oil production

Figure 1. Map of basins with assessed shale oil and shale gas formations, as of May 2013

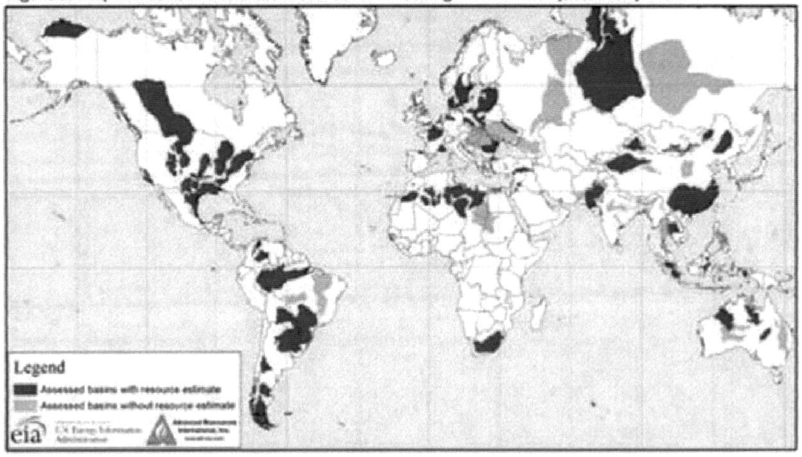

Source: United States basins from U.S. Energy Information Administration and United States Geological Survey; other basins from ARI based on data from various published studies

14 AIChE Spring Meeting, New Orleans, LA

Tight Shale Oil

- "Table 5. Top 10 countries with technically recoverable shale oil resources
- RANK COUNTRY SHALE OIL Billions of Bbls
- 1. Russia 75
- 2. US 58
- 3. China 32
- 4. Argentina 27
- 5. Libya 26
  - 6. Australia 18
- 7. Venezuela 13
- 8. Mexico 13
- 9. Pakistan 9
- 10. Canada 9
- Total Above 345

(EIA, 2013)

14 AIChE Spring Meeting, New Orleans, LA

Full Spectrum Consulting LLC

Tight oil formations don't stop at borders

14 AIChE Spring Meeting, New Orleans, LA

Full Spectrum Consulting LLC

Tight oil formations don't stop at borders

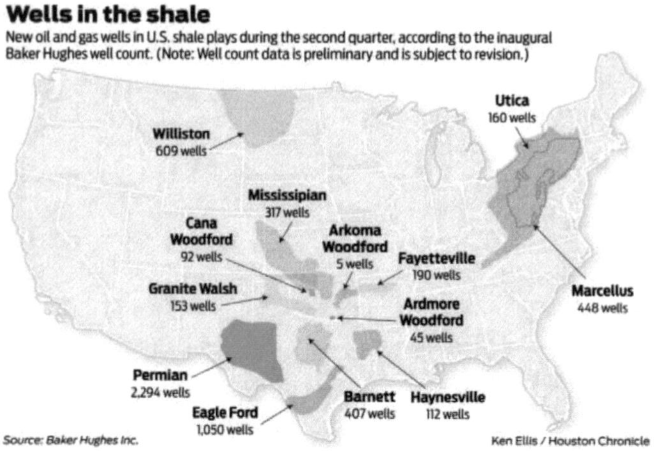

http://fuelfix.com/blog/2013/07/26/four-quarters-4092-new-wells-in-the-eagle-ford-shale/?cmpid=eefl

14 AIChE Spring Meeting, New Orleans, LA

Full Spectrum Consulting LLC

Tight oil, Eagle Ford

- "Eagle Ford's future might lie in Mexico's demand"
- "But while South Texas is well established as a top U.S. onshore field, for the long term, Pioneer and other Eagle Ford operators are watching for potential opportunities in Mexico, where President Enrique Peña Nieto unveiled a proposal last month to allow some direct foreigh investment in projects of Petroleos Mexicanos, or PEMEX."
- http://fuelfix.com/blog/2013/09/20/eagle-fords-future-might-lie-in-mexicos-demand/

14 AIChE Spring Meeting, New Orleans, LA

Full Spectrum Consulting LLC

Tight oil formations South America

- "La Luna Formation. The Maracaibo-Catatumbo Basin hosts some of the world's richest source rocks and conventional oil and gas reservoirs. The Late Cretaceous (Cenomanian-Santonian) shale of the La Luna Formation, the primary source rock in the basin22 and time-equivalent with the Eagle Ford Shale in Texas, appears to be the most prospective target for shale oil and gas exploration. The black calcareous La Luna Shale ranges from 100 to over 400 feet thick across the basin, thinning towards the south and east. 23,24
- Total organic carbon (TOC) varies across the basin, with values ranging from 3.7% to 5.7% in the northwest to 1.7% to 2% in the south and east. Maximum TOC values can reach 16.7%. A large portion of this shale-gas-prospective area includes part of Lake Maracaibo itself. ARI chose to include this submerged area because water depths are shallow (less than 100 feet) and there are numerous conventional production platforms that could provide access to shale drilling and development."
- (EIA, 2014)

Tight oil formations South America

- .4 Recent Activity
- A number of companies -- including Ecopetrol, ConocoPhillips, ExxonMobil, Nexen, and Shell -- have initiated shale oil and gas exploration programs at existing conventional oil and gas lease positions in Colombia during the past two years. Activity has been concentrated in the Middle Magdalena Valley Basin, close to the Bogota market. More than 12 vertical and horizontal shale exploration wells were planned for 2012, including several re-entries.
- State-owned Ecopetrol S.A., which controls about one-third of the oil and gas licenses in Colombia, first announced its shale exploration program in early 2011 and drilled the La Luna-1 stratigraphic test in the MMVB later that year (results not disclosed). Ecopetrol already has been drilling horizontal wells in the MMVB for non-shale targets during the past several years, providing a good foundation for future horizontal shale development in the basin.12
- Canacol holds three conventional exploration licenses in Colombia, which the company estimates have a total 260,000 gross acres with shale oil potential. The company has disclosed a Mean Estimate of 2.9 billion barrels of recoverable resource potential within their lease position. In recent months Canacol has signed separate joint-venture agreements with ConocoPhillips, ExxonMobil, and Shell to conduct shale exploration within Canacol's acreage. These companies plan to drill a total of 19 shale exploration wells at an estimated cost of $123 million. ConocoPhillips expects to drill its first exploration well to test the La Luna Shale in the second quarter of 2013.13 Canacol continues to review the shale potential of two of its other blocks.
- Nexen was one of the first companies to report exploring for shale gas in Colombia. The company reports it holds several shale blocks in Colombia for a total 1.5 million acres with shale gas potential.14 In late 2011 Nexen began drilling the first of four planned shale gas wells. These wells, located in Sueva and Chiquinquira blocks in the Sabana de Bogota high savannah plateau of the Eastern Cordillera mountain range, reportedly target the La Luna Formation. No further details are available.
- Sintana Energy has reported that its third-party consultant estimated 210 million bbl of prospective recoverable resources in shale formations at the company's VMM-37 block in the MMVB, which cover 44,000 acres (Mean Estimate). Sintana estimated initial horizontal well costs at about $13 million.
- (EIA, 2014)

Crude by Rail, including tight oil, railbit-half the diluent of pipeline dilbit, some heated rail cars transporting undiluted bitumen

Total U.S. Plus Canada Quarterly Originated Crude-by-Rail Carloads

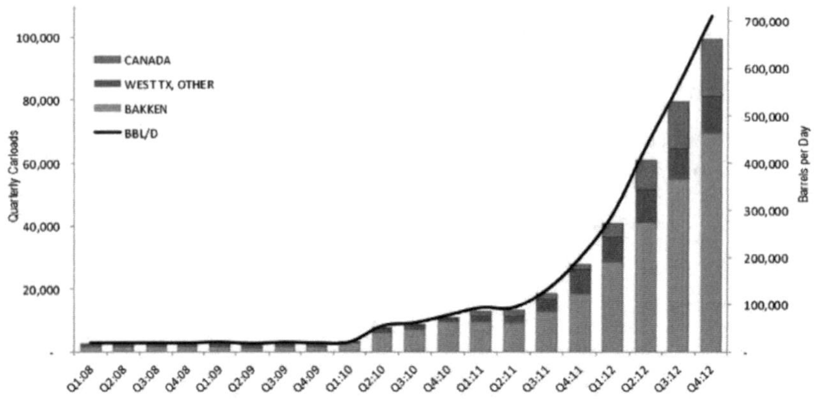

(Source: Association of American Railroads (AAR), Surface Transportation Board (STB), Raymond James research)

14 AIChE Spring Meeting, New Orleans, LA

Full Spectrum Consulting LLC

For all projects dilution has been selected, not partial upgrading

- Availability of Tight Oil for Diluent is Increasing
- Partial Upgrading had extensive evaluations
- Partial Upgrading has no commercial installation
- Partial Upgrading capital cost unknown, billions?
- Partial Upgrading reliability unknown
- Partial Upgrading history of plugging
- Partial Upgrading schemes abandoned after plugging
- Partial Upgrading history of unstable products
- Partial Upgrading history of overstating value
- Given choices, dilution standard is selected

14 AIChE Spring Meeting, New Orleans, LA

Full Spectrum Consulting LLC

CCU-UOP patent figure

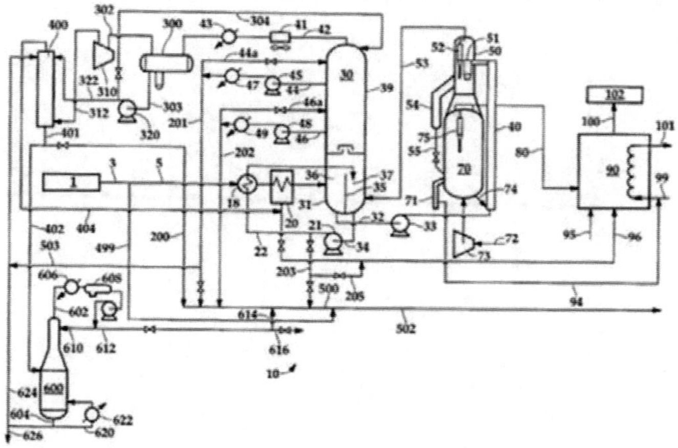

Viscositor Patent figure

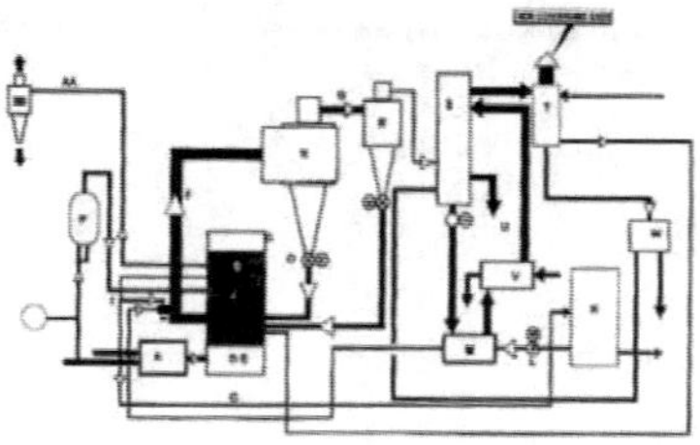

14 AIChE Spring Meeting, New Orleans, LA

Full Spectrum Consulting LLC

HTL patent figure

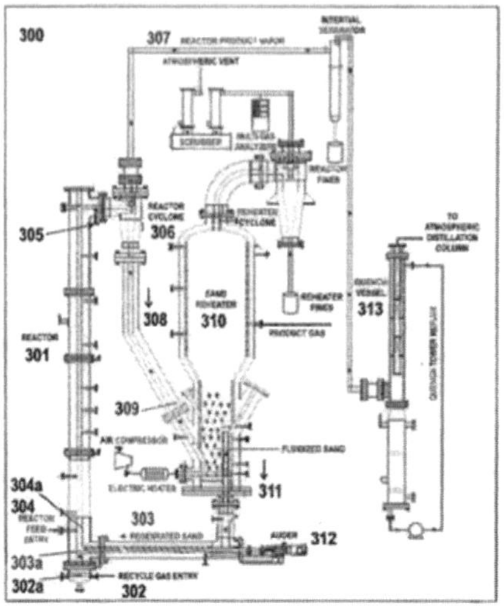

Pavel et al., 2012, USPTO Patent Application

14AIChE Spring Meeting, New Orleans, LA

Full Spectrum Consulting LLC

HTL – FTF Patent Figure

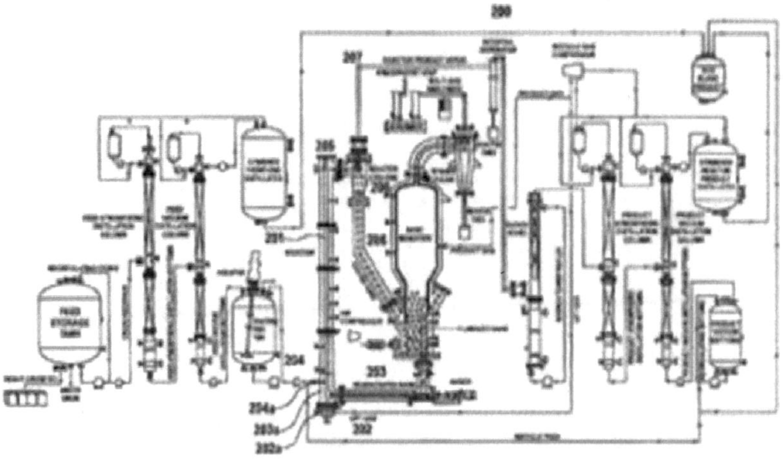

Pavel et al., 2012, USPTO Patent Application

14AIChE Spring Meeting, New Orleans, LA

Full Spectrum Consulting LLC

Visbreaker patent figure

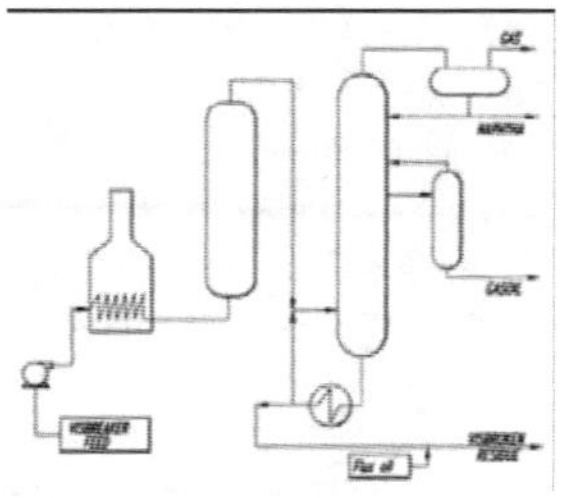

IyQ Patent figure

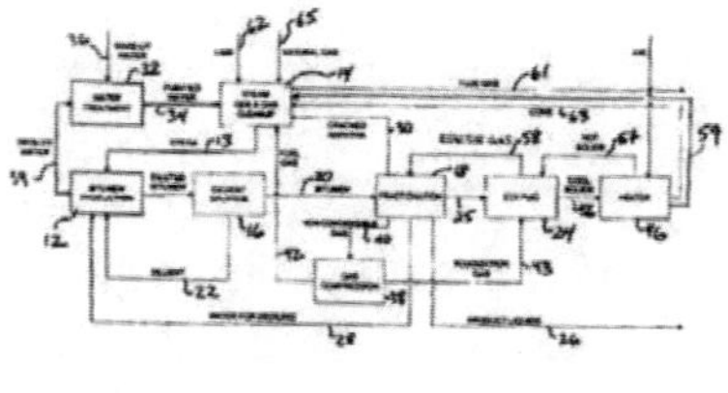

FIGURE 1

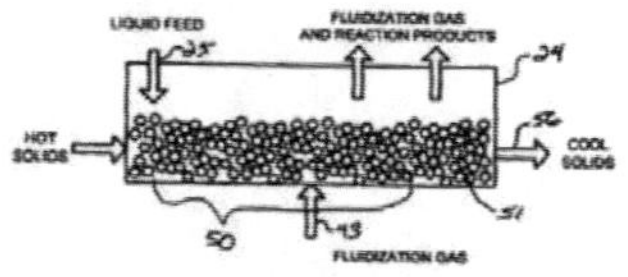

FIGURE 2

Heavy Oil Production and Upgrading HSE Considerations
HSE Considerations and Emissions

- Next sections address some of the exported Heavy Oil and Bitumen HSE considerations in press and journals:

- Caprock Integrity. Project Risk Identified for 5 of 155 projects, 3% of 221 in-situ projects affected.
- Flawed emission factors calculations vastly wrong . Actual emissions 4 to 9780 times estimated or permitted.
- LIDAR and DIAL for accurate emission measure, used in Europe, demonstrated in Canada and U.S.
- Petcoke, solid waste and toxic particulate distribution of refining waste. Petcoke not burned, it's only export.
- Distribution from oil sands production and upgraders of
- Oil sands Heavy Metal Emissions, particulates less than 2.5 microns, PM2.5s, downstream and downwind
- Oil sands PACs, Polycyclic Aromatic Compounds PM2.5 Emissions downstream and downwind
- Oil sands Mercury Emissions 7350 sq mile radius bulls eye radiating from oilsands and upgraders
- History of Heavy Metal, PAH, PM2.5 emissions, contamination and denial by government and emitters.
- Canadian Scientists Muzzled and Scripted by Government
- Air Emissions , Alberta allows carcinogenic, toxic, hazardous poisonous, recent example Peace River Baytex
- Government manipulation, misrepresentation of zero health effects associated with oil sands.
- Physician intimidation: Fort Chipewyan, Peace River, etc.; Alberta MD testimony in U.S.Senate
- TransCanada labeling environmental groups and individuals as Terrorists
- Canadian measurements, the quest for actual numbers
- Who Killed the Tar sands report, and We got that deleted
- Public Health Costs Omission and Denial
- RCMP Spying on Indigenous and Environmental Groups
- Keystone schedule
- Nebraska court, Health costs from exploration to finished product consumption, etc..

Full Spectrum Consulting LLC

14AIChE

Caprock Integrity? What's Causing Mysterious Bitumen Seepages in Alberta?

- What's Causing Mysterious Bitumen Seepages in Alberta?
- Spills show fragility of so-called in-situ, or steam-assisted, production, say experts.

14AIChE Spring Meeting, New Orleans, LA

Full Spectrum Consulting LLC

Next Oil Sands Threat: Cracking Caprock Risks of steam-assisted bitumen recovery are too little discussed, experts say.

- Short Caprock History:
- Texaco/1980s/Fort McMurray
-
- Imperial/1988/Cold Lake
- Total/2006/JoslynFrench multinational Total blasted a 300-metre crater
- CNRL/2009/Primrose EastAt Primrose East, Canadian Natural Resources Ltd.
- Devon/2010/Jackfish:In 2010,
- Imperial Oil/2013/Cold Lake
- CRNL/2013/Primrose/Cold Lake Seeping bitumen

14AIChE Spring Meeting, New Orleans, LA //www.resilience.org/print/2013-10-23/next-oil-sands-threat-cracking-caprock

Full Spectrum Consulting LLC

Next Oil Sands Threat: Cracking Caprock Risks of steam-assisted bitumen recovery are too little discussed, experts say.

- "A recent blow-out at Canadian Natural Resources Ltd.'s Primrose facility in northern Alberta sheds light on a serious but little discussed topic in the oil sands industry: caprock integrity.The blow-out allowed more than 10,000 barrels of steamed bitumen to seep into the boreal forest through ground fissures as long as 159 metres, putting groundwater at risk.
- …
- As a consequence of assorted fractures, well casing failures and leaks, Imperial is now investigating levels of benzene, toluene and ethylbenzene that exceed Canadian Drinking Water Guidelines in local groundwater. Imperial also reports numerous "bitumen in shale" incidents, where steamed bitumen finds its way into groundwater aquifers at depths of 153 metres or nearly 400 metres away from the target formation.
- Steam has mobilized arsenic in the region and forced an elaborate groundwater monitoring program. "

14AIChE Spring Meeting, New Orleans, LA

Full Spectrum Consulting LLC

Caprock Integrity, Shallow In-situ Regulatory Approach Pending

- "Regulatory Approach for Shallow Thermal In Situ Oil Sands Applications in the Wabiskaw-McMurray Deposit of the Athabasca Oil Sands Area
- Jan 28, 2014
- The Alberta Energy Regulator (AER) is currently completing a thorough technical review of the factors that affect reservoir containment of steam-assisted gravity drainage (SAGD) projects and will be consulting with stakeholders to develop formal regulatory requirements. Until those requirements are issued, the AER will defer decisions on applications for thermal oil sands projects that meet the following criteria:
- 1. Fall within a designated shallow thermal area of the Wabiskaw-McMurray Deposit in the Athabasca Oil Sands Area (see attached map).
- The AER believes that the risk of steam and reservoir fluids being released at surface is greater if reservoir containment is compromised in this area due to the shallow nature of the resource."

See more at AER website

14AIChE Spring Meeting, New Orleans, LA

Full Spectrum Consulting LLC

Caprock Integrity, Shallow In-situ Individual Project Elevated Risk Identified

Attachment Designated Shallow Thermal Area for Applications in the Wabiskaw-McMurray Deposit of the Athabasca Oil Sands Area

The designated shallow thermal area below represents the part of Alberta's in situ oil sands resources that is in the Athabasca Wabiskaw-McMurray Deposit within and next to the surface mineable area. The designated shallow thermal area is delineated where the Clearwater shale is shallower than 150 metres at its base or is absent, and where the net bitumen pay in the Athabasca Wabiskaw-McMurray Deposit is greater than zero.

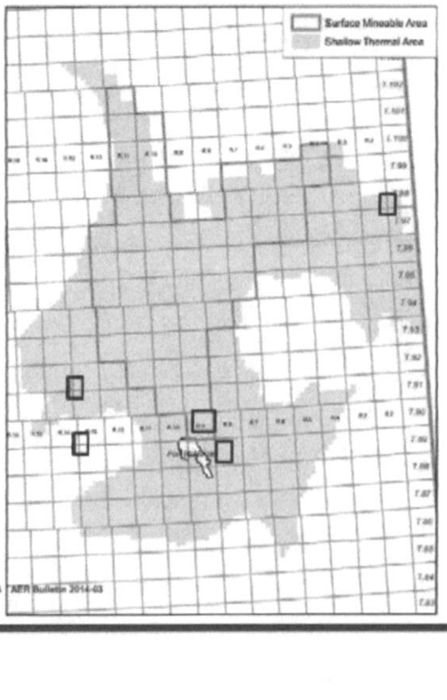

Only affects 5 projects out of the 221:
47 projects producing
17 projects under construction
38 projects with regulatory approval
70 projects under regulatory review
49 projects announced/disclosed
221Total above projects

Affects 5 projects AER ranked by risk. Out of 221 projects, only 3% of projects affected due to shallow location, 97% projects are not affected.

- **1) Producing Projects: 932,650 bpd**
- **2) 20 Projects Under Construction: 419,905 bpd**
- **3) 35 Projects with Regulator Approval: 809,650 bpd**
- **4) 55 Projects Under Regulator Review: 1,381,270 bpd**
- **5) 45 Projects Announced/Disclosed: 1,389,500 bpd**

Rough Map Location
1. T.90, R.09 Ivanhoe Energy-Tamarack
2. T.89, R.08 Value Creation-AdvancedTriStar
3. T.98, R.02 SilverWillow-Auden
4. T.89, R.15 Grizzly-Thickwood
5. T.91. R.14 Southern Pacific-MacKay SAGD Phase 2

14AIChE Spring Meeting, New Orleans, LA

Full Spectrum Consulting LLC

Caprock Integrity, Shallow In-situ

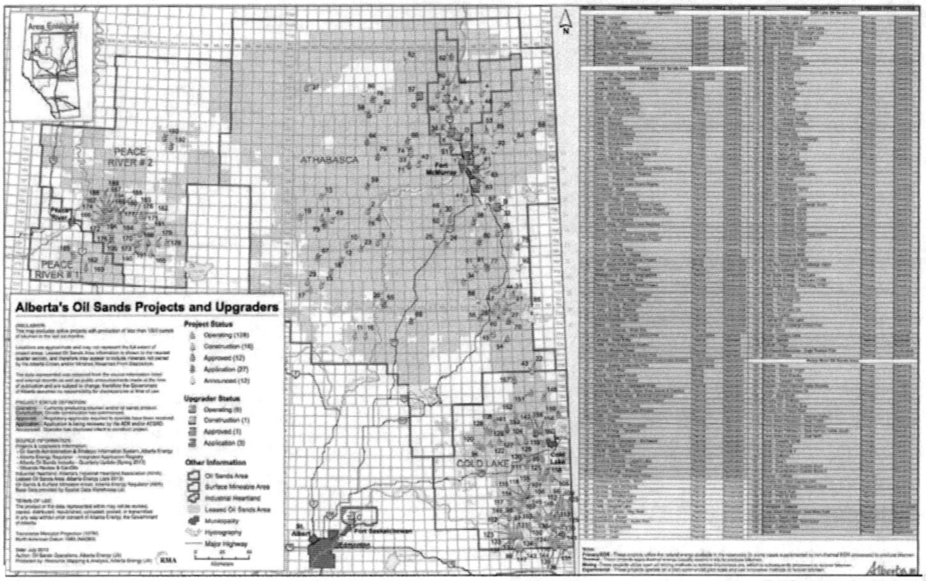

14AIChE Spring Meeting, New Orleans, LA

Full Spectrum Consulting LLC

Caprock Integrity: Shallow Thermal In-situ delayed projects ranked by AER with risk shown in circles

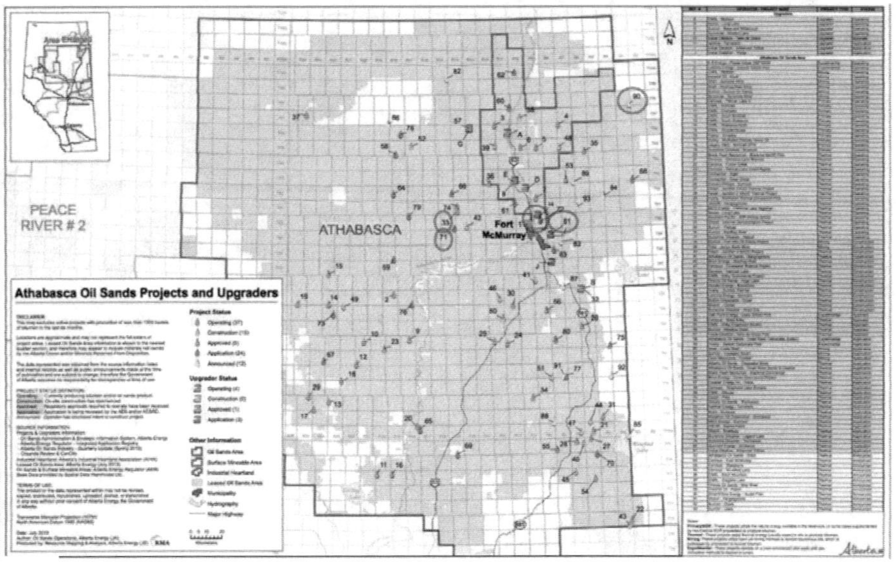

14AIChE Spring Meeting, New Orleans, LA

Full Spectrum Consulting LLC

Caprock Integrity, Shallow In-situ Regulatory Approach Pending

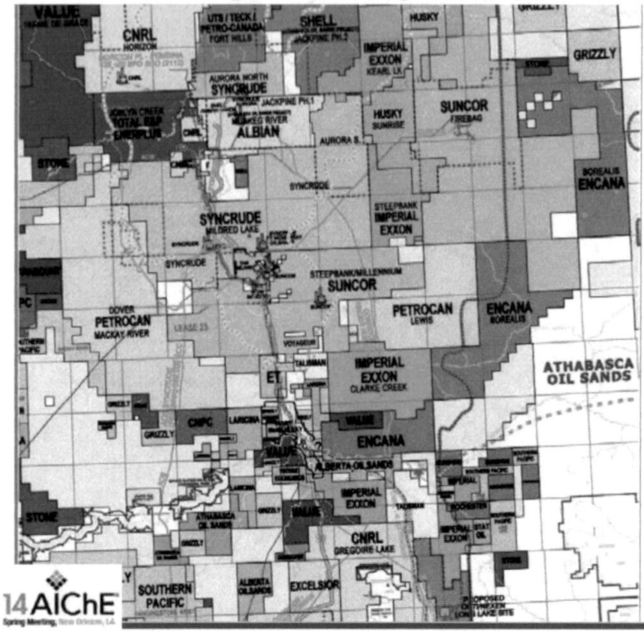

Only affects 5 insitu projects out of the 221 including:
47 operating
17 under construction
38 with regulatory approval
70 in regulatory review
49 projects announced
221 total projects
(OSDG, 2013.04)

<u>"… each project is being assessed individually based on its risk of reservoir containment problems."</u>

Affects 5 AER risk ranked projects out of 155 projects, only 0.3% of projects, 99.7% projects are not affected.

Full Spectrum Consulting LLC

14 AIChE Spring Meeting, New Orleans, LA

Next Oil Sands Threat: Cracking Caprock Risks of steam-assisted bitumen recovery are too little discussed, experts say.

- Short Caprock History:
- Texaco/1980s/Fort McMurray
-
- Imperial/1988/Cold Lake
- Total/2006/JoslynFrench multinational Total blasted a 300-metre crater
- CNRL/2009/Primrose EastAt Primrose East, Canadian Natural Resources Ltd.
- Devon/2010/Jackfish:In 2010,
- Imperial Oil/2013/Cold Lake
- CRNL/2013/Primrose/Cold Lake Seeping bitumen

14 AIChE Spring Meeting, New Orleans, LA

//www.resilience.org/print/2013-10-23/next-oil-sands-threat-cracking-caprock

Full Spectrum Consulting LLC

Next Oil Sands Threat: Cracking Caprock CRNL/2013/Primrose/Cold Lake Seeping bitumen

- "As a scientist, Timoney finds the lack of transparency dangerous, especially to the regulatory process.
- "This is one of the problems with the whole regulatory system, because ... AER just simply reports, apparently reports, what industry tells them. They don't do any checking."
- Reproducing industry figures in the name of public disclosure isn't much of a solution, he said.
- "I think the regulator has a responsibility to provide an accurate assessment of the company's activities. So if the regulator is not verifying information, it's just simply acting as a clearing house for information industry gives it, it's not doing its job. It's not acceptable.""
- He added: "It's a problem that's only gotten worse over the years, in the sense that now AER is basically a non-governmental entity. It's funded by industry. It's not an agent of the crown so we don't have the same sort of access to information we would if they were a government agency. So the AER can basically do whatever it wants to do and the public doesn't have any recourse. It's unbelievable, really, when you think about it."
- http://www.resilience.org/print/2013-10-23/next-oil-sands-threat-cracking-caprock
- Timoney, K., Lee, P. (2014.02.04). CNRL's Persistent 2013-2014 Bitumen Releases near Cold Lake, Alberta: Facts, Unanswered Questions, and Implications. Treeline Ecological Research and Global Forest Watch Canada. http://www.globalforestwatch.ca/pubs/2014Releases/02CNRLRelease/CNRL_Release_Bull

14AIChE

Full Spectrum Consulting LLC

Further Production and Upgrading

HSE Considerations: Caprock Integrity

- "Regulator defers thermal oilsands projects to adjust rules Applications on hold for five thermal in situ oilsands proposals. .... .
- "The bottom line on 3-D seismic is the fact that we need to have that information in front of us to ensure their projects can be completed safely," said Barter, who added **each project is being assessed individually based on its risk of reservoir containment problems."**

14AIChE Spring Meeting, New Orleans, LA

Full Spectrum Consulting LLC

Oil Sands Emissions and Health Risks

- Export of oil sands production technology with associated emissions need accurate calculation for downstream and downwind human impact
- Exporting untested partial upgrading technology to other countries with associated emissions must be accurately calculated for downstream and downwind population exposure.
- Use actual numbers, current emission factors are deceptive and misrepresentative with low bias estimates and proven to be 4 to 9870 times lower than actual emissions.

14 AIChE Spring Meeting, New Orleans, LA

Full Spectrum Consulting LLC

Houston mayor letter to EPA to correct emission factor errors

This Request for Correction is filed to ensure that the emission factors used by EPA to make air quality decisions that impact public health are accurate, reliable and unbiased, as required by law. The emission factors currently used by EPA significantly undercount emissions from petroleum refineries and chemical manufacturing plants.[1] Several studies have shown that actual emissions of volatile organic compounds (VOCs) and hazardous air pollutants (HAPs) from refineries and chemical manufacturing plants can be *100 times greater* than reported emissions based on EPA's emission factors, equations, and estimates.

VOCs and HAPs are air pollutants that have significant public health impacts, some of which are found in high concentrations in the air Houstonians breathe. VOCs contribute to the formation of ozone, and include toxic air pollutants like benzene, 1,3-butadiene,

and toluene. HAPs are certain toxic air pollutants that Congress has identified as presenting "a threat of adverse human health effects."[2] Certain HAPs, like benzene and 1,3 butadiene, are also VOCs. This request to correct the data quality errors in the emission factors pertaining to refineries and chemical manufacturing plants is of particular urgency in light of the recently revised ozone standard and the significant amount of toxic emissions that communities situated near these facilities are exposed to. Without accurate emissions data, local, state, and federal governments are not able to make air quality management decisions that are protective of public health. Residents of Houston and its environs are particularly adversely impacted by the inaccurate inventories because of the large concentration of refining and chemical manufacturing facilities in close proximity to one another and to the more than 5 million residents of this area. Indeed, even the current under-counted inventories reported in the TRI show that Harris County, in which Houston is located, has the highest level of industrial benzene emissions of any county in the U.S. and is among the counties with the highest emissions of HAPs from industrial sources.[3]

14 AIChE Spring Meeting, New Orleans, LA

Full Spectrum Consulting LLC

Houston mayor letter to EPA to correct emission factor errors

- ***EPA's Own Admissions and Scientific Studies Using Remote Sensing Technologies and Airborne Measurements Confirm that the Emission Factors are Inaccurate.*** One study found that measured emissions from a refinery were *100 times greater* than emission estimates based on emission factors.[21]
- ***The Emission Factors Fail to Account for Emissions Generated During Startup, Shutdown, and Malfunctions ("SSM[22]"), and Increased Emissions that Result from Poor Maintenance of Equipment.*** These uncounted emissions can be significant. The flare emissions produced during just one SSM event can *exceed* annual average emissions.[23]
- ***The Emission Factors Fail to Account for Environmental Variables that Significantly Impact Emissions.*** Emissions from storage tanks and flares increase significantly as wind speed increases. A study of one refinery using remote sensing technology estimated annual VOC emissions from tankage at 5,000 tons as opposed to the 153 tons estimated using emission factors.[24]
- ***The Emission Factors Do Not Account for Emissions from Delayed Coker Units.*** EPA has acknowledged, and other scientific studies have confirmed, that delayed coker units can be one of the largest sources of benzene emissions at refineries.[25]
- ***EPA Authorizes the Use of Emission Factors that EPA's own quality system determines are unreliable and likely to be inaccurate.*** EPA uses emission factors that EPA rates as "below average," "poor," or "unratable" quality even though EPA knows the data generated by the emission factor is likely to be inaccurate and unreliable.
- ***The Emission Factors Are so Fundamentally Flawed they Produce Emissions Data that is of No Utility to Decision Makers.*** The inaccuracy, unreliability, and bias of the emission factors undermines the EPA and other agencies in making health-protective decisions based on this data.

HSE Priority Pollutants, toxic, hazardous, poisonous, carcinogenic

check application, especially thermal cracking emissions, particulates

1. Acenaphthene
2. Acrolein
3. Acrylonitrile
4. Benzene
5. Benzidine
6. Carbon tetrachloride
7. Chlorobenzene
8. 1,2,4-trichlorobenzene
9. Hexachlorobenzene
10. 1,2-dichloroethane
11. 1,1,1-trichloreothane
12. Hexachloroethane
13. 1,1-dichloroethane
14. 1,1,2-trichloroethane
15. 1,1,2,2-tetrachloroethane
16. Chloroethane
17. REMOVED
18. Bis(2-chloroethyl) ether
19. 2-chloroethyl vinyl ethers
20. 2-chloronaphthalene
21. 2,4,6-trichlorophenol
22. Parachlorometa cresol
23. Chloroform
24. 2-chlorophenol
25. 1,2-dichlorobenzene
26. 1,3-dichlorobenzene
27. 1,4-dichlorobenzene
28. 3,3-dichlorobenzidine
29. 1,1-dichloroethylene
30. 1,2-trans-dichloroethylene
31. 2,4-dichlorophenol
32. 1,2-dichloropropane
33. 1,3-dichloropropylene
34. 2,4-dimethylphenol
35. 2,4-dinitrotoluene
36. 2,6-dinitrotoluene
37. 1,2-diphenylhydrazine
38. Ethylbenzene
39. Fluoranthene
40. 4-chlorophenyl phenyl ether
41. 4-bromophenyl phenyl ether
42. Bis(2-chloroisopropyl) ether
43. Bis(2-chloroethoxy) methane
44. Methylene chloride
45. Methyl chloride
46. Methyl bromide
47. Bromoform
48. Dichlorobromomethane
49. REMOVED
50. REMOVED
51. Chlorodibromomethane
52. Hexachlorobutadiene
53. Hexachlorocyclopentadiene
54. Isophorone
55. Naphthalene
56. Nitrobenzene
57. 2-nitrophenol
58. 4-nitrophenol
59. 2,4-dinitrophenol
60. 4,6-dinitro-o-cresol
61. N-nitrosodimethylamine
62. N-nitrosodiphenylamine
63. N-nitrosodi-n-propylamine
64. Pentachlorophenol
65. Phenol
66. Bis(2-ethylhexyl) phthalate
67. Butyl benzyl phthalate
68. Di-N-Butyl Phthalate
69. Di-n-octyl phthalate
70. Diethyl Phthalate
71. Dimethyl phthalate
72. benzo(a) anthracene
73. Benzo(a)pyrene
74. Benzo(b) fluoranthene
75. Benzo(k) fluoranthene
76. Chrysene
77. Acenaphthylene
78. Anthracene
79. Benzo(ghi) perylene
80. Fluorene
81. Phenanthrene
82. Dibenzo(,h) anthracene
83. Indeno (1,2,3-cd) pyrene
84. Pyrene
85. Tetrachloroethylene
86. Toluene
87. Trichloroethylene
88. Vinyl chloride
89. Aldrin
90. Dieldrin
91. Chlordane
92. 4,4-DDT
93. 4,4-DDE
94. 4,4-DDD
95. Alpha-endosulfan
96. Beta-endosulfan
97. Endosulfan sulfate
98. Endrin
99. Endrin aldehyde
100. Heptachlor
101. Heptachlor epoxide
102. Alpha-BHC
103. Beta-BHC
104. Gamma-BHC
105. Delta-BHC
106. PCB-1242 (Arochlor 1242)
107. PCB-1254 (Arochlor 1254)
108. PCB-1221 (Arochlor 1221)
109. PCB-1232 (Arochlor 1232)
110. PCB-1248 (Arochlor 1248)
111. PCB-1260 (Arochlor 1260)
112. PCB-1016 (Arochlor 1016)
113. Toxaphene
114. Antimony
115. Arsenic
116. Asbestos
117. Beryllium
118. Cadmium
119. Chromium
120. Copper
121. Cyanide, Total
122. Lead
123. Mercury
124. Nickel
125. Selenium
126. Silver
127. Thallium
128. Zinc
129. 2,3,7,8-TCDD

Full Spectrum Consulting LLC

Houston Industrial, Shipping, Offshore, etc. Emissions
What you don't know, can't see, and can't smell can kill you
Houston Cancer Map by Census Tract; Port Arthur on right side

AIR TOXICS, ONLINE

How You, Too, Can Browse Interactive Maps of Houston's Cancer Hotspots for Fun and Panic

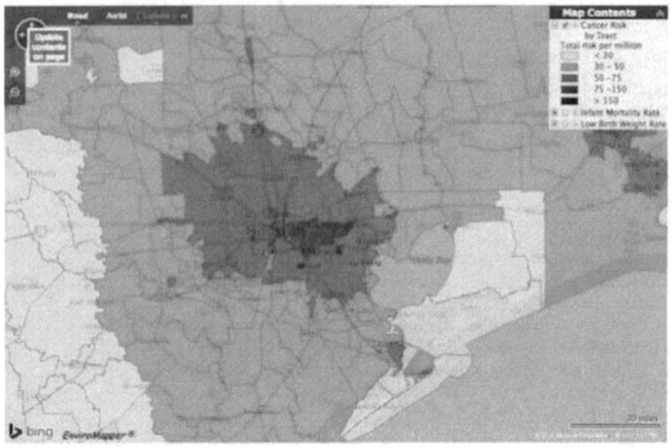

11 Comments

02/25/14 4:00pm

Air Pollution, Air Quality, Hazards, Maps, Online Resources, Public Health

Like Share 26

Tweet 7

8+1 0

Submit

tumblr. +

Pin it

- (EPA, 2014)

14 AIChE Spring Meeting, New Orleans, LA

Full Spectrum Consulting LLC

Houston Industrial, Shipping, Offshore, etc. Emissions
What you don't know, can't see, and can't smell can kill you
Houston Cancer Map by Census Tract

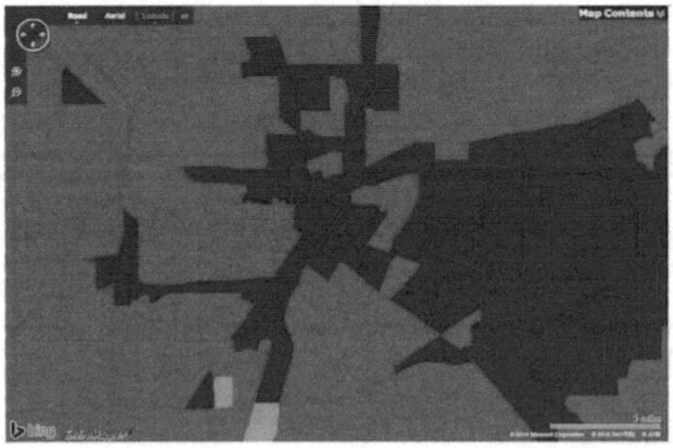

Notably, Houston's max dark blue zone as mapped by the EPA site extends not just over the sorta-bluish waters of the **Houston Ship Channel,** but across Downtown as well, and even into the steadily beating heart of the **Texas Medical Center.** The numbers for this danger zone, 75-150, represent the risk (per million persons) of someone developing any type of cancer over a lifetime — above and beyond, that is, the risk of anyone else not exposed to the same measured toxins in that area.

14 AIChE Spring Meeting, New Orleans, LA

Full Spectrum Consulting LLC

Particulate Matter less than 2.5 microns,
PM2.5s, are known as invisible killers.

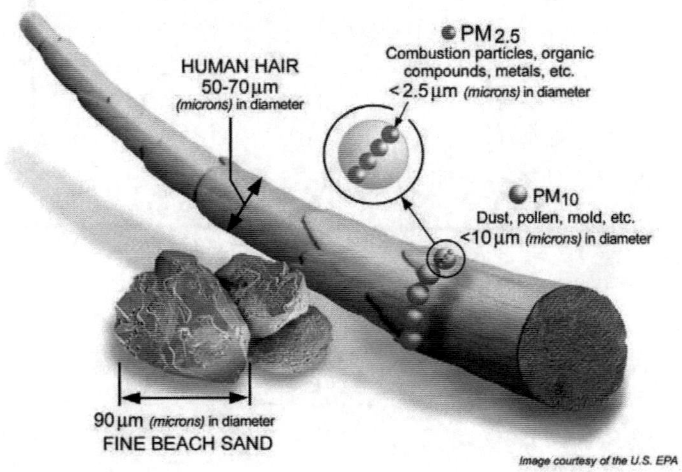

- (EPA, 2014)

14AIChE Spring Meeting, New Orleans, LA

Full Spectrum Consulting LLC

Particulate matter less than 2.5 microns, PM2.5s, are known as invisible killers.

There is no safe level of PM2.5s for any air breather. Premature death from PM2.5s cuts 12 years off life.

PM2.5s are easily measured and seen in the air, often labeled as haze. Higher PM2.5s, lower visibility.

Particulate emissions, ozone impact on Houston public health:
Out of hospital cardiac arrests: 90% fatal, each dot a life

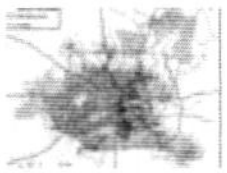

- Cardiac Arrests match and correlate to PM2.5s and Ozone

14AIChE Spring Meeting, New Orleans, LA

Full Spectrum Consulting LLC

Particulate emissions, ozone impact on Houston public health: each dot a cardiac arrest, 90% fatal
Out of hospital cardiac arrests: 90% fatal, each dot a life; I knew 3 dots

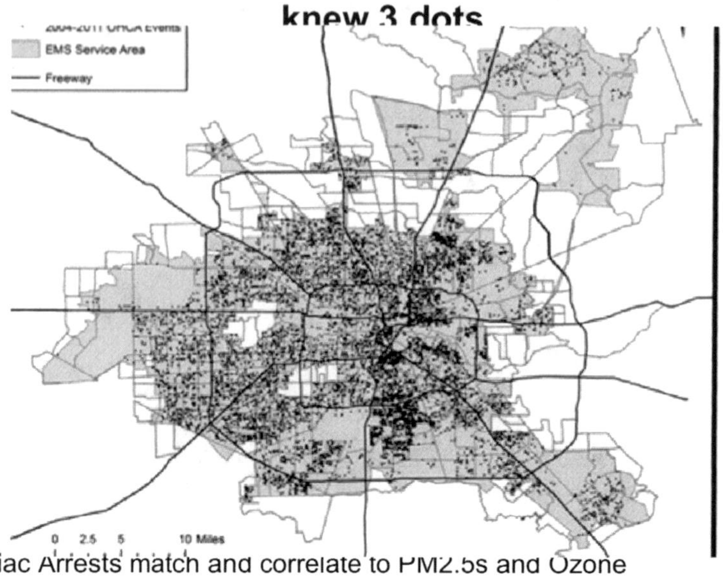

- Cardiac Arrests match and correlate to PM2.5s and Ozone
- (Ensor, Raun, 2013)

14 AIChE

Full Spectrum Consulting LLC

Particulate matter less than 2.5 microns, PM2.5s, are known as invisible killers.

There is no safe level of PM2.5s for any air breather.
Premature death from PM2.5s cuts 12 years off life.

Particulate emissions, ozone impact on Houston public health:
Out of hospital cardiac arrests: 90% fatal, each dot a life

Personally at Funerals of 3 of the dots:

1. High School football coach, massive funeral, college players
2. Father of five children all headed to college.
3. Colleague from API Chapter 17.2 committee

- Cardiac Arrests match and correlate to PM2.5s and Ozone

14 AIChE

Full Spectrum Consulting LLC

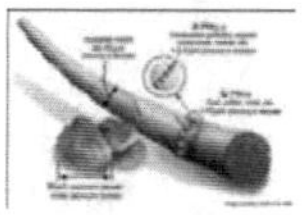

Particulate Matter less than 2.5 microns, PM2.5s cause premature deaths, avg 12 yrs early

Researchers estimate over two million deaths annually from air ...
https://www.iop.org/news/13/jul/page_60518.html - Institute of Physics
Jul 12, 2013 - Over two million **deaths** occur **each year** as a direct result of ... caused **each year** by human-caused increases in fine **particulate matter** (**PM2.5**) ...

WHO | Air quality and health
www.who.int/mediacentre/factsheets/.../en/ - World Health Organization
Almost half of these **deaths** are due to pneumonia in children under 5 years ... Urban outdoor air pollution is estimated to cause 1.3 million **deaths** worldwide **per year**. ... The 2005 AQG set for the first time a guideline value for **particulate matter** ...

[PDF] **Overview Fact Sheet (PDF) - US Environmental Protection Agency**
www.epa.gov/.../decfsov... - United States Environmental Protection Agency
Dec 14, 2012 - FOR PARTICLE POLLUTION (**PARTICULATE MATTER**) ... fine particle pollution by strengthening the **annual** health National ... pollution, also known as fine **particulate matter** (**PM2.5**), can cause premature **death** and harmful.

The Global Toll of Fine Particulate Matter : Image of the Day
earthobservatory.nasa.gov/IOTD/view.php?id... - NASA Earth Observatory
Sep 19, 2013 - In most **cases**, the most toxic pollution lingers for a few days or even ... of **deaths** per 1,000 square kilometers (386 square miles) **per year** due ...

Air Pollution Causes 2 Million Deaths Yearly | LiveScience
www.livescience.com/38125-air-pollution-global-de... - LiveScience
by Rachael Rettner - in 297 Google+ circles
Jul 11, 2013 - The study estimated that 2.1 million **deaths each year** are linked with fine **particulate matter**, tiny particles that can get deep into the lungs and ...

[PDF] **Stronger Relationship Between Particulate Matter (PM) and ...**
www.arb.ca.gov/research/.../march06.pdf - California Air Resources Board
found **a** stronger effect of **particulate matter** on premature **death**. Before ... In addition, U.S. EPA has estimated **a** value of life at $7.9 million in the **year** 2005.

[PDF] **Estimate of Premature Deaths Associated with Fine Particle Poll...**
www.arb.ca.gov/.../pm-report_2010.pdf - California Air Resources Board
Aug 31, 2010 - Premature **Deaths** in California Associated with Current **PM2.5** .... California by Attainment of the **Annual**-Average Federal **PM2.5** NAAQS ...

Death by Particles: The Link Between Air Pollution and Fatal ...
www.ncbi.nlm.nih.gov/... - National Center for Biotechnology Information
Death by Particles: The Link Between Air Pollution and Fatal Coronary Heart ... to air pollution—especially **particulate matter** (PM)—with mortality resulting from **a** ... by **a** team of epidemiologists at Loma Linda University, is part of the **22-year** ...

14 AIChE Spring Meeting, New Orleans, LA

Full Spectrum Consulting LLC

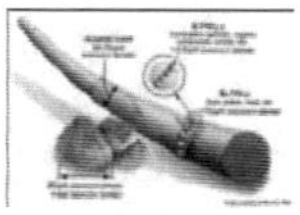

Particulate Matter less than 2.5 microns, a very bad air day in Houston

14 AIChE Spring Meeting, New Orleans, LA

Full Spectrum Consulting LLC

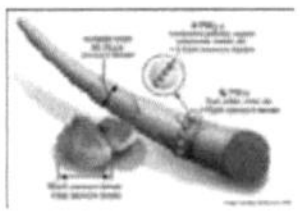

Particulate Matter less than 2.5 microns, a bad air day in Houston

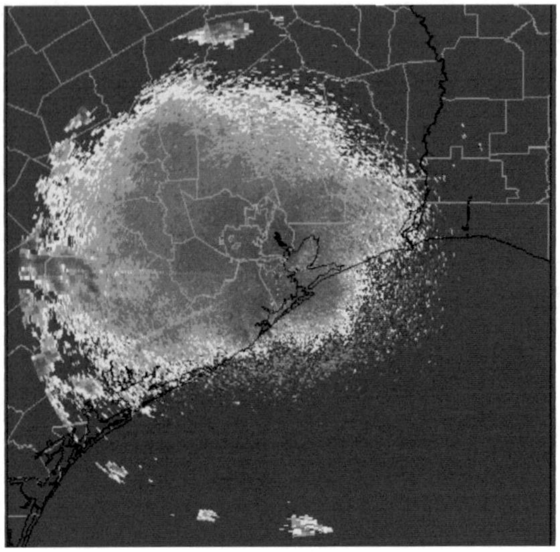

14AIChE Spring Meeting, New Orleans, LA

Full Spectrum Consulting LLC

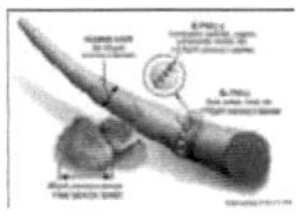

Particulate Matter less than 2.5 microns, "good day" by TCEQ misleading AQI, still PM2.5s contributing to early deaths

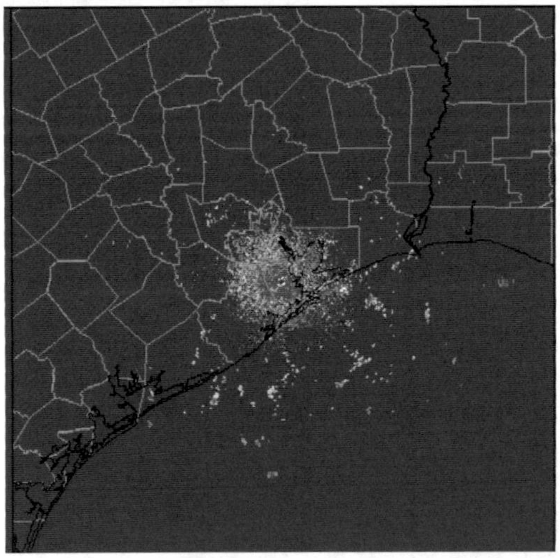

14AIChE Spring Meeting, New Orleans, LA

Full Spectrum Consulting LLC

Particulate Matter less than 2.5 microns, PM2.5s cause children pain, cost billions

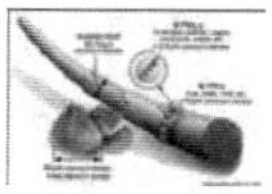

PM2.5s are also associated with Otitis Media, the ear infection that parents know from the whimpers and screams of their children in the middle of night.

AIChE

Full Spectrum Consulting LLC

Channel Industrial Point Sources combine to create ozone plume and clouds. This one heading northwest to the Woodlands, usual path is to west below I-610.

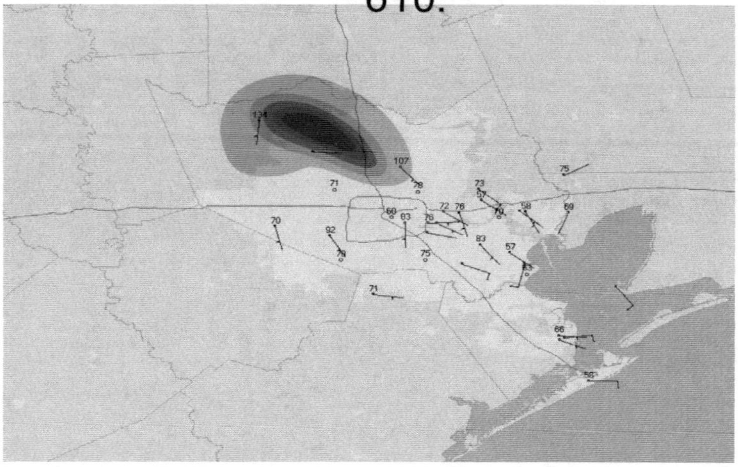

Thousands of ozone track charts were removed from the TCEQ, Texas Commission of Environmental Quality Website, This one from a saved file page.

- http://www.tceq.state.tx.us/assets/public/compliance/monops/air/sigevents/01/011002hou-ani.html

DIAL Measurements of Fugitive Emissions

- DIAL Measurements of Fugitive Emissions from Natural Gas Plants and the Comparison with Emission Factor Estimates

- Allan K. Chambers, Mel Strosher
- Alberta Research Council Inc., 250 Karl Clark Rd., Edmonton, AB, Canada T6N 1E4 allan.chambers@arc.ab.ca
- Tony Wootton, Jan Moncrieff and Philip McCready
- Spectrasyne Ltd., Suite 22, Worting House, Church Lane, Basingstoke, Hampshire, RG23 8PX, U.K. info@spectrasyne.ltd.uk

- http://www.epa.gov/ttnchie1/conference/ei15/session14/chambers.pdf

DIAL Measurements of Fugitive Emissions Used at Canadian Gas Processing Plants

- ABSTRACT

- Natural gas processing is a major industry in Alberta, Canada, and a significant source of fugitive emissions of both methane and volatile organic hydrocarbon (VOCs). This project investigated fugitive emissions at natural gas processing plants in Alberta using two complementary optical measurement methods. At five gas plants, the fugitive emissions of methane and hydrocarbons ethane and larger (C2+ hydrocarbons) were measured and quantified using Differential Absorption Lidar (DIAL). The DIAL was also used to measure emissions from process flares at two of the gas plants. At two of the plants, a gas leak imaging camera was used to locate individual hydrocarbon leaks.

- For the five gas plants surveyed in Alberta, DIAL measured methane emissions ranged from 100 to 146 kg/hr and C2+ hydrocarbon emissions ranged from 38 to 342 kg/hr. Compressors and condensate storage tanks were two significant emission sources at all of the gas plants. Process flares operating on pilot were typically responsible for 10 to 15% of the total methane emissions.

- At two gas plants the DIAL measured emissions of methane, VOCs and benzene were compared with values calculated using emission factor methods. Measured emissions of methane and VOCs were four to eight times higher than the emission factor estimates. The largest differences between measured values and estimates were for the flares and storage tanks. DIAL measured values gave a more realistic evaluation of revenue lost as fugitives than the industry accepted estimation methods, leading to an increased incentive to improve leak detection and repair.

- http://www.epa.gov/ttnchie1/conference/ei15/session14/chambers.pdf

DIAL Measurements of Fugitive Benzene Emissions

Figure 3 Concentration map of benzene emissions.

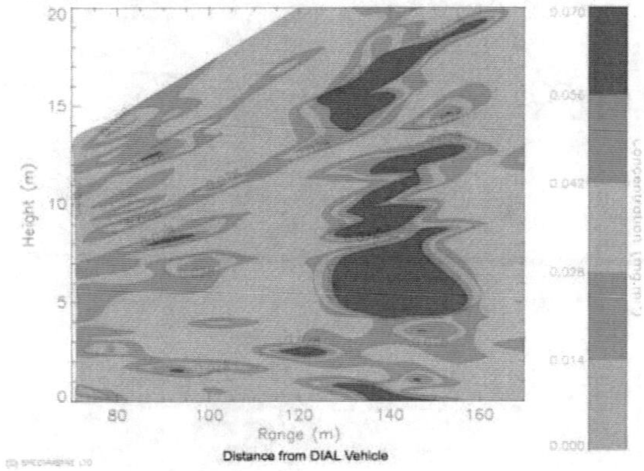

- http://www.epa.gov/ttnchie1/conference/ei15/session14/chambers.pdf

Typical Distribution of Refinery VOC Emissions Based on DIAL Measurements

- Typical location of emissions from a refinery based on a report from Spectrasyne who has completed over 30 refinery studies.
- Results will vary significantly depending on refinery design.
- All calculations using emission factors are wrong.

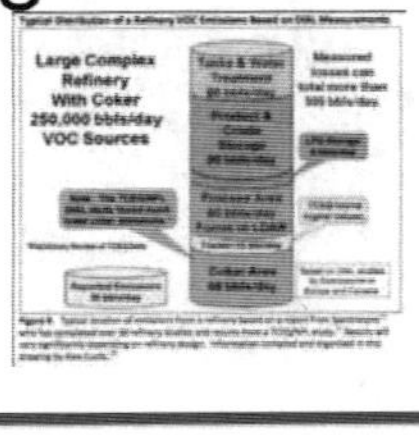

Typical Distribution of Refinery VOC Estimate versus Actual Emissions Based on DIAL Measurements

Typical Distribution of a Refinery VOC Emissions Based on DIAL Measurements

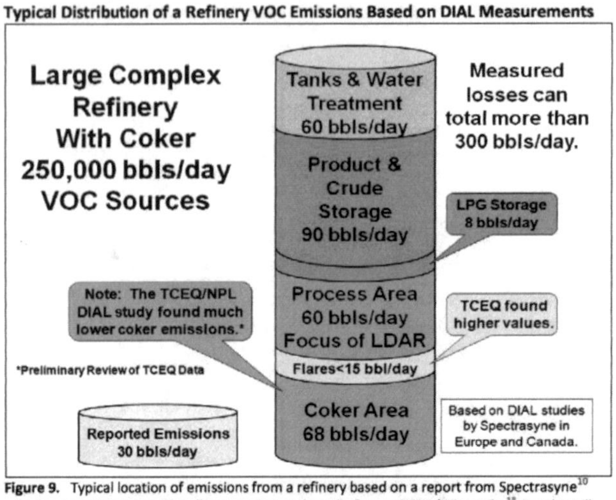

Figure 9. Typical location of emissions from a refinery based on a report from Spectrasyne[10] who has completed over 30 refinery studies and results from a TCEQ/NPL study.[13] Results will vary significantly depending on refinery design. Information compiled and organized in this drawing by Alex Cuclis.[40]

- http://www.epa.gov/ttn/chief/conference/ei20/session7/acuclis.pdf

Typical Distribution of Refinery VOC Estimate versus Actual Emissions Based on DIAL Measurements

- **Examples of carcinogenic benzene emissions include:**

- **Refinery Delayed Coker, measured up to 9780 times "Texas Flexible permit."**
- Measured....3.3 to 48.9 lbs/hr
- Texas Flexible Permit....0.005 lbs/hr

- **Chemical plant pyrolysis gasoline storage tank, measured up to 18 times permit**
- Measured....1.2 to 32.5 lbs/hr
- Texas Flexible Permit....1.83 lbs/hr

- **Benzene emissions also appeared to originate from an unpermitted tank**

- **Refinery Wastewater Treatment Basins measured up to 13 times permit.**
- Measured....2.6 to 10.3 lbs/hr
- Flexible Permit...0.82 lbs/hr

“Alberta Hides Dirty Truth as US Demands Tar Sands Facts”

- Alberta Hides Dirty Truth as US Demands Tar Sands Facts for Keystone XL, and Investment Decisions.
- “Potential buyers of tar sands oil want to know its true carbon footprint, but industry won't come clean.”

- (Nikiforuk, 2010)

14 AIChE Spring Meeting, New Orleans, LA

Full Spectrum Consulting LLC

Tar sands result in heavy metals distribution.

- Tar sands result in heavy metals distribution.
- Oil sands development contributes elements toxic at low concentrations to the Athabasca River and its tributaries

- http://thetyee.ca/News/2010/08/30/TarSandsStudy/ accessed 12.21.2014. and
- http://www.pnas.org/content/107/37/16178.full.pdf?with-ds=yes

14 AIChE Spring Meeting, New Orleans, LA

Full Spectrum Consulting LLC

Tar sands result in heavy metals distribution downstream and downwind of production and upgrading.

- "We show that the oil sands industry releases the 13 elements considered priority pollutants (PPE) under the US Environmental Protection Agency's Clean Water Act, via air and water, to the Athabasca River and its watershed. In the 2008 snowpack, all PPE except selenium were greater near oil sands developments than at more remote sites. Bitumen upgraders and local oil sands development were sources of airborne emissions. Concentrations of mercury, nickel, and thallium in winter and all 13 PPE in summer were greater in tributaries with watersheds more disturbed by development than in less disturbed watersheds. In the Athabasca River during summer, concentrations of all PPE were greater near developed areas than upstream of development. At sites down- stream of development and within the Athabasca Delta, concentrations of all PPE except beryllium and selenium remained greater than upstream of development. Concentrations of some PPE at one location in Lake Athabasca near Fort Chipewyan were also greater than concentration in the Athabasca River upstream of development. Canada's or Alberta's guidelines for the protection of aquatic life were exceeded for seven PPE—cadmium, copper, lead, mercury, nickel, silver, and zinc—in melted snow and/or water collected near or downstream of development. ...
- Conclusions. Contrary to claims made by industry and government in the popular press, the oil sands industry substantially increases loadings of toxic PPE to the AR and its tributaries via air and water pathways. This increase confirms the serious defects of RAMP (11–13), which has not detected such patterns in the AR watershed. Detailed long-term monitoring is essential to distinguish the sources of these contaminants and control their po- tential impacts on environmental and human health (13). A robust monitoring program to measure exposure and health of fish, wildlife, and humans should be implemented in the region affected by oil sands development (38, 39)"
- http://thetyee.ca/News/2010/08/30/TarSandsStudy/ accessed 12.21.2014.
- http://www.pnas.org/content/107/37/16178.full.pdf?with-ds=yes

Tar sands result in fish killing toxins, heavy metals poisoning the fish.

- Despite repeated government claims that the world's largest energy project doesn't contaminate the Athabasca River, a new scientific study released today shows that air pollution from the oil sands industry combined with extensive watershed destruction has released a highly toxic brew of heavy metals into northern waterways.
- The study, published in the prestigious Proceedings of the National Academy of Sciences (PNAS),
- The heavy metals, rated as priority pollutants by the U.S. Environmental Protection Agency, include mercury, arsenic, beryllium, copper, cadmium, thallium, lead, nickel, zinc and silver. All are toxic. Arsenic is a known human carcinogen; cadmium can severely destroy the kidneys and other organs; and thallium is so poisonous that it pops up in Agatha Christie mysteries as a murder weapon.
- The study directly contradicts claims by the Alberta government and an industry-run monitoring program that pollution in the river basically comes from the erosion of natural outcrops of metal-rich bitumen along the region's river banks. In particular, Preston McEachern, head of science and innovation with Alberta Environment, has repeatedly alleged that "contamination in area soils and rivers is natural and poses no serious health risk."
- But the chemical analysis of water and snow samples taken from a total of 47 sites shows that industry "substantially increases loadings of toxic priority pollutants to the Athabasca River."

Tar sands result in PAH distribution

- Tar sands result in Polycyclic Aromatic Hydrocarbons, PAH, distribution.
- Oil sands development contributes polycyclic aromatic compounds to the Athabasca River and its tributaries

Tar sands result in Mercury distribution.

- Tar sands result in Mercury distribution.
- Researchers find 7,300-sq-mile ring of mercury around tar sands in Canada

Tar sands result in Mercury distribution.

- ""Bitumen in the tar sands being excavated to produce oil is the likely culprit of the mercury deposits"
- Scientists have found a more than 7,300-square-mile ring of land and water contaminated by mercury surrounding the tar sands in Alberta, where energy companies are producing oil and shipping it throughout Canada and the U.S.
- Government scientists are preparing to publish a report that found levels of mercury are up to 16 times higher around the tar-sand operations — principally due to the excavation and transportation of bitumen in the sands by oil and gas companies.
- The revelations add to growing concerns over the environmental impact of mining the tar sands. Many environmentalists charge that extracting oil from the sands will lead to an increase in carbon emissions, the destruction of the land, water contamination and health problems for Canadians. The debate over the tar sands crossed over into the United States when energy company TransCanada proposed building the Keystone XL pipeline to transport crude oil to the southeastern U.S. for refining and distribution.
- Kirk and her colleagues' research shows that the development of the tar sands may be responsible for spreading mercury — which can cause nervous-system damage — far beyond the areas where drilling and transportation are taking place.
- The research suggests that the tar-sand development has created a ring of mercury contamination, with areas close to the sands showing much higher levels of mercury than before development."

14AIChE

Full Spectrum Consulting LLC

History of environmental contamination by oil sands extraction

- History of environmental contamination by oil sands extraction
- Legacy of a half century of Athabasca oil sands development recorded by lake ecosystems

- Kureka, J., Hodson, P., Kirk, J., Muir, D., Wang, X.,Evans, M., and Smola, J. (2012) History of environmental contamination by oil sands extraction. Legacy of a half century of Athabasca oil sands development recorded by lake ecosystems. .http://www.ncbi.nlm.nih.gov/pmc/articles/PMC3562832/pdf/pnas.201221660.pdf http://www.pnas.org/content/early/2013/01/02/1217675110 and http://www.pnas.org/content/suppl/2013/01/02/1217675110.DCSupplemental/pnas.201217675SI.pdf http://www.pnas.org/content/110/5/1761.full.pdf+html?sid=63c119e4-bce1-4c05-88ce-651ec586e525 . Accessed 12.21.2014 .

14AIChE Spring Meeting, New Orleans, LA

Full Spectrum Consulting LLC

'Constant Denial': Report on Tar Sands Is Scathing

- The water footprint of Canada's greatest engineering project is enormous. According to Natural Resources Canada, the tar sands mining industry, which produces 1.3 million barrels of bitumen a day, now uses 12 barrels of freshwater to produce one barrel of the asphalt-like heavy crude.
- The industry has also constructed some of the world's largest dams along the Athabasca River to contain 170 square kilometres of toxic mining waste. That's enough sludge to fill a 10-by-10 metre canal across the U.S.-Canadian border from sea to sea. According to public testimony to the parliamentary committee, at least two of these toxic lakes are leaking into groundwater.
- Unlike the mines, which will impact waterways and wetlands over a 4,000 kilometre area, the steam plants could affect aquifers over an area the size of Florida. These energy-intensive operations, which melt deep deposits of bitumen with steam, typically consume 3.5 barrels of groundwater to extract one barrel of bitumen. Some operations, such as Opti-Nexen, use up to six barrels.
- 'There was constant denial'
- "What surprised me during the hearings was the defensive posturing of government. There was constant denial that there could be anything wrong," says the report's author, Francis Scarpaleggia, the federal MP for the Montreal West Island riding of Lac Saint-Louis.

14 AIChE Spring Meeting, New Orleans, LA — Full Spectrum Consulting LLC

'We Got that Deleted': Canada's Oil Sands Lobby Twisting Washington's Arm

- **“'We Got that Deleted': Canada's Oil Sands Lobby Twisting Washington's Arm**
- US politicians bend to foreign-backed pressure to soften climate bill.”
- (Dembicki, 2010)

14 AIChE Spring Meeting, New Orleans, LA — Full Spectrum Consulting LLC

'We Got that Deleted': Canada's Oil Sands Lobby Twisting Washington's Arm

- "As U.S. senators debate some of the most sweeping climate change laws in American history, a powerful lobbying effort led by Canadian officials and huge oil firms may be winning big concessions.
- Their goal is to keep Alberta's generous "oil sands" reserves flowing south through pipelines. It's a crucial and stable energy supply for the U.S., but one often lambasted for high carbon emissions, strip-mined landscapes and giant ponds of toxic sludge.
- Any attempt to close the spigot is fought by an informal coalition that includes Albertan diplomats, Big Oil lobbyists and Canadian cabinet ministers.
- A key architect of that effort, Tom Corcoran, whose lobby group has some of the biggest oil companies in the world as members, details later in this story how past battles have been waged and won, and why he can say: "We've been successful.""
- ExxonMobil 60 MM, Koch Bros 120 MM, etc. 220 MM.

14AIChE Spring Meeting, New Orleans, LA

Full Spectrum Consulting LLC

PETCOKE

- The Canadian tar sands have been called the "most environmentally destructive project on earth", with good reason. Extracting tar sands bitumen from under the boreal forests of Alberta, Canada requires huge amounts of energy and water. It has cleared vast tracts of forest, left scars on the land that are visible from space and threatened the health and livelihoods of indigenous First Nations communities across the region.
- it is a well established fact that full exploitation of the tar sands is a grave threat to the climate. emissions from tar sands extraction and upgrading are between 3.2 and 4.5 times higher than the equivalent emissions from conventional oil produced in north america. on a lifecycle basis, the average gallon of tar sands bitumen derived fuel has between 14 and 37 percent more greenhouse gas emissions than the average gallon of fuel from conventional oil.
- But as bad as these impacts already are, existing analyses of the impacts of tar sands fail to account for a byproduct of the process that is a major source of climate change causing carbon emissions: petroleum coke - known as petcoke. Petcoke is the coal hiding in north america's tar sands oil boom.
- **Petcoke is like coal, but dirtier.** Petcoke looks and acts like coal, but it has even higher carbon emissions than already carbon-intensive coal.
- on a per-unit of energy basis petcoke emits 5 to 10 percent more
- carbon dioxide than coal.
- a ton of petcoke yields on average 53.6 percent more co2 than a
- ton of coal.
- the proven tar sands reserves of canada will yield roughly 5 billion
- tons of petcoke – enough to fully fuel 111 U.S. coal plants to 2050.
- Because it is considered a refinery byproduct, petcoke emissions are
- not included in most assessments of the climate impact of tar sands or conventional oil production and consumption. thus the climate impact of oil production is being consistently undercounted.
- **Petcoke in the tar sands is turning American refineries into coal factories.**
  - there is 24 percent more co2 embedded in a barrel of tar sands bitumen than in a barrel of light oil.
  - 15 to 30 percent of a barrel of tar sands bitumen can end up as petcoke, depending on the upgrading and refining process used.
  - of 134 operating U.S. refineries in 2012, 59 are equipped to produce petcoke.
  - U.S. refineries produced over 61.5 million tons of petcoke in 2011 - enough to fuel 50 average U.S. coal plants each year.
  - Koch carbon, Oxbow, "the" dealer of petcoke, most to gain from Keystone pipeline

14AIChE Spring Meeting, New Orleans, LA

Full Spectrum Consulting LLC

PETCOKE FROM BITUMEN

- **keystone xl will fuel five coal plants and thus emit 13% more Co2**
- **than the u.S. State Department has previously considered.**
- nine of the refineries close to the southern terminus of Keystone Xl have nearly 30 percent of U.S. petcoke production capacity, over 50,000 tons a day.
- the petcoke produced from the Keystone Xl pipeline would fuel 5 coal plants and produce 16.6 million metric tons of co2 each year.
- these petcoke emissions have been excluded from State department emissions estimates for the Keystone Xl pipeline. including these emissions raises the total annual emissions of the pipeline by 13% above the State department's calculations.

- **Cheap petcoke helps the coal industry.**
  - as a refinery byproduct, petcoke is "priced to move", selling at roughly a 25 percent discount to conventional coal.
  - rising petcoke production associated with tar sands and heavy oil production is helping to make coal fired power generation dirtier and cheaper - globally.
  - From January 2011 to September 2012, the United States exported over 8.6 million tons of petcoke to china, most of which was likely burnt in coal-fired power plants.
  - **"Petkoch": the largest global petcoke trader in the world is Florida based oxbow Corporation, owned by William koch – the brother of Charles and David koch.**
  - oxbow carbon has donated $4.25 million to goP super
  - Pacs, making it the one of the largest corporate donors
  - to super Pacs.
  - oxbow also spent over $1.3 million on lobbyists in
  - ^012.

A Black Mound of Canadian Oil Waste Is Rising Over Detroit

Fabrizio Costantini for The New York Times

Petroleum coke, a waste byproduct of refining oil sands oil, is piling up along the Detroit River.

14AIChE Spring Meeting, New Orleans, LA
Full Spectrum Consulting LLC

Detroit's mountain of oilsands waste caused some alarm for Windsor, Ont., residents across the river over the weekend when black clouds of petroleum coke were seen filling the sky.

A YouTube video uploaded July 27 shows black clouds rising up from the Detroit River, where Koch Carbon, a company owned by libertarian-leaning billionaire brothers Charles and David Koch, has been keeping a large pile of the low-quality fuel. A storm rolled in while a ship was being loaded with inventory over the weekend, resulting in the unnerving sight captured on video.

Petcoke dust clouds seen in Windsor, Ont. (Youtube)

The so-called petcoke comes from a Detroit refinery that processes crude oil originating in Alberta's oil sands. At its peak, it stood three storeys tall and was visible from across the river on Canadian soil.

The New York Times first reported on the petcoke in May, ominously headlining its story "A Black Mound of Canadian Oil Waste Is Rising Over Detroit."

14AIChE Spring Meeting, New Orleans, LA

Full Spectrum Consulting LLC

Illinois expands crackdown on petroleum coke.

- "Already targeted as a source of lung-damaging air pollution, a company that has been piling petroleum coke on Chicago's Southeast Side now faces new accusations that it illegally allowed contaminated runoff to spill into the Calumet River.
- In a lawsuit, Illinois Attorney General Lisa Madigan urged a Cook County judge to cite KCBX Terminals for violating water quality and open dumping laws by failing to prevent petroleum coke and coal from washing into the river at the company's site.
- more…

14AIChE Spring Meeting, New Orleans, LA

Full Spectrum Consulting LLC

http://priceofoil.org/content/uploads/2013/01/
OCI.Petcoke.FINALSCREEN.pdf

Alberta doctor tells U.S.: Canada is 'lying' about tar sands' health effects

- American Senators told that oil sands are linked to a huge spike in cancer, despite Canadian government claims ...
- Alberta doctor tells U.S.: Canada is 'lying' about tar sands' health effects
- American Senators told that oil sands are linked to a huge spike in cancer, despite Canadian government claims
- Mychaylo Prystupa Posted: Feb 26th, 2014
- Facebook Twitter del.icio.us Buzz Up! Google Digg Reddit StumbleUpon
- Share on Tumblr Send ArticlePrint Article Read More:News oil sands
- Dr John O'Connor speaks to US Senators about Tar Sands health impacts and cancer Alberta family doctor Dr. John O'Connor speaking to US Senators in Washington on tar sands' health impact
- A northern Alberta doctor warned U.S. Senators on what he says have been the devastating health impacts of the tar sands on families – effects, he says, that have been willfully "ignored" by the Canadian and Alberta governments.
- "I appeal to you to keep up the pressure – this is an ongoing tragedy. A total disgrace," said Dr. John O'Connor, Wednesday in Washington, D.C.
- He sighted statistics for rare cancers – of the bile duct for example – that have shot up 400 times for what is considered normal for a tiny community, such as Fort Chipewyan – which is downstream, to the north of the oil sands.
- "These are published, peer-reviewed studies that indicate that the government of Alberta and Canada have been lying, misrepresenting the impact of industry on the environment," said O'Connor.
- The Alberta government has long denied cancer links with the province's multi-billion-dollar crown energy jewel. It states on its website that there is "insufficient evidence to link the incidence of cancer in Fort Chipewyan to oil sands operations" and rates of cancer are "within the expected range."
- O'Connor finds that hard to believe.
- "All of the scientific studies that have accumulated, it's almost like they don't exist," he said.
- A new study for example, shows Leukaemia and non-Hodgkin's lymphoma have spiked in the last 10 years -- especially among men who live downwind of pollution plumes from the oil, gas and tar sands facilities east of Edmonton.
- The family physician was invited to brief two U.S. Senators who are against the proposed TransCanada Keystone XL pipeline, that if approved, would carry bitumen from Alberta to Texas

Oil sands public health impact

- "Alberta MD: Canada 'Lying' About Tar Sands Health Impacts
- Washington, DC, February 27, 2014 (ENS) – The governments of Alberta and Canada have been "lying," to cover up the health impacts of the Canadian tar sands industry on northern Alberta communities, a Canadian medical doctor told U.S. senators at a news conference Wednesday on Capitol Hill.
- Dr. John O'Connor, a family physician who serves mostly aboriginal patients between Fort McKay and Fort McMurray, appealed to the lawmakers "to keep up the pressure. This is an ongoing tragedy. A total disgrace."
- Dr. O'Connor was invited to brief two U.S. Senators on the health effects of extracting, refining, transporting and burning the world's largest deposits of bitumen, or extremely heavy crude oil, located in northeastern Alberta. …

Text

14 AIChE Spring Meeting, New Orleans, LA

Full Spectrum Consulting LLC

Oil sands public health impact

- "Alberta MD: Canada 'Lying' About Tar Sands Health Impacts
- "Saying the health impacts of tar sands oil are being ignored, the two senators Wednesday called on Secretary of State John Kerry and the Obama Administration to conduct "an immediate and comprehensive study" of the public health risks to communities from the proposed Keystone XL pipeline that would carry diluted bitumen from Alberta, across the U.S.-Canada border to refineries on the Texas Gulf Coast."
- "Health miseries follow tar sands – from extraction to transport to refining to waste disposal," Senator Boxer told reporters at the news conference."
- http://ens-newswire.com/2014/02/27/alberta-md-canada-lying-about-tar-sands-health-impacts/

14 AIChE Spring Meeting, New Orleans, LA

Full Spectrum Consulting LLC

Oil sands deadly public health impact

- "Alberta MD: Canada 'Lying' About Tar Sands Health Impacts
- "Keith Wilson, lawyer for Peace River area residents who left their homes due to bitumen fumes from a Baytex oil extraction facility, said he's "not surprised" legislators in Washington are taking an interest in related health issues. "
- "Alberta has been talking about its commitment to "responsible energy development" but the words and actions don't line up, he said. A month after the Alberta Energy Regulator held an inquiry into Baytex's operation, the company is still operating without proper pollution control equipment, he said."

Health impacts downwind and downstream of oil sands and upgraders

- "The most recent person in Fort Chipewyan stricken with cholangiocarcinoma, Chadi's illness was raised in the Senate by Fort McMurray doctor John O'Connor during a briefing about the health impacts of the oilsands.
- John Chadi of Fort Chipewyan was diagnosed with cholangiocarcinoma, a rare form of bile duct cancer, the day after he was elected to municipal council in Fort McMurray.
- In the last decade, a handful of residents in the community of 1,100 have been stricken with cholangiocarcinoma — a disease that normally occurs in fewer than one in 200,000 people.Doctors have given Chadi a year to live but he is not giving up hope."

Flawed Keystone: Tar Sands Toxins Underestimated

- **Tar Sands Toxins with Keystone XL Link Underestimated**

- "New University of Toronto-Scarborough study published Monday says there's another reason to be concerned about oil production in the tar sands: The Canadian government may have underestimated emissions of carcinogens known as polycyclic aromatic hydrocarbons, or PAHs, from the Alberta tar sands, and they may be a major hazard to both human and ecosystem health.

- "Official estimates of PAH emissions from the Alberta tar sands have been used by the Canadian government to approve new tar sands development, and estimates for PAH concentrations in air, water and food in the region may also be too low, leading to an underestimation of PAH risk to human health, the study says."

- PAHs, which include phenanthrene, pyrene and benzo(a)pyrene, "are among the most toxic hydrocarbons," Blais said. "They're some of the worst things out there."The study, conducted by a team co-led by University of Toronto-Scarborough environmental chemistry professor Frank Wania, used computer models to examine the plausibility of official Canadian inventories of PAHs against actual measured PAH concentrations found in the Athabasca Oil Sands Region where Alberta's tar sands are produced near the northern end of the proposed Keystone XL Pipeline.

- "Blowing dust could also be a source of PAHs, he said.Because sources of these emissions may not have been considered in official inventories, the risk PAHs pose to human health and the environment may be much greater than previously thought, Wania said."

- "That's important because PAHs are a health issue for anyone living near the tar sands or the Keystone XL pipeline if it were to leak, he said."

- (Magill, B. (2014.02.03) . http://www.climatecentral.org/news/tar-sands-toxins-with-keystone-xl-link-underestimated-17028

14 AIChE

Full Spectrum Consulting LLC

Flawed Keystone XL Report

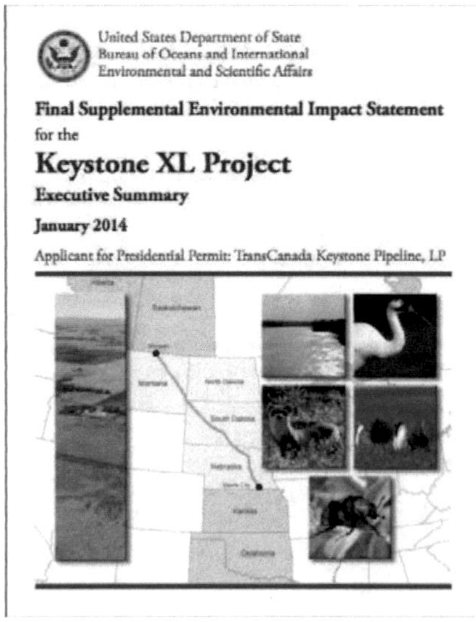
United States Department of State
Bureau of Oceans and International
Environmental and Scientific Affairs

Final Supplemental Environmental Impact Statement
for the
Keystone XL Project
Executive Summary
January 2014
Applicant for Presidential Permit: TransCanada Keystone Pipeline, LP

14 AIChE
Spring Meeting, New Orleans, LA

Full Spectrum
Consulting LLC

Flawed Keystone Final SEIS

- U.S. Department of State: Keystone XL Pipeline Project Final Supplemental Environmental Impact Statement (SEIS)
- There are serious problems that discount validity of the SEIS that have not been published including, but not limited to,
- Oil Sands
- Flawed Carbon Calculations from Emission Factors,
- Unaccounted Carbon Emissions, PAHs, CH4, etc.
- Unaccounted Global Warming Contributions, PM2.5 soot, CH4
- Undeclared Conflict of Interest from 2009 studies, and
- Catastrophic Human Costs of Pollution from production through upgrading, to refining, e.g., petcoke pm2.5s, etc.

14 AIChE Spring Meeting, New Orleans, LA

Full Spectrum Consulting LLC

Flawed Keystone Final SEIS Report

- There is concern that the Keystone Final SEIS report is founded on
- reduction of environmental requirements and oversight,
- the vicious assault on science,
- heavy muzzling and scripting of scientists,
- intimidating of physicians in downwind and downstream communities,
- Canadian government has declared U.S. will be competition (Keystone to be export pipeline of crude or Koch Carbon petcoke to Asia tankers).
- limited First Nations consultation,
- minimizing disclosure, and using the
- RCMP to spy on environmental, indigenous, and pipeline opposition
- TransCanada training U.S. police to deal with protesters and
- branding pipeline protesters and environmental advocates as Terrorists.

14 AIChE Spring Meeting, New Orleans, LA

Full Spectrum Consulting LLC

Boxer warns Keystone will cause cancer

- "The Environmental Impact Statement was woefully inadequate when it came to exploring human impacts of the pipeline," Boxer said Wednesday at a press conference held with Sen. Sheldon Whitehouse (D-R.I.) and anti-Keystone XL activists and environmentalists.
- She said Keystone would add pollutants to the air that will increase the likelihood of people getting cancer or heart disease.
- "I do believe the public health impacts are something that average people can really relate to because they know cancer is the second leading cause of death in this country — heart disease is number one — and all of this filthy air contributes to both of those," she said.
- Barron-Lopez, L. (2014.02.26) Boxer warns Keystone will cause cancer. The Hill, E2 Wire. http://thehill.com/blogs/e2-wire/e2-wire/199329-boxer-pivots-keystone-xl-talk-to-health-impacts#ixzz2vRcfcZA3 . Accessed 2014.02.26.

RCMP, intelligence agency accused of spying on pipeline opponents

- RCMP, intelligence agency accused of spying on pipeline opponents
- www.theglobeandmail.com › News › British Columbia
- The Globe and Mail
- Feb 6, 2014 - RCMP, intelligence agency accused of spying on pipeline opponents Add to ... Mark Hume. Vancouver — The Globe and Mail. Published ...

RCMP, intelligence agency accused of spying on pipeline opponents

- Environmental and aboriginal groups opposed to Enbridge Inc.'s proposed pipeline across British Columbia say federal law enforcement agencies are spying on them and they want the "intimidating and anti-democratic" practices stopped.

- In a news conference on Thursday, the B.C. Civil Liberties Association announced it has filed complaints with the agencies overseeing the RCMP and the Canadian Security Intelligence Service.
- The BCCLA said it appears that covert agents monitored meetings and other activities of the pipeline opponents who participated in the National Energy Board hearings into the proposed Northern Gateway project, and that CSIS and the RCMP then shared their intelligence with the oil industry and the NEB.

- Josh Paterson, executive director of the B.C. Civil Liberties Association, said the complaints to the Security Intelligence Review Committee, which oversees CSIS, and the Commission for Public Complaints Against the RCMP, are based on documents released by the media last year.
- Hume, M., (2014.02.06) RCMP, intelligence agency accused of spying on pipeline opponents. Vancouver — The Globe and Mail. http://www.theglobeandmail.com/news/british-columbia/csis-rcmp-accused-of-spying-on-pipeline-opponents/article16726444/

RCMP, intelligence agency accused of spying on pipeline opponents

- "Environmental and aboriginal groups opposed to Enbridge Inc.'s proposed pipeline across British Columbia say federal law enforcement agencies are spying on them and they want the "intimidating and anti-democratic" practices stopped."

- "In a news conference on Thursday, the B.C. Civil Liberties Association announced it has filed complaints with the agencies overseeing the RCMP and the Canadian Security Intelligence Service."

- Hume, M., (2014.02.06) RCMP, intelligence agency accused of spying on pipeline opponents. Vancouver — The Globe and Mail. http://www.theglobeandmail.com/news/british-columbia/csis-rcmp-accused-of-spying-on-pipeline-opponents/article16726444

RCMP, intelligence agency accused of spying on pipeline opponents, OK by Oliver

- "In a statement sent by e-mail, Natural Resources Minister Joe Oliver said he did not direct any such security measures, but he understood why police might be interested in protest groups.
- "The safety of Canadians is a priority for our government. As part of its commitment to safety, the National Energy Board may conduct a security assessment prior to a regulatory hearing," he said. "The NEB will often work with the RCMP to protect the safety of everyone involved. Neither I, nor any member of my department, gave instructions to any federal agency in this matter."
- An Enbridge spokesman said the company will not comment as the complaints are directed at CSIS and the RCMP.
- Tehara Mufti, a spokesperson for CSIS, said the agency does not comment on specific complaints, but "what I can say is that CSIS investigates – and advises government on – threats to national security, and that does not include peaceful protest and dissent."
- Hume, M., (2014.02.06) RCMP, intelligence agency accused of spying on pipeline opponents Vancouver — The Globe and Mail. http://www.theglobeandmail.com/news/british-columbia/csis-rcmp-accused-of-spying-on-pipeline-opponents/article16726444/

RCMP, intelligence agency accused of spying on pipeline opponents

- "The BCCLA said the documents show the RCMP had informants at environmental meetings, including one in a church basement in Kelowna at which people hand-painted signs.
- "It's hard to believe this is Canada," said Grand Chief Stewart Phillip, president of the Union of B.C. Indian Chiefs. "We take great exception to being spied up by the RCMP, by CSIS...clearly our democratic rights have been affected. It's a sad day for this country."
- "Will Horter, executive director of the Dogwood Initiative, said spying on people participating in a federal pipeline review is "anti-democratic, it's very likely illegal and, most disturbingly, it's a complete abuse of power."
- Hume, M., (2014.02.06) RCMP, intelligence agency accused of spying on pipeline opponents. Vancouver — The Globe and Mail. http://www.theglobeandmail.com/news/british-columbia/csis-rcmp-accused-of-spying-on-pipeline-opponents/article16726444/

TransCanada Caught Training U.S. Police to Treat Peaceful Anti-Keystone XL U.S. Activists as 'Terrorists'

- 'HOUSTON - June 12 - In the midst of recent national controversy surrounding government surveillance of the public, a recent Freedom of Information Act request to the Nebraska State Patrol has exposed evidence that
- "TransCanada provided trainings to federal agents and local Nebraska police to suppress nonviolent activists protesting the Keystone XL tar sands pipeline by arresting them on "anti-terrorism statutes."
- "The presentation slides, obtained by grassroots landowner advocacy group Bold Nebraska, target Tar Sands Blockade activists by name."
- Common Dreams (2013.06.12) https://www.commondreams.org/newswire/2013/06/12-1

14AIChE

Full Spectrum Consulting LLC

TransCanada Caught Training Police to Treat Peaceful Anti-Keystone XL Activists as 'Terrorists' (1)

- TransCanada does not specify any limitation of the label of terrorists. If individual in any country is anti-Keystone in any way they are terrorists. Review who would be TransCanada labeled "Terrorists"
- U.S. Citizens, Activists w/ photos in TransCanada training slides (2)
- Over 2 million U.S. Citizens that submitted anti-Keystone comments to the State Department in due process review.
- U.S. Senators, U.S. Representatives not paid by, or voting for Keystone
- U.S. Scientists, Celebrities, Actors, Actresses, Parents, Children
- 10 million U.S. Citizens for Clean Air and Water (Earth Day, 1970)
- Others [see RCMP]
- (1) Common Dreams (2013.06.12) https://www.commondreams.org/newswire/2013/06/12-1
- (2) TransCanada powerpoint http://www.tarsandsblockade.org/transcanadapolice/

14AIChE

Full Spectrum Consulting LLC

TransCanada caught with 60 miles of pipeline that failed integrity inspections, some three per mile

- "Grassroots resistance to Keystone XL is growing in Texas and Oklahoma where TransCanada is currently digging up freshly laid sections of the pipeline that failed integrity inspections. The East Texas Observer reports that at least 70 "anomalies," including dents and shoddy welds in the pipe, were identified in a 60 mile span, and in some areas these imperfections occur at a rate of three per mile."

- "A discovery like this presentation reveals that TransCanada has no problem lying to authorities and intentionally misleading the public in pursuit of its own private gain," says Seifert. "If TransCanada officials cannot be trusted to tell the truth about the peaceful nature of Keystone XL protestors, why should we believe they've told the truth about the integrity of their pipeline and the real threats it poses to the hundreds of rivers and creeks that it crosses?"

- Common Dreams (2013.06.12) https://www.commondreams.org/newswire/2013/06/12-1

14AIChE

Full Spectrum Consulting LLC

CBC — REGULATORS 'BURIED' REPORT CRITICAL OF TRANSCANADA OVERSIGHT:

- "CBC News reports that regulators "effectively buried" a report on a TransCanada natural gas pipeline that exploded in 2009 and destroyed five acres of woods. "In an early 2011 draft report about the incident, the National Energy Board criticized TransCanada, the operator of the line owned by its subsidiary NOVA Gas Transmission, for 'inadequate' field inspections and 'ineffective' management.

- Final reports are typically published by the investigative bodies, either the NEB or the Transportation Safety Board, but this report wasn't released until this January when the CBC obtained it through an access-to-information request. The NEB said the delay was caused by an 'administrative error' when an employee left without transferring the file over."

- (Guillen, A., 2014) Keystone XL public comment period begins, Politico. http://www.politico.com/morningenergy/0214/morningenergy12919.html and CBC: http://bit.ly/1e06faj)

14AIChE Spring Meeting, New Orleans, LA

Full Spectrum Consulting LLC

Harper govt makes moves to silence Canada's leading environmental groups

- "In both the US and Canada, activism against tar sands, pipelines and climate change has soared in recent years. But while President Obama has encouraged citizens to "speak up", Canadian Prime Minister Stephen Harper's administration has tried to silence critics."

- Bagley, K. (2014) Harper Govt Makes Moves to Silence Canada's Leading Environmental Groups. Inside Climate News. http://insideclimatenews.org/news/20140214/harper-govt-makes-moves-silence-canadas-leading-environmental-groups. Accessed 2014.02.15.

14AIChE Spring Meeting, New Orleans, LA

Full Spectrum Consulting LLC

Harper Govt Makes Moves to Silence Canada's Leading Environmental Groups

- "The audits are the latest in a string of attacks on environmentalists and scientists in recent years by a Harper government that has made oil sands expansion a centerpiece of its economic agenda. Since Harper took office in 2006, production of Alberta's carbon-intensive tar sands crude has doubled, and Canada has set a goal to at least double it again in the years ahead.

- Simultaneously, the conservative administration has
- cut funding for climate science research,
- shuttered environmental research labs and
- buried government reports highlighting the dangerous climate and environmental implications of runaway oil sands development. It has also
- muzzled its own scientists from speaking about the issue,
- cancelled thousands of environmental assessments for pipelines, oil sands mining and other development projects, and
- gutted 70 environmental protection laws in 2013 alone.
- In 2012, the government listed environmentalists as a terrorist threat.
- Amid the erosion of environmental law in Canada, across the border in the United States the Obama administration has sought to apply federal law to combating climate change. The centerpiece of this work is an historic regulation under the Clean Air Act to cut carbon pollution from the nation's coal plants.
- (Bagley, 2014)

14AIChE

Full Spectrum Consulting LLC

7 Ways Canadian Environmental Groups Are Being Attacked as "Terrorists"

- "Media campaigns smear environmentalists as extremists.
- Canada's first counter-terrorism program targets environmentalists
- Harper threatens the tax status of environmental charities
- New counter-terrorism unit formed to protect the oil sands
- Canada sends diplomats to scare Fortune 500 corporations
- Government trainings identify environmentalists as "terrorists"
- "Under which terrorist group do animal rights activists and environmental extremists fall?"
- "Terrorism" reports that blur the line between legal, illegal conduct"
- Potter, W., (2012). 7 Ways Canadian Environmental Groups Are Being Attacked as "Terrorists". Green is the New Red. http://www.greenisthenewred.com/blog/7-ways-canada-environmental-groups-labeled-terrorists/6374/ Accessed 2013.12.21.

14 AIChE Spring Meeting, New Orleans, LA

Full Spectrum Consulting LLC

Washington to discuss health effects of Keystone XL

- "Washington will play host to a new discussion regarding the proposed Keystone XL pipeline on Wednesday, and this time the health concerns of Northern Alberta communities next to oil sands operations will be front and centre. …
- More than a decade ago, Dr. O'Connor raised red flags regarding health concerns in the oil sands region. At one point, he faced allegations of causing "undue alarm" among residents of Fort Chipewyan, but the Alberta College of Physicians and Surgeons closed that file in 2009. The same year, the Alberta Cancer Board released a report that found elevated cancer cases in Fort Chipewyan over the 1995 to 2006 study period. Wrangling between the province, Ottawa, the industry and First Nations followed, and an independent study examining whether health issues are linked to oil sands pollution has never been completed. [usual obfuscation and delay tactic on a national level]
- Dr. O'Connor continues to work as a physician in First Nations communities located downstream from production, and has advised landowners in the Peace River region who believe their health has suffered as a result of emissions from oil sands operations there.
- "The pipeline going ahead is going to open the floodgates in terms of development. I'm just fearful for what will happen five to 10 years down the road," Dr. O'Connor said in an interview Tuesday.
- When the State Department released its final environmental impact statement Jan. 31 – which concluded Keystone would be unlikely to alter greenhouse gas emissions – Ms. Boxer released a statement that said she isn't "satisfied with any analysis that does not accurately document what is really happening on the ground when it comes to the extraction, transport, refining, and waste disposal of dirty, filthy tar sands oil.
- "My biggest concerns continue to be the serious health impacts on communities and the dangerous carbon pollution that comes from tar sands oil.""
- Cryderman, K. (2014.02.25) Washington to discuss health effects of Keystone XL.. The Globe and Mail. http://www.theglobeandmail.com/report-on-business/industry-news/energy-and-resources/washington-to-discuss-health-effects-of-keystone-xl/article17103397/

14 AIChE Spring Meeting, New Orleans, LA

Full Spectrum Consulting LLC

Oliver to Obama:
"Quit ragging [sic] the puck, Mr. President."

- News Canada. OTTAWA - "Natural Resources Minister Joe Oliver used a hockey term Sunday to urge President Barack Obama to finally score with the Keystone XL pipeline after years of missing the net. ... "I certainly hope he won't drag the puck. We feel that the reasons to go ahead are very strong, that the environmental issues have been dealt with," Oliver said on CTV's Question Period."
- Dunn, M., (2014) "Quit ragging [sic] the puck, Mr. President." News Canada. http://www.camrosecanadian.com/2014/02/02/quit-ragging-the-puck-mr-president-oliver after http://politwitter.ca/page/news/user/joeoliver1. Accessed 02.02.2014.

14 AIChE Spring Meeting, New Orleans, LA

Full Spectrum Consulting LLC

Oliver to Obama:
"the environmental issues have been dealt with"

- **Oliver's, Canada's ways of dealing with environmental issues?**
- **"7 Ways Canadian Environmental Groups Are Being Attacked as "Terrorists".**
- Media campaigns smear environmentalists as extremists.
- Canada's first counter-terrorism program targets environmentalists
- Harper threatens the tax status of environmental charities
- New counter-terrorism unit formed to protect the oil sands
- Canada sends diplomats to scare Fortune 500 corporations
- Government trainings identify environmentalists as "terrorists"
- ""Under which terrorist group do animal rights activists and environmental extremists fall?""
- "Terrorism" reports that blur the line between legal and illegal conduct."

- Potter, W., (2012). 7 Ways Canadian Environmental Groups Are Being Attacked as "Terrorists". Green is the New Red (2013). http://www.greenisthenewred.com/blog/7-ways-canada-environmental-groups-labeled-terrorists/6374/ Accessed 2013.12.21.

14 AIChE Spring Meeting, New Orleans, LA

Full Spectrum Consulting LLC

Keystone XL Nebraska Court Delay

- "OTTAWA—The politics of pipeline development heated up another notch after Prime Minister Stephen Harper's much-touted project to carry oilsands-derived crude to the United States suffered its latest setback.

- A Nebraska court decision delaying approval of the $7-billion Keystone XL line is raising the stakes for the controversial Northern Gateway line through British Columbia and other projects to carry petroleum from Alberta's oilsands to export terminals."

, A., Penty, R., TransCanada Nebraska Ruling Seen as Delaying Keystone. Bloomberg. www.bloomberg.com/news/2014-02-19/transcanada-pipeline-route-through-nebraska-ruled-...html . Accessed 2014.02.19.

14AIChE Spring Meeting, New Orleans, LA

Full Spectrum Consulting LLC

Keystone XL Nebraska Court Delay

- TransCanada Corp. (TRP)'s Keystone XL pipeline route through Nebraska was ruled illegal by a state court judge in a decision seen as setting back the project by as long as a year as U.S. officials consider approving it. ...

- "Because the governor's actions of Jan. 22, 2013, in approving the Keystone XL Pipeline route were predicated on an unconstitutional statute, the court also finds the governor's actions in that regard must be declared null and void," she said. ...

- Harris, A., Penty, R., TransCanada Nebraska Ruling Seen as Delaying Keystone. Bloomberg. http://www.bloomberg.com/news/2014-02-19/transcanada-pipeline-route-through-nebraska-ruled-invalid.html . Accessed 2014.02.19.

14AIChE Spring Meeting, New Orleans, LA

Full Spectrum Consulting LLC

TransCanada Nebraska Ruling Seen as Delaying Keystone

- Canadian Prime Minister Stephen Harper has called Keystone XL a "no-brainer" for the U.S. and last month criticized Obama for "punting" the decision on the line. While the project remains on hold, Harper's government has redoubled efforts to get new oil conduits built to Canada's Atlantic and Pacific coasts, by making faster regulatory reviews law and increasing discussions with aboriginal groups along pipeline paths.

- Obama met today in Toluca, Mexico, with Harper and Mexican President Enrique Pena Nieto. Obama gave no ground on speeding up the review of the Keystone XL after the meeting.

- "Keystone will proceed along the path that's already been set forth," Obama said at a press conference after the meeting. He said he and Harper discussed "a shared interest in dealing with greenhouse gas emissions. This is something we have to deal with."

- Harper said he and Obama "had an exchange" on the Keystone XL project. "We'll continue on that discussion," he said.

- Harris, A., Penty, R., TransCanada Nebraska Ruling Seen as Delaying Keystone. Bloomberg. http://www.bloomberg.com/news/2014-02-19/transcanada-pipeline-route-through-nebraska-ruled-invalid.html . Accessed 2014.02.19.

East & West Pipeline approvals appear to make Keystone unnecessary

- **"Two giant oil pipelines proposed to speed "doubling" of tar sands.**

- "In a dizzying week of oil announcements, two new giant west-to-east pipelines passed key milestones. If built, the pipelines would rapidly expand Alberta's oil sands, cause massive environmental impacts, and trigger thousands of new jobs, according to several observers.

- "Cancer concerns have also been raised downstream of the oil sands - especially among First Nations - but the industry and Alberta government deny any significant link." [See Alberta MD: Canada 'Lying' About Tar Sands Health Impacts]

- Vancouver Observer, Canada. 7 March 2014. http://www.vancouverobserver.com/news/two-giant-oil-pipelines-proposed-speed-doubling-tar-sands

Flawed Keystone Final SEIS

- U.S. Department of State: Keystone XL Pipeline Project Final Supplemental Environmental Impact Statement (SEIS)
- There are serious problems that discount validity of the SEIS that have not been published including, but not limited to,
- Oil Sands
- Flawed Carbon Calculations from Emission Factors,
- Unaccounted Carbon Emissions, PAHs, CH4, Carcinogens, etc.
- Unaccounted Global Warming Contributions, PM2.5 soot, CH4
- Undeclared Conflict of Interest from 2009 studies, and
- Catastrophic Human Costs of Pollution from production through upgrading, to refining, e.g., petcoke pm2.5s, etc.

Flawed Keystone Final SEIS Report

- There is concern that the Keystone Final SEIS report is founded on
- reduction of environmental requirements and oversight,
- the vicious assault on science,
- heavy muzzling and scripting of scientists,
- intimidating of physicians in downwind and downstream communities,
- Canadian government has declared U.S. will be competition (Keystone to be export pipeline of crude or Koch Carbon petcoke to Asia tankers).
- limited First Nations consultation,
- minimizing disclosure, and using the
- RCMP to spy on environmental, indigenous, and pipeline opposition
- TransCanada training U.S. police to deal with protesters and branding pipeline protesters and environmental advocates as terrorists.

14AIChE Spring Meeting, New Orleans, LA
Full Spectrum Consulting LLC

Thank You

14AIChE Spring Meeting, New Orleans, LA

Full Spectrum Consulting LLC

Thermal Cracking Partial Upgrading Product: Unstable, Incompatible, is not SCO, it is Self-InCompatible Partial Upgraded Thermal Cracked Unstable Product SICPUTCUP

Stephen Pavel
Full Spectrum Consulting LLC

14AIChE Spring Meeting, New Orleans, LA

2014 AIChE Spring Meeting & 10th Global Congress on Process Safety,
2014.04.01, 10:30am, Fountain, Hilton NOLA Riverside. Paper (1065b)

Full Spectrum Consulting LLC

Getting Bitumen and Extra Heavy Oil to Market: Tight Oil Synergy, Blending, and Partial Upgrading Alternative Challenges Including Thermal Cracking Product Dienes, Gums, Self-Incompatible Asphaltenes, and Pricing Review

Stephen Pavel
Full Spectrum Consulting LLC

14AIChE Spring Meeting, New Orleans, LA

2014 AIChE Spring Meeting & 10th Global Congress on Process Safety, 2014.04.02, 2:00pm, Belle Chase, Hilton NOLA Riverside. Paper (147b)

Full Spectrum Consulting LLC

- Processing Shale Oil and Oil Sands Crudes
- Panelists include
- Ken Bryden, WR Grace;
- Andrew Sloley, CH2MHILL;
- John Kulach, UOP;
- Stephen Pavel, Full Spectrum Consulting LLC.

- 2014 AIChE Spring Meeting & 10th Global Congress on Process Safety,
- 2014.04.01, 1:30 pm, Belle Chase, Hilton NOLA Riverside. Paper (147b)
- Description: 78 Processing Shale Oil and Oil Sands Crudes - Panel
- Sponsor: Topical 7: 17th Topical on Refinery Processing
- Co-sponsor: Fuels and Petrochemicals Division (16)

14AIChE Spring Meeting, New Orleans, LA

Full Spectrum Consulting LLC

Why Climate Deniers Have No Scientific Credibility: Only 1 of 9,136 Recent Peer-Reviewed Authors Rejects Global Warming

"I have brought my previous study (see here and here) up-to-date by reviewing peer-reviewed articles in scientific journals over the period from Nov. 12, 2012 through December 31, 2013.

I found 2,258 articles, written by a total of 9,136 authors. (Download the chart above here.)

Only one article, by a single author in the Herald of the Russian Academy of Sciences, rejected man-made global warming. I discuss that article here.

My previous study, of the peer-reviewed literature from 1991 through Nov. 12, 2012, found 13,950 articles on "global warming" or "global climate change." Of those, I judged that only 24 explicitly rejected the theory of man-made global warming. "

Powell, J., (2014.01.08) Why Climate Deniers Have No Scientific Credibility: Only 1 of 9,136 Recent Peer-Reviewed Authors Rejects Global Warming. http://www.desmogblog.com/2014/01/08/why-climate-deniers-have-no-scientific-credibility-only-1-9136-study-authors-rejects-global-warming . Accessed 2014.01.21 .

GRACE
Catalysts Technologies

Processing Shale Oils in FCC: Challenges and Opportunities

Kenneth Bryden, E. Thomas Habib, Jr., Olivia Topete

14 AIChE Spring Meeting, New Orleans, LA

Presented at Session 147: Shale and Heavy Oil Processing
AIChE Spring Meeting, April 2, 2014

Outline

Crude Oil Trends and Shale Oil Properties

Challenges Associated with Processing Unconventional Crudes

Mitigating Risks with Catalyst Technology

Shale Oil is a Game Changer for North America

Novel drilling techniques, horizontal drilling, and hydraulic fracturing have resulted in large amounts of oil from shale formations

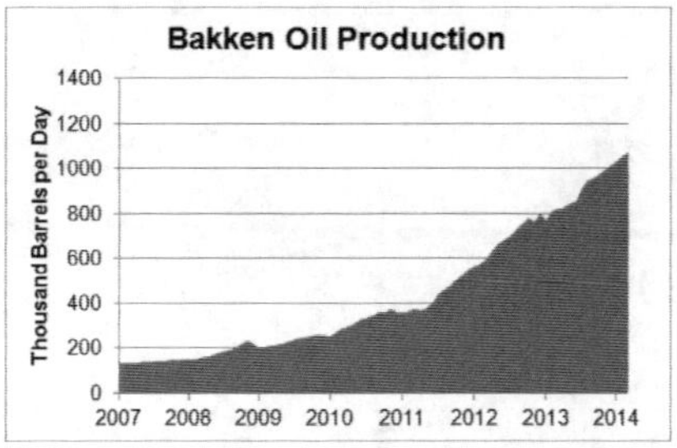

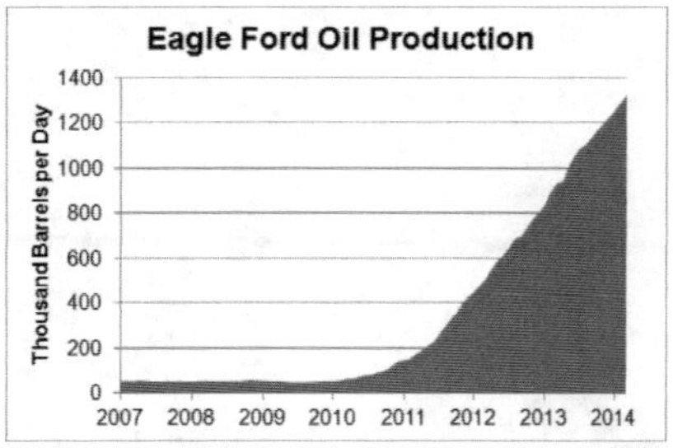

Source: March 2014 U.S. Energy Information Administration Drilling Productivity Report

Merrill Lynch estimates that by 2017, refiners will likely be "saturated with light oil."

The 2014 BP Energy Outlook projects that tight oil will be 7% of world oil supply in 2035

3

GRACE
Talent | Technology | Trust

Grace's Shale Oil Program

Understand market environment → **Identify challenges and opportunities** → **Develop catalyst solutions**

Coordinated effort between R&D and technical service, supported by Grace's industry-leading pilot plants, analytical resources, and manufacturing capabilities

4

GRACE
Talent | Technology | Trust

Shale Oils are Light and Sweet

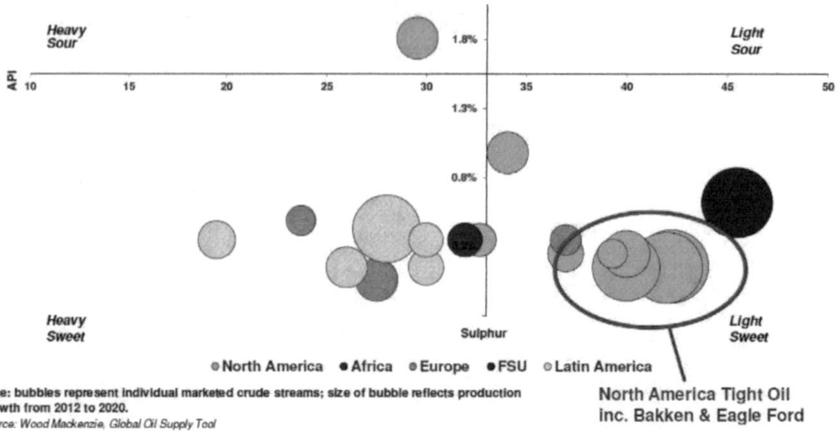

Note: bubbles represent individual marketed crude streams; size of bubble reflects production growth from 2012 to 2020.
Source: Wood Mackenzie, Global Oil Supply Tool

5

GRACE
Talent | Technology | Trust

Sources of FCC Feed

Typically the heavier portions of whole crude are fed to the FCC unit

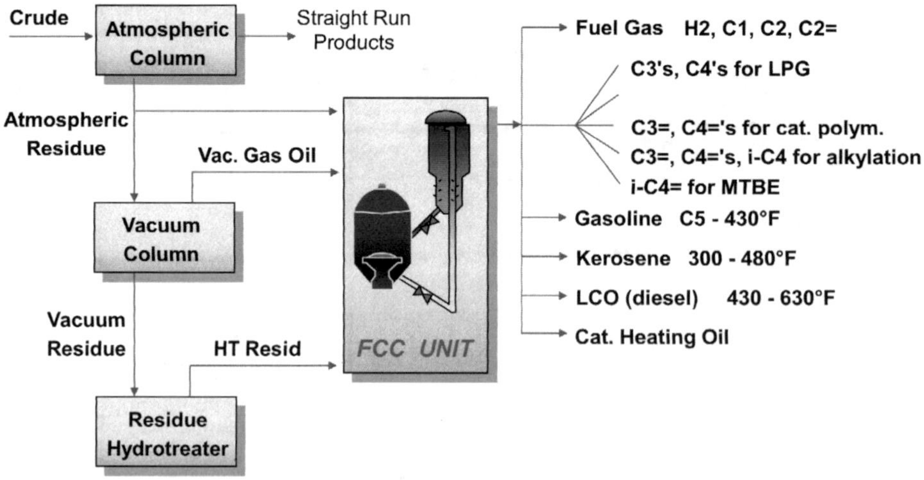

6

GRACE
Talent | Technology | Trust

Properties of a raw Bakken crude sample

		Bakken sample used in this work	Published assay data Bakken [1]	WTI [1]	LLS [1]	"Typical" Eagle Ford [2, 3]
API Gravity	Degrees	41.9	>41	40.0	35.8	47.0
Sulfur	wt%	0.19	<0.2	0.33	0.36	0.11
Distillation Yield		Weight %	Volume%	Volume%	Volume%	Volume%
Light Ends	C1-C4	1	3	1	2	1
Naphtha	C5-330°F	32	30	30	17	34
Kerosene	330-450°F	14	15	15	14	15
Diesel	450-680°F	25	25	24	34	23
Vacuum Gas Oil	680-1000°F	23	22	23	25	20
Vacuum Residue	1000+°F	5	5	7	8	7
Total		100	100	100	100	100

[1] D. Hill, "North Dakota Refining Capacity Study Final Technical Report," DOE Award No.: DE-FE0000516, January 5, 2011.
[2] Platts Methodology and Specifications Guide, "The Eagle Ford Marker: Rationale and Methodology," October 2012.
[3]"Effects Of Possible Changes In Crude Oil Slate On The U.S. Refining Sector's CO_2 Emissions," prepared for the International Council On Clean Transportation by MathPro Inc., March 29, 2013.

Bakken is light, sweet, and has a low proportion of FCC feed

7

GRACE
Talent | Technology | Trust

Properties of gasoline and LCO range of Bakken

Whole crude sample was distilled into gasoline range (C5 - 430°F) and LCO (430-650°F) ranges and characterized

Gasoline fraction properties	RON	60.6
	MON	57.6
LCO Fraction Properties	Aniline Point (°F)	155.9
	API Gravity	37.6
	Diesel Index	58.6

RON and MON were calculated via Grace's proprietary G-Con® octane calculation software

8

GRACE
Talent | Technology | Trust

Shale Oil Derived Feeds: Light, Sweet, and Paraffinic

Property	Eagle Ford Condensate Splitter Bottoms	HVGO Derived from 85% Eagle Ford	430°F+ Distillation of Whole Bakken Crude	650°F+ Distillation of Whole Bakken Crude	Mid-Continent VGO
API Gravity (degrees)	36.6	30.0	28.6	23.0	24.7
CCR (wt%)	0.15	0.17	1.34	2.27	0.32
Hydrogen (wt%)	13.7	13.4	13.1	12.7	12.9
Sulfur (wt%)	0.08	0.83	0.30	0.43	0.35
Percent Boiling >1000°F	10.7	13.1	14.5	23.6	16.5
D2287 SimDis (°F)					
10%	519	715	470	658	691
50%	735	862	699	844	848
90%	1006	1015	1057	1135	1045

9

GRACE
Talent | Technology | Trust

High Resolution Mass Spectroscopy of Shale Oil Derived Feeds

	Eagle Ford Condensate Splitter Bottoms	HVGO Derived from 85% Eagle Ford	650°F+ Distillation of Whole Bakken Crude	Mid-Continent VGO
	Avg. Wt%	Avg. Wt%	Avg. Wt%	Avg. Wt%
Paraffins	44.4	23.1	12.4	12.2
Cycloparaffins	43.2	38.6	46.8	42.5
Total Saturates	**87.6**	**61.7**	**59.2**	**54.7**
Monoaromatics	5.1	11.3	21.0	20.4
Diaromatics	4.5	7.1	8.5	14.4
Triaromatics	1.7	6.5	7.4	7.0
Tetraaromatics	0.0	1.9	1.4	0.1
Total Aromatics	**11.3**	**26.8**	**38.3**	**41.9**
Total Thiophenic Compounds	1.1	11.5	2.4	3.4

10

GRACE
Talent | Technology | Trust

Shale Oil derived feeds have high crackability

Yields at C/O = 6	Eagle Ford Condensate Splitter Bottoms	HVGO Derived from 85% Eagle Ford	430°F+ Distillation of Bakken	650°F+ Distillation of Bakken	Mid Continent VGO
Conversion (wt %)	86.3	83.3	71.7	74.4	74.5
H2 Yield (wt%)	0.05	0.09	0.06	0.08	0.04
C1's+C2's (wt%)	1.0	1.3	1.2	1.5	1.3
Total C3 (wt%)	8.1	6.7	5.1	5.2	5.1
C3= (wt%)	7.0	5.8	4.4	4.4	4.4
Total C4's (wt%)	17.3	14.3	10.8	10.7	10.8
LPG (wt%)	25.4	21.0	16.0	15.9	15.9
Gasoline (C5-430°F) (wt%)	58.4	58.6	52.1	52.8	54.3
RON (G-Con)	89.1	90.8	89.2	90.1	90.3
MON (G-Con)	78.8	79.9	78.9	79.6	79.5
LCO (430-700°F) (wt%)	11.5	12.2	24.6	19.6	19.1
Bottoms (700°F+) (wt%)	2.2	4.5	3.7	6.0	6.4
Coke (wt%)	1.3	2.5	2.7	4.1	2.9

Paraffinicity results in high gasoline yield

11

GRACE
Talent | Technology | Trust

Processing Straight Run Shale Oil

Normally you wouldn't charge whole crude oil to a FCC, but doing so has been considered by refiners running predominantly shale oil.

- Shale oil is low in 650°F+ and high in 650°F-, so a refiner running 100% shale oil can be at maximum distillation and light cut processing capacity while short on FCC feed.

As a model case whole light sweet Bakken crude was cracked in a circulating riser pilot plant over a lab-deactivated FCC Catalyst at three riser outlet temperatures 900°F, 935°F, 970°F and compared to a mid-continent VGO

- What is the effect of process conditions on yields and product quality?

12

GRACE
Talent | Technology | Trust

Grace DCR™ Circulating FCC Pilot Plant

A description of capabilities of the pilot plant can be found in:

"Flexible Pilot Plant Technology for Evaluation of Unconventional Feedstocks and Processes AM-13-04," 2013 American Fuel and Petrochemical Manufacturers Annual Meeting, San Antonio, Texas

GRACE
Talent | Technology | Trust

Processing Straight Run Shale Oil – *Yield Structure*

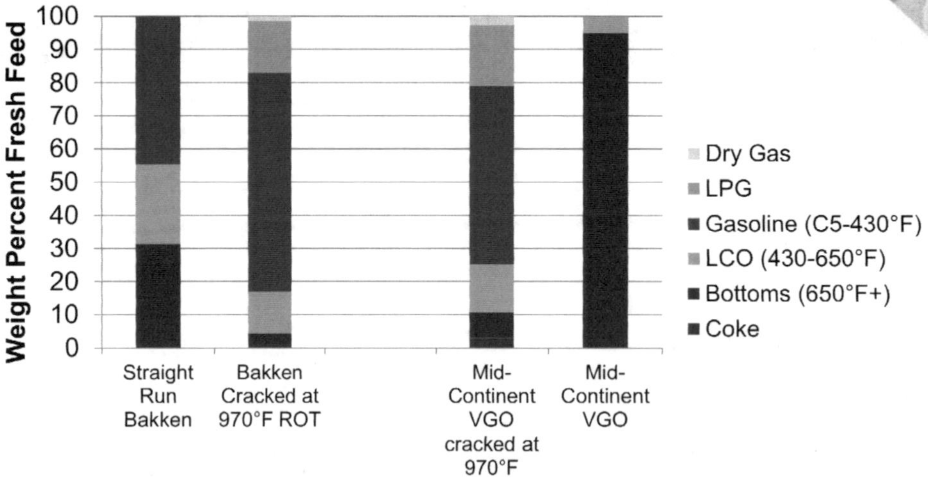

What happens when you charge whole crude to an FCC?

GRACE
Talent | Technology | Trust

Product Yields as a Function of ROT and Feed

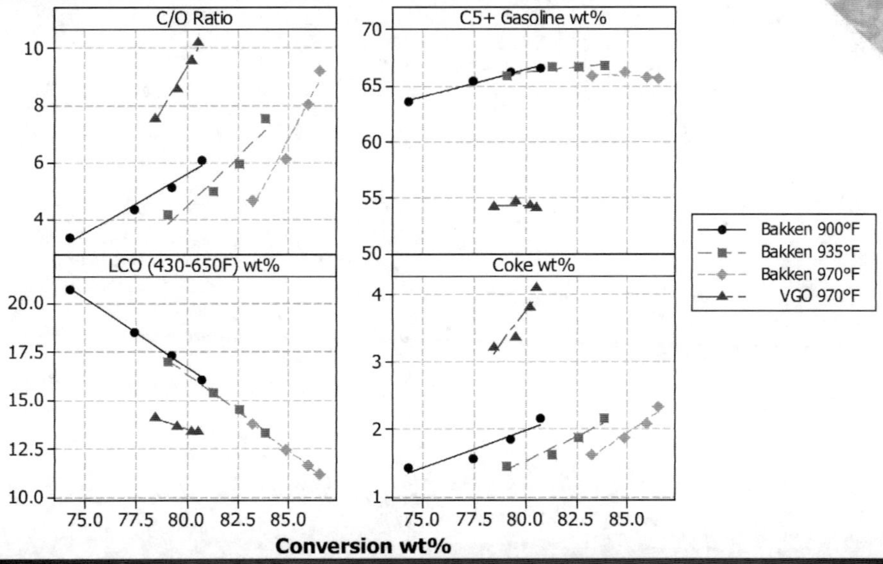

As expected, lowering riser outlet temperature increases LCO on whole shale oil.

GRACE
Talent | Technology | Trust

Gasoline Properties a Function of ROT

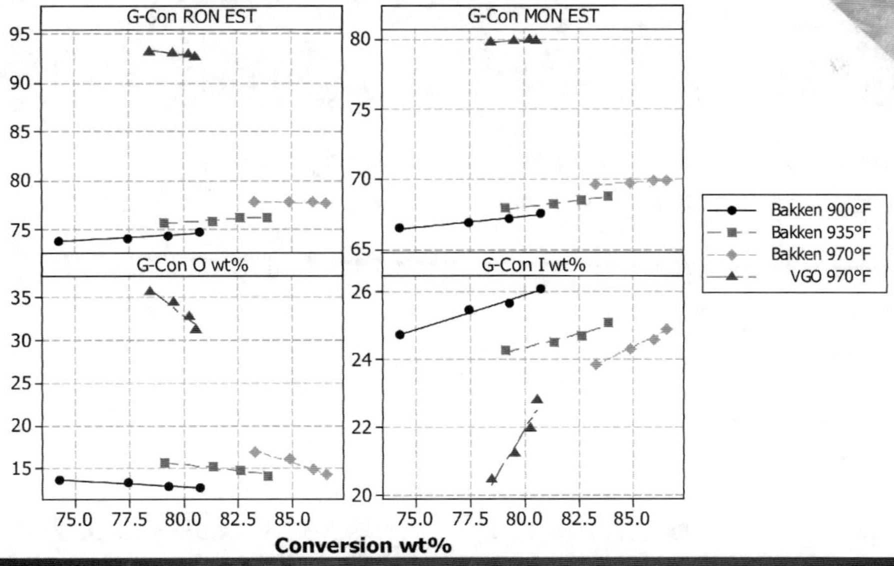

Cracking straight run crude produces a paraffinic, low-octane gasoline.

GRACE
Talent | Technology | Trust

Impact on LCO Yield and Quality

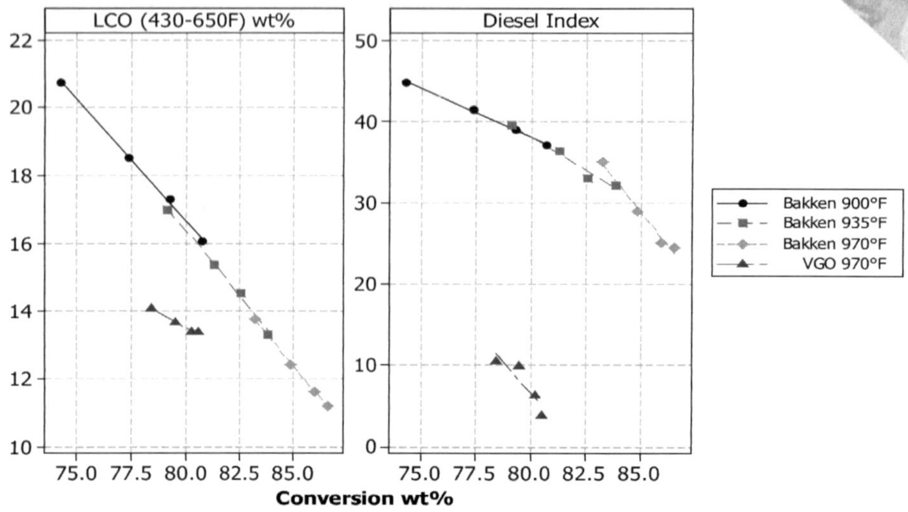

DCR™ pilot plant total liquid product distilled to get LCO cut and measure aniline point and API gravity.

GRACE
Talent | Technology | Trust

Key Findings Whole Shale Oil Processing

Whole shale oil responds to similarly to shifts in operating conditions

However, product yield slate (volume and quality) is substantially different

- **Good quality LCO from the FCC**
- **Large amounts of low octane gasoline**

For additional details see: "*Processing Shale Oils in FCC: Challenges and Opportunities,*" Hydrocarbon Processing, September 2013

GRACE
Talent | Technology | Trust

Metals in Shale Oils

Property (ppm)	Whole Bakken Crude	650°F+ Distillation of Bakken Crude	Eagle Ford Condensate Splitter Bottoms	Mid-Continent VGO
Feed Barium	0.2	0.1	0.8	<0.01
Feed Calcium	0.5	1.2	5.4	<0.1
Feed Iron	7.5	7.8	8.6	<0.1
Feed Magnesium	0.2	0.2	0.3	<0.04
Feed Nickel	0.4	1.9	0.2	<0.04
Feed Potassium	0.4	0.3	0.0	<0.04
Feed Sodium	8.7	3.9	3.1	<0.06
Feed Vanadium	0.1	0.5	0.9	<0.03

Fe and Ca in Bakken and other tight oil derived crudes are high and unpredictable

Grace analyses confirm other industry publications of tight oil metals contaminants

Tight oils have been proven to introduce an unknown assortment of metals

19

GRACE
Talent | Technology | Trust

Outline

Crude Oil Trends and Tight Oil Properties

Challenges Associated with Processing Unconventional Crudes

Mitigating Risks with Catalyst Technology

20

GRACE
Talent | Technology | Trust

Challenges of Unconventional Feeds

- Rapidly changing feed quality
- Excess light ends in refinery
- Reduced FCC feed from crude unit
- Unexpected or unconventional metals
- Difficulty maintaining FCC heat balance
- Refinery gasoline octane pool debits from paraffinic feed and unconstrained downstream units (alky)

21

GRACE
Talent | Technology | Trust

Ways to address challenges

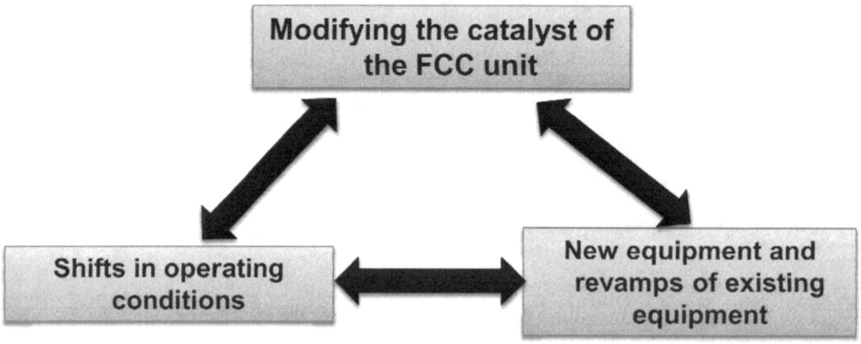

22

GRACE
Talent | Technology | Trust

Historical examples of catalyst advances that have addressed changing refinery needs

Challenge	Catalytic Solution
Meeting market demand for increased gasoline production	Zeolite-containing catalysts giving higher activity and better coke selectivity
Achieving high temperature catalyst regeneration	Combustion promoter additives
Reducing SOx emissions from the regenerator	SOx transfer additives
Processing High Boiling Point Resid Feeds	Coke selective catalysts and metals traps
Octane shortage from phase out of tetraethyl lead	Ultrastable Zeolite Y catalysts
Increased demand for light olefins	ZSM-5 based catalysts and additives

23

GRACE
Talent | Technology | Trust

Mitigating Risks of Unconventional Feeds with Technology

Challenge ➡	Consequence ➡	Catalyst Solution
Fe poisoning	Loss of bottoms cracking and conversion	Employ a high porosity matrix
Unpredictable swings in contaminant metals	Loss of surface area leads to lower MAT and conversion	Utilize traps for Ni and V with high stability zeolites
FCC heat balance	Low regenerator temps and circulation constraints	Increase catalyst activity
Refinery imbalances	Low refinery gasoline pool octane	Boost zeolite isomerization activity in the FCC

Flexible Functionality Critical for Unconventional Feeds

24

GRACE
Talent | Technology | Trust

New Grace FCC catalyst family for unconventional feeds

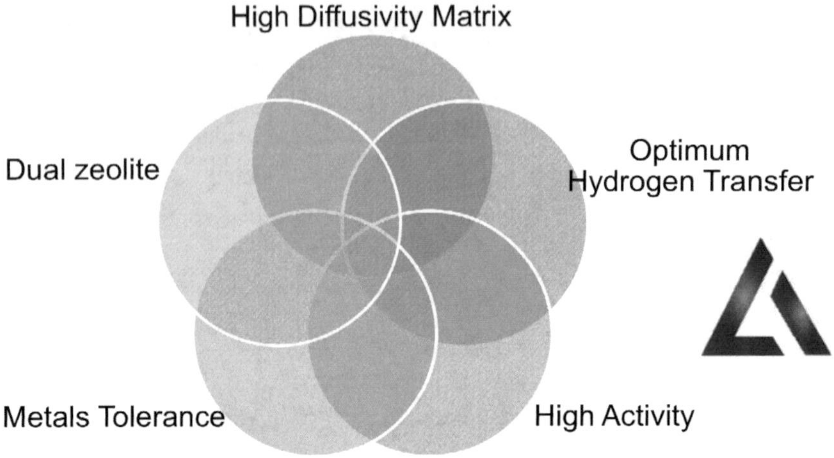

ACHIEVE™
THE ADVANTAGE
Grace FCC Catalyst

For more details see the paper "Processing Tight Oils in FCC: Issues, Opportunities and Flexible Catalytic Solutions AM-14-16," 2014 American Fuel and Petrochemical Manufacturers Annual Meeting, Orlando, Florida

25

GRACE
Talent | Technology | Trust

Outline

Crude Oil Trends and Tight Oil Properties

Challenges Associated with Processing Unconventional Crudes

Mitigating Risks with Catalyst Technology

26

GRACE
Talent | Technology | Trust

Iron and Calcium Poisoning

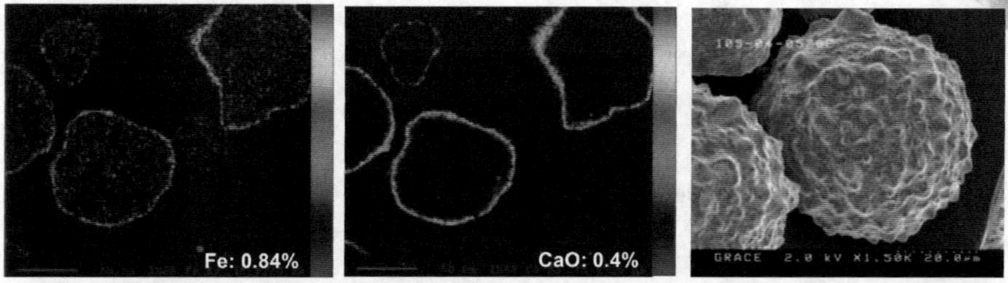

Fe poisoned Ecat: Fe forms nodules and a dense skin on the catalyst exterior

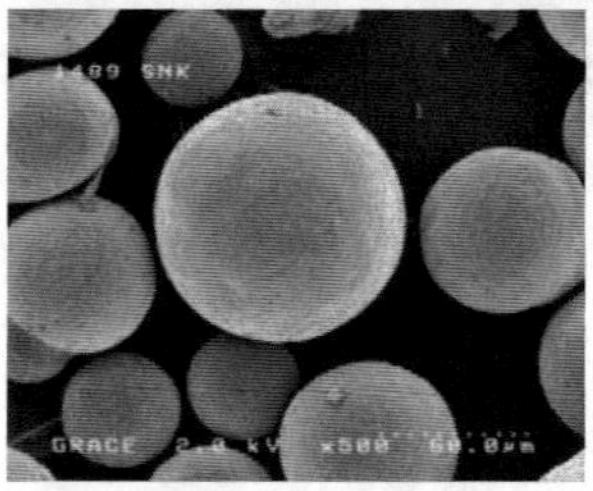

Non-Fe poisoned Ecat

Effects of Fe on FCC Unit Operations

Unit Symptoms
- **Loss of conversion**
- **Loss of bottoms cracking**

Ecat Changes
- **Fe ring visible by EPMA**
- **Nodules visible by SEM**
- **Ecat surface appear glassy**
- **ABD decreases**
- **MAT, surface area and pore volume <u>do not necessarily</u> decrease**

GRACE
Talent | Technology | Trust

Catalytic solutions for high iron feeds

High alumina matrix + optimum mesoporosity
- **Resistant to low melting point phases of Fe and Ca**
- **Unhindered diffusion to active sites**

The high diffusivity matrix of ACHIEVE® catalysts is based on technology embodied in the popular MIDAS® catalyst

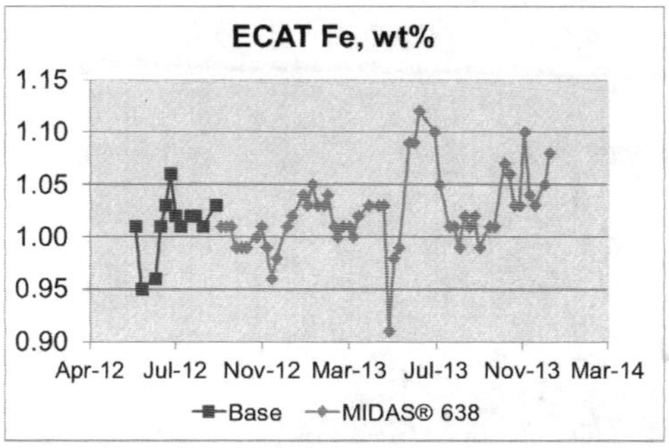

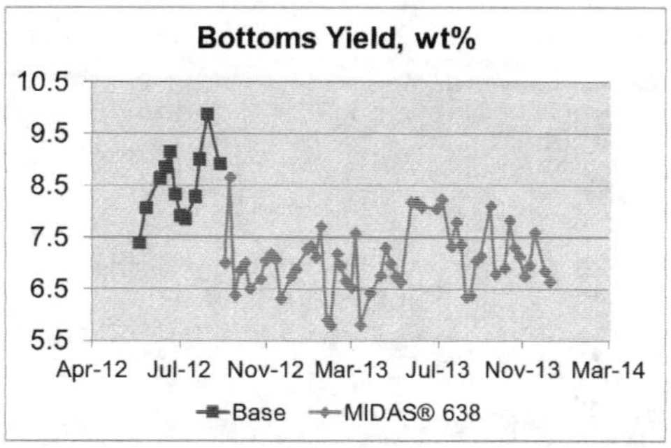

Selectivity maintained in 100% tight oil operation

GRACE
Talent | Technology | Trust

Catalytic Solutions for Refinery Imbalances

Higher than expected light ends yield could result in lower FCC severity, oftentimes leaving downstream units unconstrained

ACHIEVE® 400 catalyst delivers higher LCO yield, greater gasoline octane, and improved C4= selectivity through:

- **High Diffusivity Matrix**
- **Optimum Hydrogen Transfer**
- **Metals Tolerance**
- **Dual Zeolite Input**

Realize the yield potential within a single catalyst particle

GRACE
Talent | Technology | Trust

ACHIEVE® 400 Catalyst Improves Butylene Yield

ACHIEVE® 400 catalyst delivers superior selectivity of C4 to C3 olefin yield compared to conventional ZSM-5 additives, by preferentially cracking gasoline olefins at C7 and above

Reactant		Selectivity	Relative Reactivity	C4=/C3=
C8= $\rightarrow$	2 C4=	44%	100	1.57
	C3= + C5=	56%		
C7= $\rightarrow$	C3= + C4=	95%	12	1.0
	C2= + C5=	2%		
C6= $\rightarrow$	2 C3=	83%	1.5	0.1
	C2= + C4=	16%		

ACHIEVE® 400

ZSM-5 Additive

Buchanan, et.al., J. Catal., 158, 279 (1996)

GRACE
Talent | Technology | Trust

DCR™ Pilot Plant Evaluation – Constant Conversion Yields

	Base Midas®	Base Midas® + OlefinsMax®	ACHIEVE® 400
Cat to Oil	8.7	9.2	8.3
H_2 wt%	0.28	0.27	0.28
C2= wt%	0.8	0.8	0.8
Dry gas	2.84	2.78	2.75
C3= wt%	4.3	5.1	5.3
Total C4 wt%	9.3	10.2	10.6
iC4 wt%	1.5	1.7	1.6
nC4 wt%	0.4	0.4	0.4
Total C4= wt%	7.3	8.1	8.5
ΔC4= /ΔC3=	---	0.89	1.1
Gasoline wt%	50.8	49.1	48.7
LCO wt%	18.4	18.2	18.2
Bottoms wt%	6.6	6.7	6.7
Coke wt%	6.9	6.8	6.7
RON	93.50	93.53	94.12
MON	79.69	79.80	80.07

ACHIEVE® 400 catalyst delivers:

- Superior ΔC4= /ΔC3= selectivity
- Higher octane
- Equivalent coke and bottoms

31

GRACE
Talent | Technology | Trust

Commercial Update- ACHIEVE® 400 Catalyst

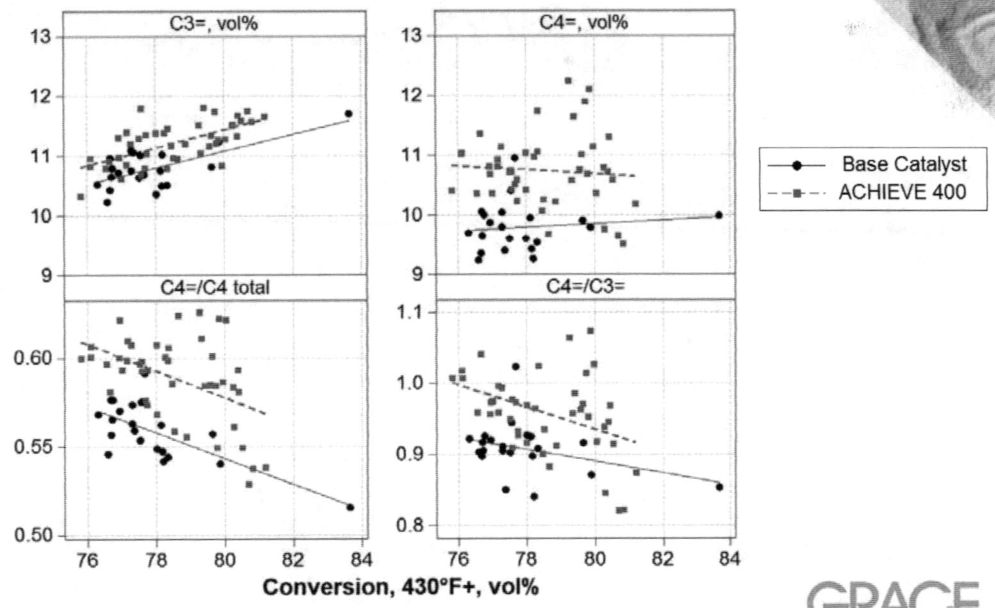

32

GRACE
Talent | Technology | Trust

Conclusions

- The tight oil boom has revitalized U.S. refining
- Though light, sweet, and abundant, tight oil feeds present numerous challenges for refiners and the FCCU
- Proper application of catalyst technology can help mitigate the challenges
- Grace's ACHIEVE® catalyst portfolio is designed to meet the needs of a market environment driven by growth of unconventional feedstocks

GRACE
Talent | Technology | Trust

Questions?

For additional information, please visit www.grace.com or contact:

Kenneth Bryden | Manager FCC Evaluations Research
Grace Catalysts Technologies
kenneth.bryden@grace.com

GRACE
Talent | Technology | Trust

352884 Shale Crude Opportunities for US Refineries

Abstract:

Refineries in the United States and throughout the world are continually optimizing feed blends, product slates and configuration to maximize value. For U.S. refiners, accessing the Shale crude production provides an additional opportunity to optimized performance. According to an independent study, about 1.8 MMb/d of incremental U.S. production will be added by 2016 and over 75% of this increase will be light crude and condensates, with much of this from the shale deposits. It is forecast that domestic light production will displace most light crude imports by 2016. This growth in domestic crude production presents new crude slate opportunities for the refining sector. In this presentation, the impact of substituting domestic shale crude for currently processed light crudes in the US refinery will be illustrated.

349043 Modifying Crude Heaters for Shale Processing

Abstract:

Refiners have started processing shale crudes in their refineries and are facing a number of problems. The atmospheric heaters which have traditionally been designed for heavy crudes preheating are suddenly processing very light unstable crudes. Several of these heaters are having severe coking issues. Furnace Improvements worked with a client and identified the issues causing the coking of the tubes. A number of modifications have been proposed to the heater to make it suitable for processing the light shale crudes and have a long run length at the same time. We will discuss the same in this presentation.

350762 Downhole Catalytic Upgradation of Heavy Oil

Abstract:

There are currently vast reserves of heavy oil/natural bitumen in the world, offering a potential boost to energy supplies in the next decade as conventional light oil reservoirs run low. However, recovering and upgrading these sources present an immense technological challenge, due primarily to their high viscosity at reservoir conditions. Currently, inefficient and expensive surface mining techniques are the main methods employed which are associated with a high environmental impact.

Technologies such as Steam Assisted Gravity Drainage (SAGD) and Cyclic Steam Stimulation (CSS) are in commercial use, however, the oil produced from these technologies needs expensive surface upgrading and steam consumption is particularly high. Toe-to-Heel Air Injection, or THAI, along with CAtalytic upgrading PRocess In-situ, or CAPRI, offers very high recovery rates along with significant upgrading of the produced oil which adds to the value of produced oil. In the present study, the various In-situ upgrading techniques and the various advantages of THAI-CAPRI are discussed in details.

Automation Solutions for Processing Light Tight Oil

Tim Olsen

Emerson Process Management

14AIChE
Spring Meeting, New Orleans, LA

EMERSON.
Process Management

LTO Crude Availability

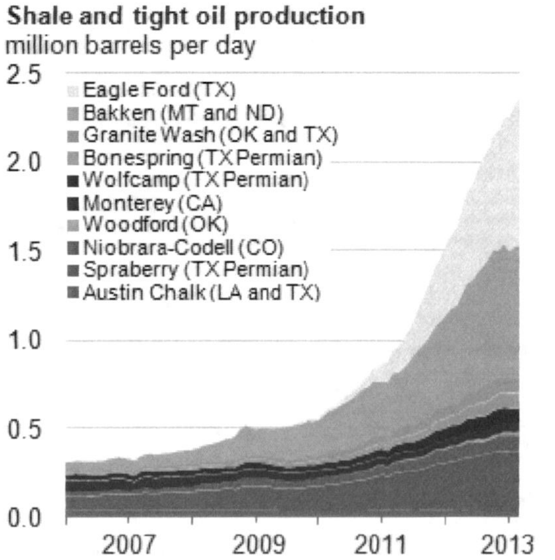

Source: U.S. Energy Information Administration based on Drilling Info and LCI Energy Insight

Did you know?

United States crude oil domestic supply has increased over one million bpd in each of the past two years, among the highest historical gains worldwide.

What does this mean?

North America refiners will need to learn how to process this tight oil.

14AIChE
Spring Meeting, New Orleans, LA

Tight Oil Challenges

- H2S, amines, and high volatility
- *Light* Tight Oil (LTO) and crude blending
- LTO crude blends with Canadian crudes
- Hot section heat exchanger and fired heater tube fouling
- Cold section heat exchanger fouling
- Variability in API Gravity
- Filterable solids and desalter operation
- Corrosion
- Downstream operations – catalyst and reactor controls

14 AIChE
Spring Meeting, New Orleans, LA

Be Aware – High RVP

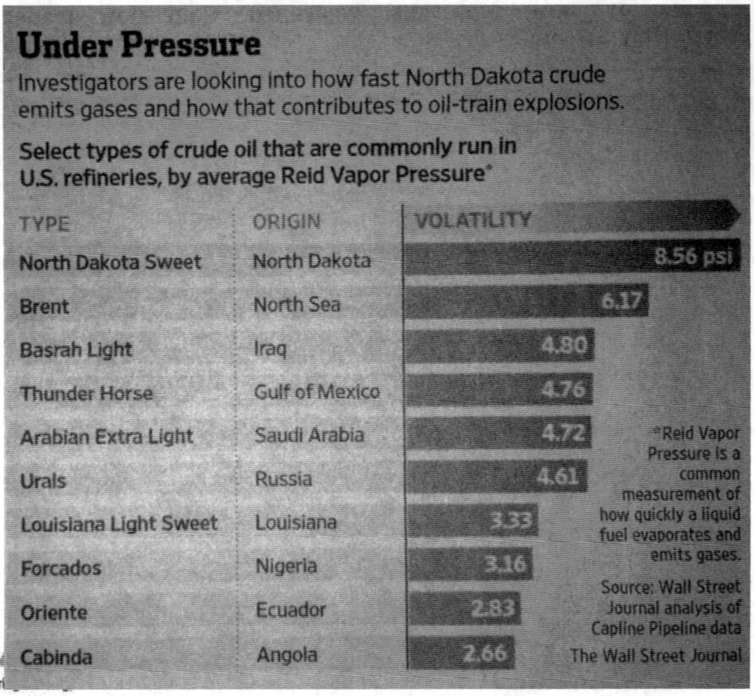

Under Pressure

Investigators are looking into how fast North Dakota crude emits gases and how that contributes to oil-train explosions.

Select types of crude oil that are commonly run in U.S. refineries, by average Reid Vapor Pressure*

TYPE	ORIGIN	VOLATILITY
North Dakota Sweet	North Dakota	8.56 psi
Brent	North Sea	6.17
Basrah Light	Iraq	4.80
Thunder Horse	Gulf of Mexico	4.76
Arabian Extra Light	Saudi Arabia	4.72
Urals	Russia	4.61
Louisiana Light Sweet	Louisiana	3.33
Forcados	Nigeria	3.16
Oriente	Ecuador	2.83
Cabinda	Angola	2.66

*Reid Vapor Pressure is a common measurement of how quickly a liquid fuel evaporates and emits gases.

Source: Wall Street Journal analysis of Capline Pipeline data

The Wall Street Journal

Bakken RVP seen as high as 12 psi

Vapor Recovery at tank unloading in the tank farm

Hydrogen Sulfide

... three pipeline companies including Enbridge Inc. warned regulators that North Dakota oil with too much hydrogen sulfide, which is toxic and flammable, was reaching terminals and putting workers at risk...

...The NTSB estimates that 69 percent of today's rail tank-car fleet has "a high incidence of tank failure during accidents," ...

North Dakota Oil Boom Seen Adding Costs for Rail Safety

By Jim Efstathiou Jr. and Angela Greiling Keane | Jan 14, 2014 1:22 PM CT | **18 Comments**
Email **Print**

(Corrects exposure level in 25th paragraph or story published Aug. 13.)

Photographer: The Canadian Press/Ryan Remiorz/AP Photo

Cleanup continues at the scene of the runaway oil train derailment and explosion in... **Read More**

Crude oil shipped by railroad from **North Dakota** is drawing fresh scrutiny from regulators concerned that the cargo is adding environmental and safety hazards, something that analysts say could raise costs.

Related

Federal Railroad Administration Letter to API

The U.S. Federal Railroad Administration is investigating whether chemicals used in hydraulic fracturing are corroding rail tank cars and increasing risks. Separately, three pipeline companies including **Enbridge Inc. (ENB)** warned regulators that North Dakota oil with too much hydrogen sulfide, which is toxic and flammable, was reaching terminals and putting workers at risk.

14AIChE
Spring Meeting, New Orleans, LA

Hydrogen Sulfide and Amines

- Although tight oil is considered low in sulfur within the oil, there is H_2S
- Amine-based H_2S Scavengers are added in the pipeline or railcars prior to transporting
- Variability in which amine and what quantity added in the field
- H_2S has been seen >100ppm when offloading at tank farm
- Paraffin waxes in the crude are trapping amines against metal in the crude unit causing local corrosion
- Automation – H_2S detection

14AIChE
Spring Meeting, New Orleans, LA

Light Tight Oil

- Light, so typically needs to be blended with other heavier crudes
  - Match downstream processing capabilities, especially bottom of the barrel processing
  - Match designed heat balance in crude unit heat exchanger train
- However, additional challenges happen when you blend crudes
  - Potential for accelerated heat exchanger fouling depending on crude blend incompatibility

14 AIChE
Spring Meeting, New Orleans, LA

Crude Blending Incompatibility

- The percentage of each blended crude used matters… 20/80 may be fine but 30/70 may not, it may cause instability and accelerated heat exchanger fouling.
- Even with the best intentions of crude blending, there may be times when the desired ratio cannot be achieved because of what is available in the crude tanks.
- Variability within and between wells (API gravity, paraffin content, etc.)

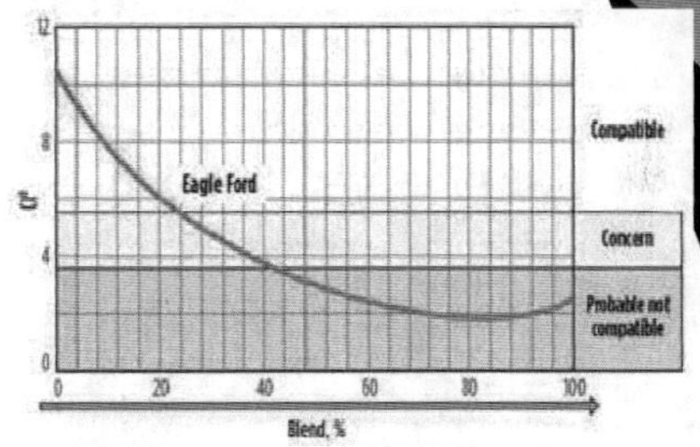

Western Canadian heavy crude is moderate in sulfur content and not paraffinic

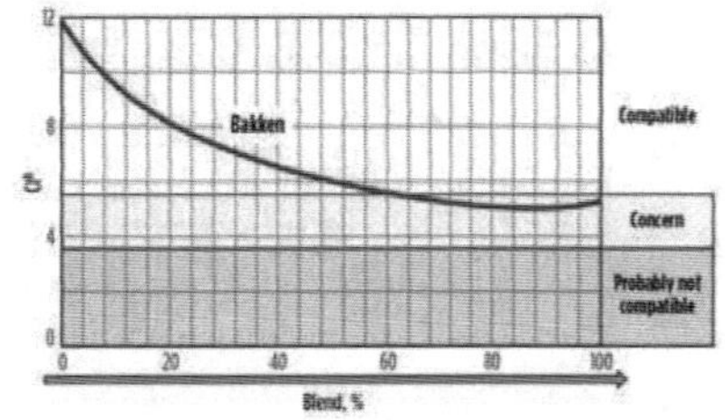

*Sayles, S.; Romero, S. KBC, (01/02/2014). Case history: Characterization of shale oils and heat exchanger performance. Hydrocarbon Processing. R.O. 2/25/2014
http://www.hydrocarbonprocessing.com/Article/3304489/Case-history-Characterization-of-shale-oils-and-heat-exchanger-performance.html

14 AIChE
Spring Meeting, New Orleans, LA

Asphaltenes, Resins, Aromatics, Paraffins

- Asphaltenes do not dissolve in crude oil but exist as a colloidal suspension. They are soluble in aromatic compounds such as xylene, but will precipitate in the presence of light paraffinic compounds such as pentane.
- Resins, on the other hand, are readily soluble in oil.
- Mixing stable crude blends and asphaltic and paraffinic oil, creates the potential for precipitating the unstable asphaltenes by reducing the solvency capability of the base crude
- Highly asphaltenic crudes require higher concentrations of aromatics or resins and lower concentrations of saturates to maintain stability in crude blends.

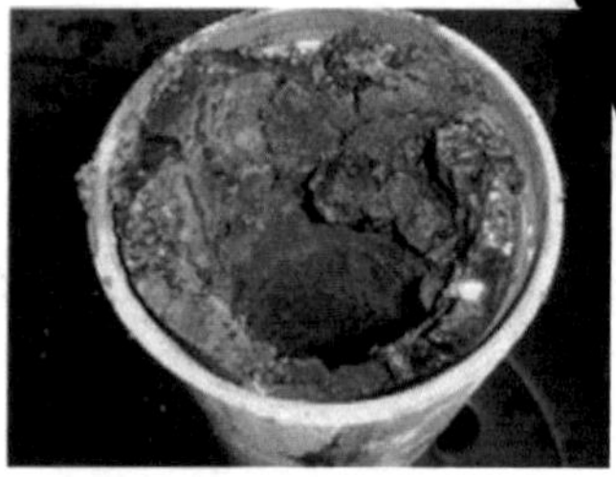

Asphaltenes clogging pipe

14 AIChE
Spring Meeting, New Orleans, LA

Blend High TAN Heavy Crude and LTO

- Canada sent an average of more than 2.5 million barrels of oil per day to the United States in 2013
  - Up 10 percent from the previous year
- Sulfur helps mitigate naphthenic acid corrosion
- Low sulfur in Shale Oil can increase risk
- Corrosion reported even in low acid concentrations
- The risk is dependent upon the ratio of sulfur to acid number, the lower the ratio the greater the risk of naphthenic acid corrosion.

14 AIChE
Spring Meeting, New Orleans, LA

Cold Section of Crude Unit Heat Exchanger Train

- Typically not automated monitoring and minimal manual monitoring for fouling
- Paraffin waxes fouling and trapping amines against walls of pipes and vessels

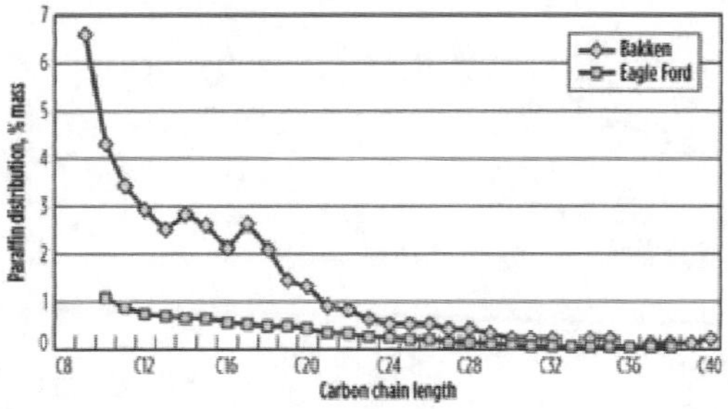

Hydrocarbon Processing - Innovative solutions for processing shale oils, *07.01.2013* | **Sandu, C.**, Baker Hughes, Sugar Land, Texas; **Wright, B.**, Baker Hughes, Sugar Land, Texas

14 AIChE
Spring Meeting, New Orleans, LA

Heat Exchanger Monitoring

- Difficult to maintain history or conduct analysis
- Difficult to take measurements
  - Hot process fluids
  - Stacked exchangers
  - Other access constraints
- Not aware of accelerated fouling
- Time consuming to take measurements manually

Up to 50% of the heat delivered to the crude unit comes from the heat exchanger train before the fired heater

Add more temperature measurements and analysis software for automated online monitoring

14 AIChE
Spring Meeting, New Orleans, LA

Fouling not limited to Heat Exchangers

- Some refiners are also experiencing fouling in the tubes of fired heaters.

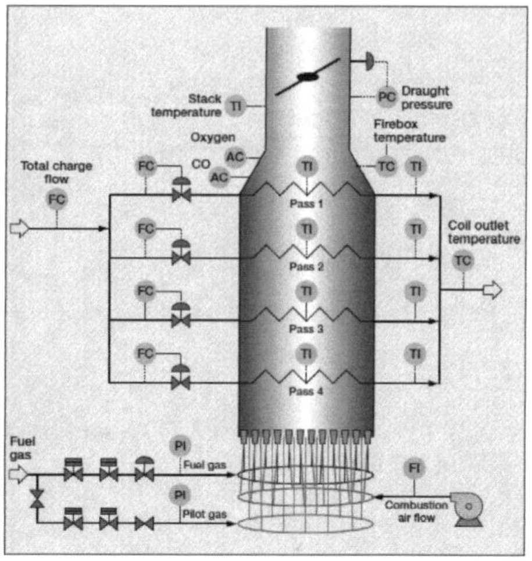

14 AIChE
Spring Meeting, New Orleans, LA

Also, Variability in API Gravity

Flint Hills Processing Larger Eagle Ford Crude Volumes

Wichita, Kansas based Flint Hills Resources LP recently reported that it will add the capability to ship up to 200,000 bpd of crude oil and condensate from its terminal in Ingleside, Texas, which is integrated into the Port of Corpus Christi navigation network. According to recent comments from Brad Urban, senior vice president of crude oil for Flint Hills Resources, "With this approval we will be able to proceed with permitting and complete our plans to modify and integrate these assets into our crude terminal that is already receiving Eagle Ford production. Flint Hills Resources expects to begin outbound waterborne shipments of Eagle Ford from the retrofitted assets by the middle of 2012."

"We believe the Eagle Ford Shale play will produce a large volume of crude oil for a long time – and in amounts that will exceed what local refiners can use," Urban said. "Due to the proximity to Eagle Ford production, we believe Corpus Christi and Ingleside are the best locations for shipping Eagle Ford crude to other markets on the Gulf Coast." The API gravity of Eagle Ford crude oil is roughly between 30-50 °API. ■

Valero Refining Facility Replacing Foreign Crude with Domestic Source

A report from Reuters dated February 10 quoted Valero Energy Corp. Chief Executive Bill Klesse as saying that Eagle Ford shale crude from south Texas was replacing foreign crude at the company's 93,000 bpd Three Rivers, Texas refinery. The refinery, located 74 miles south of San Antonio, is running 27,000 bpd of Eagle Ford crude currently, and is expected to run 40,000 bpd of the oil by May, Klesse said in a recent conference call. Within the year, the refinery will be running 60,000 bpd of Eagle Ford. ■

- Crude tank level indication
- Mass flow measurement rather than volume

14 AIChE
Spring Meeting, New Orleans, LA

Filterable Solids in Tight Oil

- Filterable solids seen >200 pounds per thousand barrels
- Some refiners are injecting chemicals directly into the desalter
- Desalters are expected to do more with tight oils
- Level interface detection will be even more important with more chemicals and solids

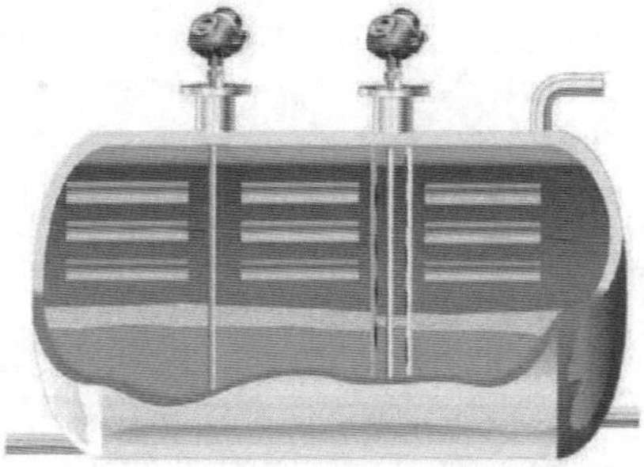

Desalter interface detection and level Measurement – guided wave radar

14 AIChE Spring Meeting, New Orleans, LA

Corrosion Monitoring

- As mentioned earlier, paraffin waxes trapping amines against wall of pipes and vessels
- Crude unit fractionator top section
  - Typically, pH measurement in the top section
  - Corrosion coupons or probes (intrusive)
- Fired Heater outlet
  - Electric field signature (non-intrusive)

Non-intrusive, continuous corrosion monitoring

14 AIChE Spring Meeting, New Orleans, LA

Downstream Operations - Reaction

- Processing with tight oils have resulted in deficiencies in product qualities (e.g. cold flow properties), however...
- Catalyst manufacturers are already changing functionality of catalyst
- Keep in mind with high selective catalyst, the process controls need tight control to maximize benefit of catalyst - e.g. reduce variability in target temperature
- Control valve positioners and Lambda tuning methodology allow for opportunity to have less variability.

CEP, January 2013

14 AIChE
Spring Meeting, New Orleans, LA

Key Points – Processing LTO

- Watch closely on what is in the crude
  - BSW and filterable solids
  - H2S and Amines
- Blending, utilize mass flow due to wide variations in API Gravity (don't forget about tank levels)
  - Balance for overall plant operations
  - Heat balance in crude heat exchanger train
  - Ratio of crude blends may result in accelerated fouling or corrosion
- Monitor fouling in heat exchangers and fired heater tubes
  - Crude incompatibilities (asphaltene precipitation)
  - Wax buildup in cold section
- Monitor desalter performance
- Monitor for corrosion
  - After crude and vacuum heaters
  - Crude tower top
- Have good process control on reactor controls to maximize functionality of highly selective catalyst

14 AIChE
Spring Meeting, New Orleans, LA

Questions

Tim.Olsen@Emerson.com

14 AIChE Spring Meeting, New Orleans, LA

EMERSON. Process Management

AUTHOR INDEX

AUTHOR INDEX